Basic Chemometric Techniques in Atomic Spectroscopy

RSC Analytical Spectroscopy Monographs

Titles in the Series:

1: Flame Spectrometry in Environmental Chemical Analysis: A Practical Guide
2: Chemometrics in Analytical Spectroscopy
3: Inductively Coupled and Microwave Induced Plasma Sources for Mass Spectrometry
4: Industrial Analysis with Vibrational Spectroscopy
5: Ionization Methods in Organic Mass Spectrometry
6: Quantitative Millimetre Wavelength Spectrometry
7: Glow Discharge Optical Emission Spectroscopy: A Practical Guide
8: Chemometrics in Analytical Spectroscopy, 2nd Edition
9: Raman Spectroscopy in Archaeology and Art History
10: Basic Chemometric Techniques in Atomic Spectroscopy

How to obtain future titles on publication:
A standing order plan is available for this series. A standing order will bring delivery of each new volume immediately on publication.

For further information please contact:
Sales and Customer Care, Royal Society of Chemistry, Thomas Graham House, Science Park, Milton Road, Cambridge, CB4 0WF, UK
Telephone: +44 (0)1223 432360, Fax: +44 (0)1223 420247, Email: sales@rsc.org
Visit our website at http://www.rsc.org/Shop/Books/

Basic Chemometric Techniques in Atomic Spectroscopy

Edited by

Jose Manuel Andrade-Garda
Department of Analytical Chemistry, University of A Coruña, Galicia, Spain

RSC Publishing

RSC Analytical Spectroscopy Monographs No. 10

ISBN: 978-0-85404-159-6
ISSN: 0146-3829

A catalogue record for this book is available from the British Library

Published by The Royal Society of Chemistry,
Thomas Graham House, Science Park, Milton Road,
Cambridge CB4 0WF, UK

Registered Charity Number 207890

For further information see our web site at www.rsc.org

Preface

This book is rooted in an informal discussion with three researchers, Dr Alatzne Carlosena, Dr Mónica Felipe and Dr María Jesús Cal, after they had some problems measuring antimony in soils and sediments by electrothermal atomic absorption spectrometry. While we reviewed the results and debated possible problems, much like in a brainstorming session, I realized that some of their difficulties were highly similar to those found in molecular spectrometry (mid-IR spectroscopy, where I had some experience), namely a lack of peak reproducibility, noise, uncontrollable amounts of concomitants, possible matrix interferences, *etc.*

As many of these difficulties are currently overcome in molecular spectroscopy using multivariate regression methods (or multivariate chemometrics), I proposed that these three ladies should apply them to their spectra. The first reaction on their faces seemed something like ' . . . this crazy chemometrician guy . . . ', but after some discussions we agreed to work together and see what might be achieved. It was not easy to get the spectral raw data that we needed from our old Perkin-Elmer device and Mónica and María spent much time on this until they found a way to do it.

The number of papers we found reporting on the use of multivariate regression in atomic spectrometry was small, and we guessed that this might be because of either a lack of general awareness of the huge potential capabilities of these techniques and/or the difficulties in extracting the spectral data from the spectrometers, something trivial with most instruments dedicated to molecular spectrometry.

We obtained some good results and one morning I read an e-mail from Dr Merlin Fox (Commissioning Editor at the RSC) with a totally surprising proposal: to prepare a monograph on the subject. After reading his e-mail several times and asking him if that was true (some electronic 'spam' seemed very possible), Dr Carlosena and I contacted several good scientists in the two fields of atomic spectrometry and chemometrics. I am indebted to all authors

RSC Analytical Spectroscopy Monographs No. 10
Basic Chemometric Techniques in Atomic Spectroscopy
Edited by Jose Manuel Andrade-Garda

Published by the Royal Society of Chemistry, www.rsc.org

for their constructive words and immediate collaboration, although, maybe, it was the first one we contacted, Professor Alfredo Sanz-Medel (a worldwide-reputed atomic spectroscopist, with several international awards, including the 2007 Robert Kellner Lecture), who really fuelled us to go on. I really want to express my deep and sincere gratitude to each participant in this exciting project. You are not only skilful scientists but also nice persons, enthusiastic workers and, first of all, good friends. Joan, thanks for the nice photograph on the cover, you have a marvelous hobby. Recognition is also due to Merlin Fox for his continuous support and encouragement.

Finally, our thanks to you, the reader. It is the honest desire of the authors to hear from you. We would like to receive your feedback. It is impossible to produce a text like this that satisfies everyone, their expectations and needs. Many things were left out of the book, but if you feel that some more explanations, reviews or information are required, please do not hesitate to contact us. If a new version comes with the winds of the future, your suggestions will be greatly appreciated and, as far as possible, included (and publicly acknowledged, of course).

Jose Manuel Andrade-Garda
A Coruña, Galicia, Spain

Contents

RSC Analytical Spectroscopy Monographs No. 10
Basic Chemometric Techniques in Atomic Spectroscopy
Edited by Jose Manuel Andrade-Garda

Published by the Royal Society of Chemistry, www.rsc.org

Chapter 2 Implementing A Robust Methodology: Experimental Design and Optimization
Xavier Tomás Morer, Lucinio González-Sabaté, Laura Fernández-Ruano and María Paz Gómez-Carracedo

Chapter 3 Ordinary Multiple Linear Regression and Principal Components Regression
Joan Ferré-Baldrich and Ricard Boqué-Martí

Chapter 5 Multivariate Regression using Artificial Neural Networks
Jose Manuel Andrade-Garda, Alatzne Carlosena-Zubieta, María Paz Gómez-Carracedo and Marcos Gestal-Pose

Contributors

Jose Manuel Andrade-Garda
Department of Analytical Chemistry
University of A Coruña
A Coruña, Spain

Alatzne Carlosena-Zubieta
Department of Analytical Chemistry
University of A Coruña
A Coruña, Spain

Laura Fernández-Ruano
Department of Applied Statistics
Institut Químic de Sarrià
Universitat Ramon Llull
Barcelona, Spain

Marcos Gestal-Pose
Department of Information and
Communications Technologies
University of A Coruña
A Coruña, Spain

Lucinio González-Sabaté
Department of Applied Statistics
Institut Químic de Sarrià.
Universitat Ramon Llull
Barcelona, Spain

Rosario Pereiro-García
Department of Physical and
Analytical Chemistry
University of Oviedo
Oviedo, Spain

Ricard Boqué-Martí
Department of Analytical and
Organic Chemistry
University Rovira i Virgili
Tarragona, Spain

José Manuel Costa-Fernández
Department of Physical and
Analytical Chemistry
University of Oviedo
Oviedo, Spain

Joan Ferré-Baldrich
Department of Analytical and
Organic Chemistry
University Rovira i Virgili
Tarragona, Spain

María Paz Gómez-Carracedo
Department of Analytical Chemistry
University of A Coruña
A Coruña, Spain

Alfredo Sanz-Medel
Department of Physical and
Analytical Chemistry
University of Oviedo
Oviedo, Spain

Xavier Tomás Morer
Department of Applied Statistics
Institut Químic de Sarrià.
Universitat Ramon Llull
Barcelona, Spain

Structure of the Book

It was very clear from the beginning that the main objective of this project would be to present atomic spectroscopists with the basis of the most widely applied multivariate regression techniques. We did not want just '*another book on chemometrics*' and, instead, we challenged ourselves to present some practical material with clear links to common problems in atomic spectroscopy. Although mathematics were avoided as far as possible, a minimum amount was required to present correct explanations and to justify why some issues are tackled in one way rather than in another. We tried to keep technical explanations within the general scientific background that chemists receive in their careers. Besides, in our opinion, atomic spectroscopists should be conscious that things are changing so rapidly in Analytical Chemistry that some sound background on chemometrics is highly advisable. Please, consider it as another tool to be combined with your instruments, not as an end in itself.

In this respect, the first chapter is devoted to a general overview of the most common atomic spectroscopic techniques. The very basics of the analytical techniques are discussed and, most importantly, pros and cons are presented to the reader. Practical difficulties are referred to, their solutions depicted and, when possible, multivariate chemometric solutions pointed out.

The second chapter deals with a critical statement that any analyst and chemometrician has to remember: no good chemometric analysis can be obtained unless the original data are trustworthy. One of the key objectives of chemometrics is to obtain relevant information from the data, but this is possible if, and only if, the data are correct. To obtain reliable data, we can use a suite of dedicated chemometric tools aimed at developing good analytical methodologies. Thus, experimental design, optimization and robustness are cornerstones to assess accuracy during any method development process. Typical methodologies are introduced and discussed, along with extensive literature reviews that combined objective optimization and atomic spectrometry.

Chapter three presents the basic ideas of classical univariate calibration. These constitute the standpoint from which the natural and intuitive extension of multiple linear regression (MLR) arises. Unfortunately, this generalisation is not suited to many current laboratory tasks and, therefore, the problems associated with its use are explained in some detail. Such problems justify the use of other more advanced techniques. The explanation of what the multivariate space looks like and how principal components analysis can tackle it is the next step forward. This constitutes the root of the regression methodology presented in the following chapter.

Chapter four presents the most widely applied and, probably, most satisfactory multivariate regression method used nowadays: partial least squares. Graphical explanations of many concepts are given, along with the more formal mathematical background. Several common approaches to solve current problems are suggested, along with the golden rule that '*there is not a golden rule*'. The development of a satisfactory regression model can alleviate the typical laboratory workload (preparation of many standards, solutions with concomitants, *etc.*), but only when a strict and serious job is performed with the regression software. Iteration is the key word here, as the analyst has to iterate with his/her data and with the software capabilities. Validation is a key point that can never be stressed sufficiently strongly.

Finally, chapter five goes into a new regression paradigm: artificial neural networks. Quite different from the other regression methods presented in the book, they have gained acceptance because they can handle non-linear systems and/or noisy data. This step forward is introduced briefly and, once again, a review is presented with practical applications in the atomic spectroscopy field. Not surprisingly, most papers referred to deal with complex measurements (*e.g.*, non-linear calibration or concomitant-affected measurements in ETAAS) and/or new analytical atomic techniques (which, therefore, yield very complex data, *e.g.*, X-ray fluorescence in complex systems and several SIMS-based methodologies).

CHAPTER 1

A General Overview of Atomic Spectrometric Techniques

ALFREDO SANZ-MEDEL, ROSARIO PEREIRO AND JOSÉ MANUEL COSTA-FERNÁNDEZ

Department of Physical and Analytical Chemistry, University of Oviedo, Oviedo, Spain

1.1 Introduction: Basis of Analytical Atomic Spectrometric Techniques

Analytical atomic spectrometry comprises a considerable number of techniques based on distinct principles, with different performance characteristics and hence with varied application scopes, but in all cases providing elemental chemical information about the composition of samples. Figure 1.1 shows that these techniques can be classified into three main groups according to the type of particle detected: *optical spectrometry*, where the intensity of either non-absorbed photons (absorption) or emitted photons (emission and fluorescence) is detected as a function of photon energy (in most cases, plotted against wavelength); *mass spectrometry* (MS), where the number of atomic ions is determined as a function of their mass-to-charge ratio; and *electron spectroscopy*, where the number of electrons ejected from a given sample is measured according to their kinetic energy, which is directly related to the bonding energy of the corresponding electron in a given atom.

X-ray photoelectron spectroscopy (XPS) and Auger electron spectroscopy (AES) are the two main techniques based on electron spectroscopy. In XPS, a source of photons in the X-ray energy range is used to irradiate the sample.

RSC Analytical Spectroscopy Monographs No. 10
Basic Chemometric Techniques in Atomic Spectroscopy
Edited by Jose Manuel Andrade-Garda

Published by the Royal Society of Chemistry, www.rsc.org

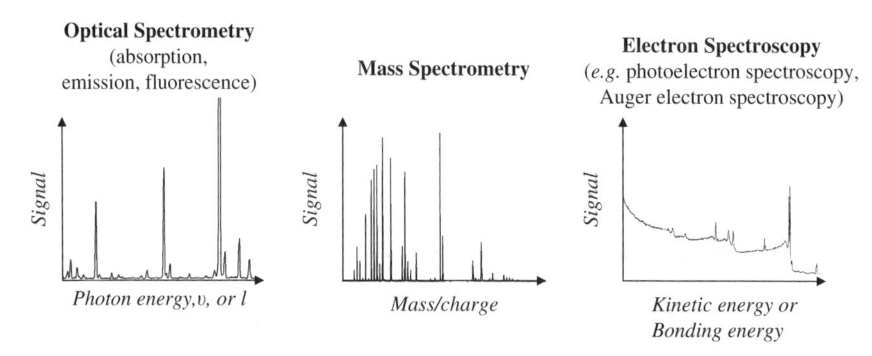

Figure 1.1 Classification of spectrometries according to the detection principle.

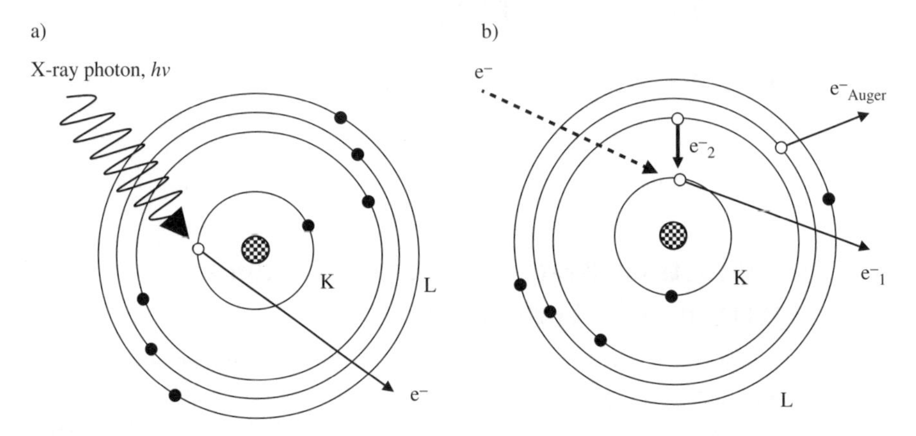

Figure 1.2 Electron spectroscopy. (a) Schematic representation of the XPS process. (b) Schematic representation of the processes for auger electron emission.

Superficial atoms emit electrons (called photoelectrons) after the direct transfer of energy from the photon to a core-level electron (see Figure 1.2a). Photoelectrons are subsequently separated according to their kinetic energy and counted. The kinetic energy will depend on the energy of the original X-ray photons (the irradiating photon source should be monochromatic) and also on the atomic and, in some cases, the molecular environment from which they come. This, in turn, provides important information about oxidation states and chemical bonds as the stronger the binding to the atom, the lower is the photoelectron kinetic energy.

In an Auger process, the kinetic energy of the emitted electron does not depend on the energy of the excitation source. AES consists of a two-step process: first, the sample is irradiated with an electron beam (or, less commonly, with X-rays), which expels an inner electron (e^{-}_{1}). In a second step, the relaxation of the excited ion takes place through the fall of a more external electron (e^{-}_{2}) to fill the 'hole', and then a third electron (e^{-}_{Auger}) uses the energy released in that movement to exit the atom (Figure 1.2b). XPS and AES are considered powerful techniques for surface analysis, with good depth and lateral resolution. However, due to their narrow range of applications in qualitative studies and the scarcity of quantitative analyses, they will not be considered further in this chapter.

The aim of this chapter is, therefore, to introduce briefly the most common quantitative atomic techniques based on both optical and mass spectrometric detection. The main emphasis will be given to conceptual explanations in order to stress the advantages and disadvantages of each technique, the increase in the complexity of the data they generate and how this can be addressed. References to chemometric tools presented in the following chapters will be given.

For these techniques, a dissolved sample is usually employed in the analysis to form a liquid spray which is delivered to an atomiser (*e.g.* a flame or electrically generated plasma). Concerning optical spectrometry, techniques based on photon absorption, photon emission and fluorescence will be described (Section 1.2), while for mass spectrometry (MS) particular attention will be paid to the use of an inductively coupled plasma (ICP) as the atomisation/ionisation source (Section 1.3). The use of on-line coupled systems to the above liquid analysis techniques such as flow injection manifolds and chromatographic systems will be dealt with in Section 1.4 because they have become commonplace in most laboratories, opening up new opportunities for sample handling and pretreatment and also to obtain element-specific molecular information.

Finally, direct solid analysis by optical and mass spectrometry will be presented in Section 1.5. This alternative is becoming more appealing nowadays and implemented in laboratories because of the many advantages brought about by eliminating the need to dissolve the sample. Techniques based on the use of atomiser/excitation/ionisation sources such as sparks, lasers and glow discharges will be briefly described in that section.

1.2 Atomic Optical Spectrometry

Routine inorganic elemental analysis is carried out nowadays mainly by atomic spectrometric techniques based on the measurement of the energy of photons. The most frequently used photons for analytical atomic spectrometry extend from the ultraviolet (UV: 190–390 nm) to the visible (Vis: 390–750 nm) regions. Here the analyte must be in the form of atoms in the gas phase so that the photons interact easily with valence electrons. It is worth noting that techniques based on the measurement of X-rays emitted after excitation of the sample with X-rays (*i.e.* X-ray fluorescence, XRF) or with energetic electrons (electron-probe X-ray micro-analysis, EPXMA) yield elemental information directly from solid samples, but they will not be explained here; instead, they will be briefly treated in Section 1.5.

The measurement of analytes in the form of gaseous atoms provides atomic spectra. Such spectra are simpler to interpret than molecular spectra (since atoms cannot rotate or vibrate as molecules do, only electronic transitions can take place when energy is absorbed). Atomic spectra consist of very narrow peaks (*e.g.* a few picometres bandwidth) providing two types of crucial analytical information: the observed wavelength (or frequency or photon energy), which allows for qualitative analysis, and the measurement of the peak height or area at a given frequency, which provides quantitative information about the particular element sought. The relative simplicity of such atomic spectra and the fairly straightforward qualitative and quantitative information have led to the enormous practical importance of atomic optical spectrometry for inorganic elemental analysis. However, it should be stressed again that to obtain such spectra the analytes must be converted into atoms which will absorb or emit photons of UV–Vis radiation and so an 'atomiser', for example a dynamic medium of high temperature where molecules containing the analyte are broken down into individual gaseous atoms, is needed.

1.2.1 Classification of Techniques: Absorption, Emission and Fluorescence

The interaction processes between UV–Vis photons and the outer electrons of the atoms of the analytes can be understood using quantum mechanics theory. In the thermodynamic equilibrium between matter and interacting electromagnetic radiation, according to the radiation laws postulated by Einstein, three basic processes between two stable energy levels 1 and 2 are possible. These processes, which can be defined by their corresponding transition probabilities, are summarised in Figure 1.3.

- *Spontaneous emission of photons*. This process refers to a spontaneous transition of the electron from the excited state 2 to the lower energy state 1 with emission of a photon of frequency $\nu_{12} = (E_2 - E_1)/h$. This process

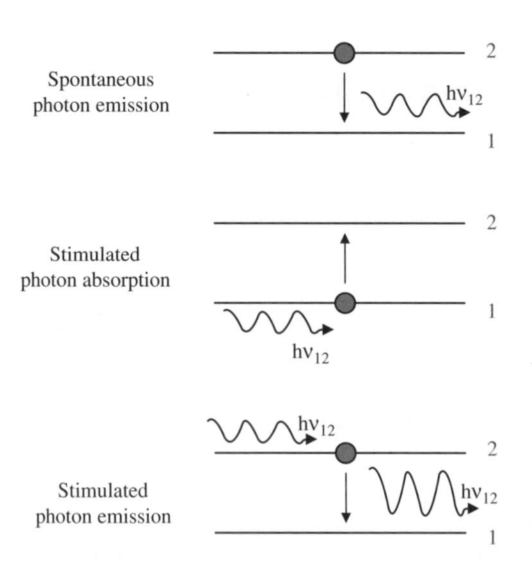

Figure 1.3 Basic interaction processes between matter and interacting electromagnetic radiation.

constitutes the photophysical basis of atomic emission spectrometry, which will be termed here optical emission spectrometry in order to use the acronym OES instead of AES because the latter acronym can be confused with that for Auger electron spectroscopy.

- *Stimulated absorption of photons.* In this case, the electronic transition takes place from state 1 to state 2 in response to the action of an external radiation of the appropriate frequency. Atomic absorption spectrometry (AAS) is based on this process. On the other hand, atomic fluorescence spectrometry (AFS) corresponds to the sequential combination of a stimulated absorption followed by spontaneous emission.
- *Stimulated emission of photons.* This process consists of electronic transitions from the excited energy level to the lower one stimulated by an external radiation of the appropriate frequency $(E_2 - E_1)/h$ and constitutes the basis of the laser (light amplification by stimulated emission of radiation) phenomenon.

Atomic lines can arise from electronic transitions in neutral atoms or in atomic ions (in general, atomic lines for a given element M are denoted M I, whereas their ionic counterparts are denoted M II). The transitions of outer electrons of an atom may be represented as vertical lines on an 'energy level' diagram, where each energy level of the outer electron possesses a given energy and is represented by a horizontal line. For example, Figure 1.4 shows the diagram for the neutral sodium atom (the wavelengths corresponding to the transitions in the diagram are expressed in ångströms, Å). The energy scale is linear in electronvolt (eV) units, assigning a value of zero to the 3s orbital. The scale extends up to about 5.2 eV, which corresponds to the energy necessary to extract the 3s electron and so to produce a sodium ion. All electronic transitions ending on the same energy level are usually called 'series', the most likely ones being those ending in the lowest possible energy level (the ground state) of the electron in the atom.

The light coming from such transitions is separated according to its frequency (or its wavelength, λ) and the intensity observed for each frequency measured electronically (*e.g.* with a photomultiplier tube). Thus, if the observed intensity of the emitted light is plotted against the frequency (or wavelength) of the corresponding transition (line), an 'atomic emission' spectrum is obtained (see Figure 1.1). Similarly, an 'atomic fluorescence' spectrum would be the plot of the measured intensity (coming from atoms excited by appropriate electromagnetic radiation) as a function of the frequency of the emitted radiation. Finally, if stimulated absorption of light in response to an electronic transition between a lower and a higher energy level is measured, a plot of 'percent absorption versus frequency of the light' can be drawn; such a plot represents an 'atomic absorption' spectrum.

The atomic lines in the spectrum appear as vertical lines or 'peaks' due to the nature of the transition involved. That is, in molecules an electronic transition is usually accompanied by simultaneous changes in the molecule vibrational and rotational energy levels; sometimes all the three energy types may change

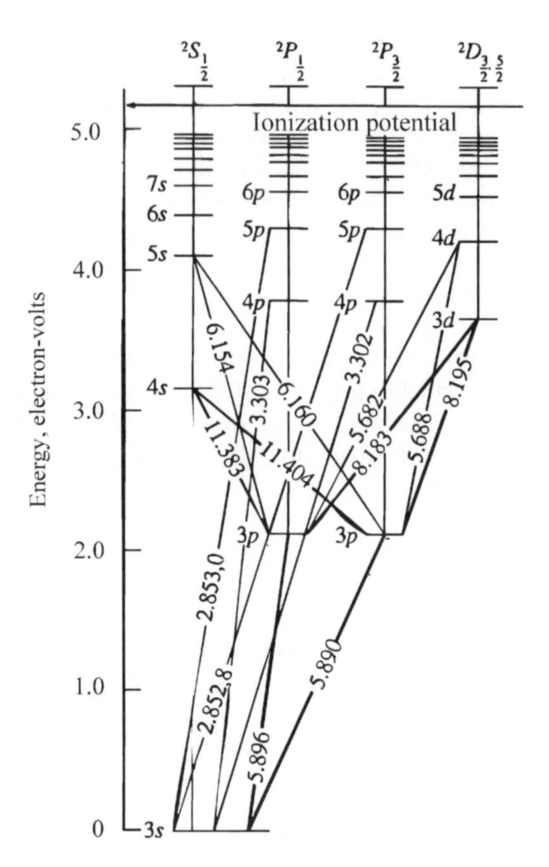

Figure 1.4 Diagram of energy levels and electronic transitions for atomic sodium.

simultaneously in an electronic transition in a molecule. The many transition possibilities allowed in this way and the solvent effect derived from the aggregation state of the sample (the 'excited' sample is in liquid form) determines that in UV–Vis molecular absorption (or emission) the corresponding 'peaks' in the spectrum are widely broadened. Typically, the half-bandwidth of an absorption 'band' in such molecular UV–Vis spectra is around 40 nm (or 400 Å), whereas in atomic 'lines' the half-bandwidth observed, as a result of pure electronic transitions, is a few hundredths of an ångström (typically 0.03–0.05 Å).

Thus, spectral interferences in atomic spectroscopy are less likely than in molecular spectroscopy analysis. In any case, even the atomic 'lines' are not completely 'monochromatic' (*i.e.* only one wavelength per transition). In fact, there are several phenomena which also bring about a certain 'broadening'. Therefore, any atomic line shows a 'profile' (distribution of intensities) as a function of wavelength (or frequency). The analytical selectivity is conditioned by the overall broadening of the lines (particularly the form of the wings of such atomic lines).

The selection of the most appropriate atomic line among all possible transitions for qualitative and quantitative purposes is critical. For most elements, the 'resonance' atomic lines (*i.e.* when the lowest energy level in the corresponding transition is the fundamental or 'ground state' level, $E_0 = 0$) are the most sensitive ones in flames and they are used in the majority of flame methods. However, with plasma sources (commonly used in combination with OES), the choice is more difficult because several emission lines from neutral atoms or atomic ions of the same element may appear useful. Often, the expected concentration range will dictate whether to use a neutral atom resonance line, an ion line or a line arising from transitions between excited atomic states. Resonance lines are useful for trace constituents, but they are susceptible to self-absorption of emitted radiation at high concentrations (this effect is due to an excess of analyte atoms in the ground-state level). Lines of lower relative intensities are often used for minor and major constituents. Moreover, the abundance of nearby, potentially interfering lines from other elements, has to be assessed carefully.

1.2.1.1 Atomic Absorption Spectrometry. Principles of Quantitative Analysis

For quantitative purposes in AAS, a magnitude called transmittance (T) which relates, for a given wavelength, the intensity (measured by the detector) of the light source (I_0) and the intensity not absorbed which has passed through the atomiser or transmitted light (I) is used:

$$T = \frac{I}{I_0} \tag{1.1}$$

The amount of light absorbed is a function of the so-called absorption coefficient (k') and of the optical pathlength in the atomiser cell (b); k' depends on the frequency of the selected analytical line and on the concentration of the analyte absorbing atoms. The general absorbance law (Lambert–Beer–Bouguer law) relates transmittance (and so measured intensities I and I_0) to k' and b through the following equation:

$$T = \mathrm{e}^{-k'b} \tag{1.2}$$

The parameter used in the analytical calibrations by AAS is absorbance (A), which is linearly related to k' (that is, at a given λ, with the atomic concentration of the analyte in the atomiser) and with the length of the optical path:

$$A = -\log T = \log 1/T = k'b \log e = 0.434k'b \tag{1.3}$$

For a given set of experimental conditions in an absorption experiment, we obtain

$$\mathrm{A} = \text{constant} \times b \times N_0 \tag{1.4}$$

N_0 being the analyte atom density (number of atoms per unit volume) in the ground state in the atomiser. The relationship between the atom concentration per unit volume ($N_T \approx N_0$) and the concentration of the analyte in the sample, C, is linear under fixed working conditions for a given line of the analyte. Therefore, we can establish a linear relationship between absorbance and C:

$$A = KbC \tag{1.5}$$

1.2.1.2 Optical Emission Spectrometry. Principles of Quantitative Analysis

Optical emission spectrometry is one of the oldest physical methods of analysis enabling multielement determinations. In this process, free atoms which are generated by thermal dissociation of the sample material are excited or ionised and excited additionally (several collisions or other processes may be responsible for delivering the required energy). The higher the temperature, the higher is the percentage of excited analyte species (at least, in general) and the higher the emission intensity. The Boltzmann equation relates the temperature (T) with the number of atoms in an energy state E_0 and an excited state E_q, provided that the source is in a so-called thermal equilibrium, as

$$n^*/n_0 = \frac{g_q}{g_0} e^{-(E_q - E_0)/k_B T} \tag{1.6}$$

where n_0 is the number of atoms in the energy level E_0, n^* the number of atoms in an energy state E_q, k_B the Boltzmann's constant, g_q and g_0 the statistical weights for each energy state (E_q and E_0) and T the temperature in Kelvin.

The flames commonly used as atomisers have temperatures in the range 2000–3000 K allowing for the analysis of elements such as Na, K and Cs by OES. The flame temperatures are not high enough to excite many other elements, so other atomisers such as spectroscopic plasmas have to be used.

Linear (straight-line) relationships can be easily achieved between the emission intensity of a given transition and the total atomic concentration of the element in the atomisation/excitation system. However, under certain conditions, spectral lines from resonance transitions can display a phenomenon called self-absorption, giving rise to non-linearity in calibration graphs at high concentrations. If changes to the experimental setup cannot correct the problem, it will cause difficulties in classical linear calibration, although they can be solved by multivariate calibration techniques (some of which are introduced in the following chapters).

1.2.1.3 Atomic Fluorescence Spectrometry. Principles of Quantitative Analysis

AFS involves the emission of photons from an atomic vapour that has been excited by photon absorption. For low absorbance signals (and thus for low analyte concentrations), the following linear relationship applies:

$$I_F = 2.3K'AI_0 \tag{1.7}$$

where I_F is the fluorescence intensity, A the absorbance and I_0 the intensity of the light excitation source. K' depends on the quantum efficiency of the fluorescence process (*i.e.* the ratio between the number of atoms emitting fluorescence and the number of excited atoms). Considering eqn 1.5:

$$I_F = 2.3K'KbCI_0 \tag{1.8}$$

$$I_F = k'CI_0 \tag{1.9}$$

Therefore, I_F depends on the concentration of analyte atoms in the atomiser and on I_0 (in fact, much of the research on AFS as an analytical technique has involved the development of stable and intense suitable light sources). Using a constant light excitation output, linear calibration graphs can be achieved for low analyte concentrations:

$$I_F = kC \tag{1.10}$$

1.2.2 A Comparative View of Basic Instrumentation

The basic components of AAS, OES and AFS instruments are illustrated by the simple schematics shown in Figure 1.5. They need an atomiser to convert the analyte contained within the sample to gaseous atoms. A device is also required to separate the electromagnetic radiations arising from the atomiser and a 'light read-out' system, which is integrated by a transducer or detector (transforming the light intensity into a measurable signal, *e.g.* an electric current), and a electronic read-out system.

In AAS (Figure 1.5a) the external energy is provided by a light source in a straight-line optical axis configuration. Figure 1.5b shows that the basic instrumental components in AFS are the same as in AAS, only the geometry of the arrangement changes as the external light source used for analyte photo-excitation has been rotated 90° (with respect to the straight-line optical axis used in absorption measurements) to minimise the collection of scattered light from the excitation source. Finally, as depicted in Figure 1.5c, OES measurements do not use any external light source since the sample is excited in the atomiser by the energy provided by a flame, a plasma (*i.e.* a hot, partially ionised gas), *etc.*

Based on the configurations in Figure 1.5, many analytical techniques have been developed employing different atomisation/excitation sources. For example, two powerful AAS techniques are widespread: one uses the flame as atomiser (FAAS) whereas the other is based on electrothermal atomisation (ETAAS) in a graphite furnace. Although the flame has limited application in OES, many other analytical emission techniques have evolved in recent decades based on different atomisation/excitation plasma sources.

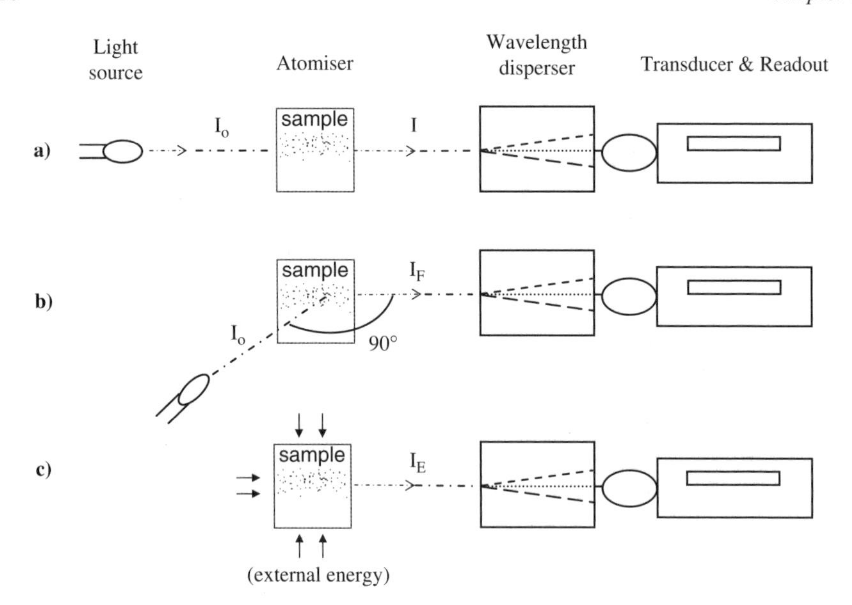

Figure 1.5 Schematics of basic components of analytical techniques based on atomic optical spectrometry. (a) Atomic absorption spectrometry; (b) atomic fluorescence spectrometry; (c) atomic emission spectrometry.

Concerning AFS, the atomiser can be a flame, plasma, electrothermal device or a special-purpose atomiser (*e.g.* a heated quartz cell). Nowadays, commercially available equipment in AFS is simple and compact, specifically configured for particular applications (*e.g.* determination of mercury, arsenic, selenium, tellurium, antimony and bismuth). Therefore, particular details about the components of the instrumentation used in AFS will not be given in this chapter.

1.2.2.1 Atomic Absorption Spectrometry. Instrumentation

Figure 1.6a shows the simplest configuration of an atomic absorption spectrometer, called a 'single-beam spectrometer'. As can be seen, the lamp, the atomiser and the wavelength selector (most wavelength selectors used in AAS are monochromators) are aligned. The selected wavelength is directed towards a detector (*e.g.* a photomultiplier tube), producing a signal proportional to the light intensity. To remove the atomiser continuum emission, the radiation source is modulated to provide a means of selectively amplifying modulated light coming from the lamp while the continuous emission from the atomiser is disregarded. Source modulation can be accomplished with a rotating chopper (mechanical modulation) located between the lamp and the atomiser or by pulsing the lamp (electronic modulation). Synchronous detection eliminates the unmodulated dc signal emitted by the atomiser and so measures only the amplified ac (modulated) signal coming from the lamp.

If a single beam is used, a blank sample containing no analyte should be measured first, setting its absorbance to zero. If the lamp intensity changes when

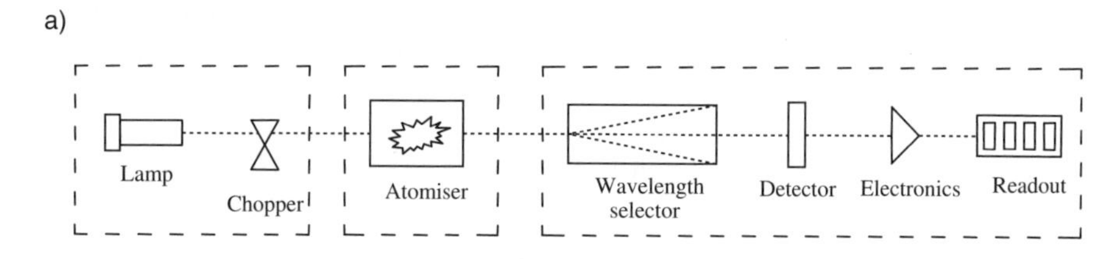

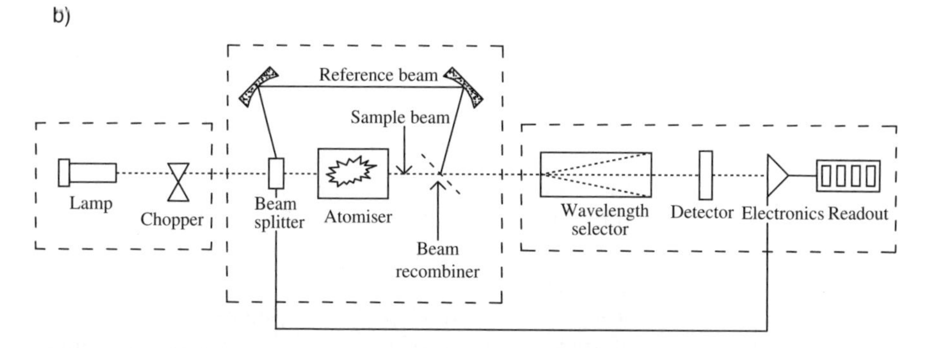

Figure 1.6 Configurations of instruments for atomic absorption spectrometry. (a) Single-beam spectrometer; (b) double-beam spectrometer.

the sample is put in place, the measured absorbance will be inaccurate. An alternative configuration is the 'double-beam spectrometer', which incorporates a beamsplitter so that part of the beam passes through the atomiser and the other part acts as a reference (Figure 1.6b), allowing for a continuous comparison between the reference beam and the light passing through the atomiser.

By far the most common lamps used in AAS emit narrow-line spectra of the element of interest. They are the hollow-cathode lamp (HCL) and the electrodeless discharge lamp (EDL). The HCL is a bright and stable line emission source commercially available for most elements. However, for some volatile elements such as As, Hg and Se, where low emission intensity and short lamp lifetimes are commonplace, EDLs are used. Boosted HCLs aimed at increasing the output from the HCL are also commercially available. Emerging alternative sources, such as diode lasers [1] or the combination of a high-intensity source emitting a continuum (a xenon short-arc lamp) and a high-resolution spectrometer with a multichannel detector [2], are also of interest.

The radiation absorbed or scattered from the light source by atoms or molecules different from the analyte will give rise to a background absorption which will add to the specific absorption of the analyte. Contributions to the background in AAS can arise from spectral interferences due to a spectral line of another element within the bandpass of the wavelength selector (such a possibility is rather uncommon in AAS; besides, such interferences are now well characterised), absorption by molecular species originating from the sample and light scattering from particles present in the atomiser. Therefore, to determine accurately the absorbance due to the analyte, subtraction of the background from the total absorbance measured in the spectrometer is necessary. In most cases instrumental means of measuring and correcting for this background absorption can be used, such as deuterium or Zeeman-based background correctors [3]. However, instrumental background correction has limitations and it is important to keep in mind that the ideal for background correction should be to be able to measure a true blank solution. Multivariate calibration techniques have very powerful resources to cope with this problem.

Two atomisers are generally used in AAS to produce atoms from a *liquid* or *dissolved* sample:

1. A flame, where the solution of the sample is aspirated. Typically, in FAAS the liquid sample is first converted into a fine spray or mist (this step is called nebulisation). Then, the spray reaches the atomiser (flame) where desolvation, volatilisation and dissociation take place to produce gaseous free atoms. Most common flames are composed of acetylene–air, with a temperature of ~2100–2400 °C, and acetylene–nitrous oxide, with a temperature of ~2600–2900 °C.
2. An electrothermal atomiser, where a drop of the liquid sample is placed in an electrically heated graphite tube which consists of a cylinder (3–5 cm in length and a few millimetres in diameter) with a tiny hole in the centre of the tube wall for sample introduction (see Figure 1.7a). Both ends of the

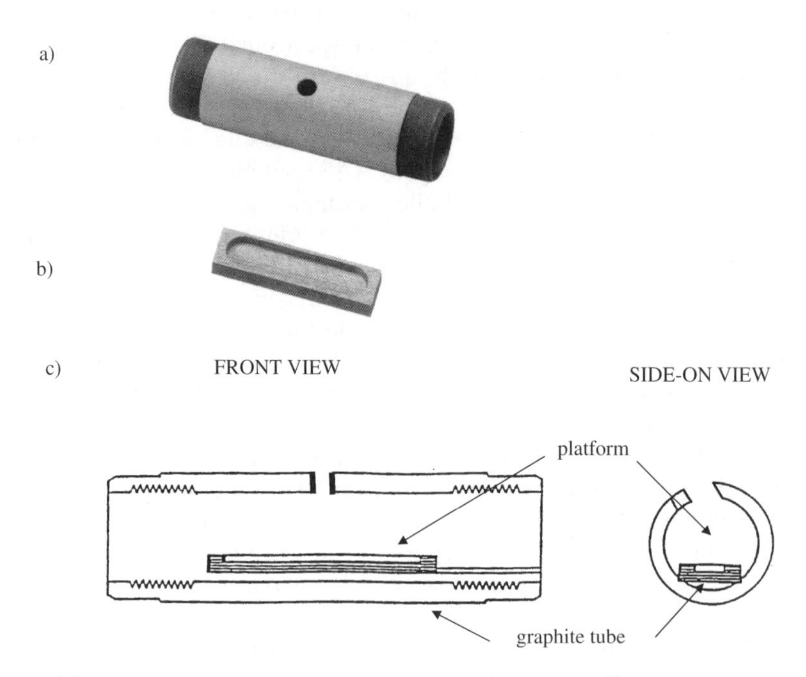

Figure 1.7 Electrothermal atomisation atomic absorption spectrometry. (a) Photograph of a graphite tube. (b) Photograph of a L'vov platform. (c) Schematic front and side-on views of a graphite tube with a L'vov platform.

tube are open to allow the light from the lamp to pass through and for the removal of the sample constituents after the analysis. Atomisers used in commercial ETAAS instruments are most commonly made of pyrolytically coated electrographite. A platform (the so-called L'vov platform) to deposit the sample within the tube is used (Figure 1.7b and c). The platform has a finite heat capacity and is heated primarily by tube radiation, so its temperature will, in principle, be lower than that of the tube wall. Hence the sample deposited on it will be volatilised later in time (relative to direct wall atomisation), at higher gas-phase temperatures which will favour isothermal atom formation, reducing interference effects from temporal non-isothermal conditions typical in wall atomisation.

An ETAAS determination starts by dispensing a known volume of sample into the furnace. The sample is then subjected to a multi-step temperature programme by increasing the electric current through the atomiser body. The heating steps include drying, pyrolysis and atomisation. Unfortunately, during the atomisation step any organic material still present in the graphite tube is pyrolysed, producing smoke which may cause severe attenuation of the light beam. Further, the presence of many salts in the tube can give rise to large background absorbance when atomised at high temperatures. The addition of appropriate chemicals known as matrix modifiers and the use of instrumental methods of background correction have been crucial to overcoming these effects; however, problems still remain for some analytes and some types of samples.

Gaseous and volatilised analytes can also be easily determined by FAAS and ETAAS. For example, the determination of several elements by the formation of covalent volatile hydrides (*e.g.* arsenic, selenium) and cold vapour generation (mercury and cadmium) is feasible with good analytical sensitivity (see Section 1.4.1.1).

Solid sampling can be implemented for both FAAS and, especially, for ETAAS (sample particle size effects are less critical in ETAAS than in nebulisation-based techniques because it offers longer residence times). The availability of commercial instrumentation supporting solids analysis using the graphite furnace has led to its successful application, in particular for fast screening analysis with high sensitivity (absence of any dilution) and minimal risk of contamination [4]. Unfortunately, some problems are associated with direct solid sampling; for example, the small sample sizes required are frequently not representative enough and also matrix-matched standards are usually needed. These inconveniences can be minimised by slurry sampling [5]; however, some limitations remain and a critical one is the need to maintain the stability of the slurry until sample injection. Further, particle size may affect the reproducibility and accuracy if it is non-homogeneous or too large. On the other hand, owing to the absence of a pretreatment step to eliminate the matrix, the molecular absorption signal is often rather high and structured. Another limitation is the probable build-up of carbonaceous residues in the atomiser, reducing its lifetime.

1.2.2.2 Atomic Emission Spectrometry. Instrumentation

Flames and plasmas can be used as atomisation/excitation sources in OES. Electrically generated plasmas produce flame-like atomisers with significantly higher temperatures and less reactive chemical environments compared with flames. The plasmas are energised with high-frequency electromagnetic fields (radiofrequency or microwave energy) or with direct current. By far the most common plasma used in combination with OES for analytical purposes is the inductively coupled plasma (ICP).

The main body of an ICP consists of a quartz torch (15–30 mm in diameter) made of three concentric tubes (see Figure 1.8) and surrounded externally by an induction coil that is connected to a radiofrequency generator commonly operating at 27 MHz. An inert gas, usually argon, flows through the tubes. The spark from a Tesla coil is used first to produce 'seed' electrons and ions in the region of the induction coil. Subsequently the plasma forms, provided that the flow patterns are adequate inside the torch, giving rise to high-frequency currents and magnetic fields inside the quartz tube. The induced current heats the support gas to a temperature of the order of 7000–8000 K and sustains the ionisation necessary for a stable plasma. Usually, an aerosol from the liquid sample is introduced through the central channel transported by an argon flow

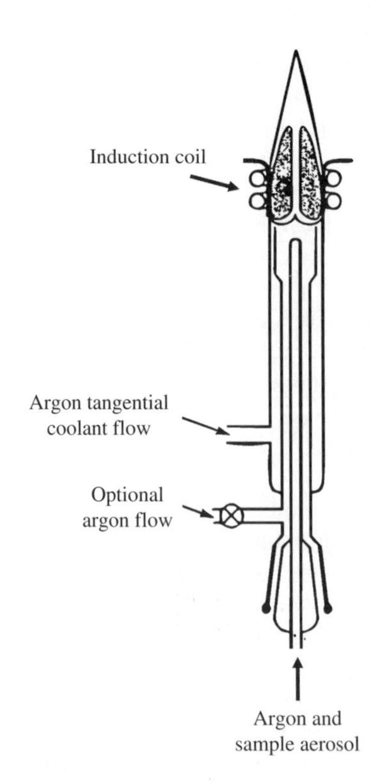

Figure 1.8 Schematic view of an ICP torch.

of about 1 l min^{-1}. A much higher Ar flow velocity (about 10 l min^{-1}) is introduced tangentially to prevent overheating. Because of efficient desolvation and volatilisation in the ICP, this atomiser/excitation source is commonly applied for the analysis of dissolved samples. The high temperatures and the relative long residence time of the atoms in the plasma (2–3 ms) lead to nearly a complete solute vaporisation and high atomisation efficiency. Accordingly, although matrix and inter-element effects should be relatively low, it has been observed that sometimes they are significant. Further, the high excitation capacity of this source gives rise to very rich spectra, so a careful assessment of potential spectral interferences is essential. On the other hand, the ICP emission frequently has an important background due to *bremsstrahlung* (*i.e.* continuous radiation produced by the deceleration of a charged particle, such as an electron, when deflected by another charged particle, such as an atomic nucleus) and to electron–ion recombination processes.

For a given ICP-OES instrument, the intensity of an analyte line is a complex function of several factors. Some adjustable parameters that affect the ICP source are the radiofrequency power coupled into the plasma (usually about 1 kW), the gas flow rates, the observation height in the lateral-viewing mode and the solution uptake rate of the nebuliser. Many of these factors interact in a complex fashion and their combined effects are different for dissimilar spectral lines. The selection of an appropriate combination of these factors is of critical importance in ICP-OES. This issue will be addressed in Chapter 2, where experimental designs and optimisation procedures will be discussed. Many examples related to ICP and other atomic spectrometric techniques will be presented.

Concerning the detection of the emitted light, the usual configuration used for signal collection is the lateral view of the plasma. However, nowadays, most instrument companies offer also, at least an ICP system based on axial viewing. For a given system, the use of axial viewing will improve the limits of detection compared with those obtained with lateral viewing, roughly by an order of magnitude. However, axial viewing has a poor reputation in terms of matrix effects and self-absorption phenomena. Chapter 2 presents several examples of the optimisation of ICP detection devices.

The availability of solid-state detectors (such as the charge-coupled detector, CCD) makes it possible to acquire simultaneously significant portions of the spectra or even the entire rich spectra obtained by ICP-OES in the UV–Vis region, thus providing a large amount of data. The commercial availability of ICP-OES instruments with these multichannel detectors has significantly renewed interest in this technique. However, some limitations, such as the degradation of the spectral resolution compared with photomultiplier-based dispersive systems, still remain.

ICP-OES has enjoyed a long-lasting success, with several companies marketing versatile and robust instruments which are being used for various research and routine applications in many laboratories worldwide. However, there is still a demand for improvement. It is expected that most future improvements will be related to more efficient data processing to take full benefit of the available emitted information. In particular, the availability of the entire UV–Vis spectra should improve the reliability of the analytical results through the use of several lines per element and through a better understanding of matrix effects [6]. Therefore, new alternatives are required for data treatment and calibration. It is the authors' opinion that some of the techniques presented in the next chapters will be of great importance. In particular, several seminal studies have applied multivariate regression to ICP data, and also pattern recognition techniques (*e.g.* principal component analysis, PCA). More details will be presented in the corresponding sections.

Examples of other plasmas which have been used in combination with OES are the following:

- *Microwave-induced plasma (MIP):* this consists of an electrodeless microwave cavity plasma. Like the ICP, an MIP is initiated by providing 'seed' electrons. The electrons oscillate in the microwave field and gain sufficient energy to ionise the support gas by collisions. Large amounts of sample or solvent vapour can result in considerable changes in plasma impedance and thus coupling efficiency. MIPs operate at 2450 MHz (this is the frequency usually available in commercial microwave generators) and at substantially lower powers than ICP devices. The low power levels do not provide plasmas of sufficient energy to get an efficient desolvation of solutions and, hence, MIPs have been used mostly with vapour-phase sample introduction (*e.g.* as detectors for gas chromatography). Sample introduction difficulties have been primarily responsible for the lower popularity of MIPs compared with ICPs.
- *Direct current plasma (DCP):* this is produced by a dc discharge between electrodes. DCPs allow the analysis of solutions. Experiments have shown that although excitation temperatures can reach 6000 K, sample volatilisation is not complete because residence times in the plasma are relatively short (this can be troublesome with samples containing materials that are difficult to volatilise). A major drawback is the contamination introduced by the electrodes.
- *Furnace atomisation plasma emission spectrometry (FAPES):* this consists of an atmospheric pressure source combining a capacitively coupled radiofrequency helium plasma formed inside a graphite tube which contains an axial powered electrode. This miniplasma has rarely been used in analytical atomic spectrometry, probably because of the small number of users and a lack of information about its applications and capabilities [7].

1.2.3 Analytical Performance Characteristics and Interferences in the Different Techniques

This section starts with a discussion of selectivity for the most extended analytical atomic techniques based on optical spectrometry. Then, aspects such as detection limits (DLs), linear ranges, precision, versatility and sample throughput will be presented. The section ends with a brief comparison of the performances of the most common optical techniques for atomic analysis of dissolved samples.

1.2.3.1 Selectivity

To understand the analytical selectivity of atomic spectroscopic methods, a basic knowledge of the different sources of interferences which may be encountered is essential. Therefore, the concept and relative magnitude of each interference will be described next and compared for the three main atomic detection modes. The following discussion is a sort of basic platform to understand and assess potential sources of error in any atomic technique we might need in our laboratory.

Spectral interferences. These interferences result from the inability of an instrument to separate a spectral line emitted by a specific analyte from light emitted by other neutral atoms or ions. These interferences are particularly serious in ICP-OES where atomic spectra are complex because of the high temperatures of the ICP. Complex spectra are most troublesome when produced by the major constituents of a sample. This is because spectral lines from other analytes tend to be overlapped by lines from the major elements. Examples of elements that produce complex line spectra are Fe, Ti, Mn, U, the lanthanides and noble metals. To some extent, spectral complexity can be overcome by the use of high-resolution spectrometers. However, in some cases the only choice is to select alternative spectral lines from the analyte or use correction procedures.

Physical (transport) interferences. This source of interference is particularly important in all nebulisation-based methods because the liquid sample must be aspirated and transported reproducibly. Changes in the solvent, viscosity, density and surface tension of the aspirated solutions will affect the final efficiency of the nebulisation and transport processes and will modify the final density of analyte atoms in the atomiser.

Chemical interferences. An important type of chemical interference in atomic spectrometry is due to the presence or formation in the atomiser of analyte refractory compounds. These interferences are probably the most serious ones when comparatively low-temperature atomisers (such as flames and graphite furnaces) are employed. The reduction of analyte atoms (which become trapped in the refractory molecule) will bring about a decrease in the analytical signal. Typical examples are phosphate interferences in determinations of Ca and Mg by flame-based methods (phosphates can form and they are only partially dissociated at normal flames temperatures). Another good illustration is the determination of elements such as Al, Si, Zr, Ta and Nb. For these 'refractory' elements, the use of hottest flames than those obtained with air–acetylene is needed, such as nitrous oxide–acetylene flames to allow the dissociation of the corresponding refractory oxides and hydroxides. Apart from 'hottest' atomisers, an alternative way to overcome these interferences is to resort to 'releasing agents' such as chemical reagents (*e.g.* organic chelating compounds, such as 8-hydroxyquinoline) that are able to form compounds with the analyte which are easily dissociated at the usual temperatures of the flame.

Another type of interference that can arise in the atomiser is called 'ionisation interferences'. Particularly when using hot atomisers, the loss of an electron from the neutral atom in metals with low ionisation energy may occur, thus reducing the free atom population (hence the sensitivity of the analyte determination, for which an atomic line is used, is reduced). These interferences can be suppressed in flames by adding a so-called 'ionisation suppressor' to the sample solution. This consists in adding another element which provides a great excess of electrons in the flame (*i.e.* another easily ionisable element). In this way, the ionisation equilibrium is forced to the recombination of the ion with the electron to form the metal atom. Well-known examples of such buffering compounds are salts of Cs and La widely used in the determination of Na, K and Ca by FAAS or flame OES.

In ICP-OES, it has been observed that analyte lines with high excitation potentials are much more susceptible to suffer matrix effects than those with low excitation potentials. The effect seems to be related to the ionisation of the matrix element in the plasma, but in fact it is a rather complicated and far from fully characterised effect [8,9]. Therefore, calibration strategies must be carefully designed to avoid problems of varying sensitivity resulting from matrix effects. A possible approach may be to combine experimental designs and multivariate calibration, in much the same way as in the case study presented in the multivariate calibration chapters.

Temperature variations in the atomiser. Variations in the temperature of the atomiser will change the population of atoms giving rise to atomic absorption, but they affect particularly the population of excited atoms, essential for OES measurements.

Light scattering and unspecific absorptions. Both of these problems occur only in AAS and AFS. When part of the light coming from the lamp is scattered by small particles in the atomiser (*e.g.* droplets or refractory solid particles) or absorbed unspecifically (*e.g.* by undissociated molecules existing in the flame), important analytical errors would be derived if no adequate corrections were made. Scattered and dispersed radiation decrease the lamp intensity and create false analytical signals. Fortunately, both sources of 'false signals' can be easily distinguished from the 'specific' analyte signals which do occur at the analytical line only (and not outside it in the spectrum) and this basic differential feature can be used for correction.

1.2.3.2 Detection Limits, Linear Ranges and Precision

Detection limits (DLs) are the most common figure used to compare analytical techniques (or methods) from the detection power point of view. DLs depend on the analyte and, for some techniques, also on the matrix. Further, DLs depend on the quality of the instrumentation (which is being continuously improved). Therefore, here just some approximate ranges will be given. DLs obtained by AAS are of the order of $mg\,l^{-1}$ (ppm). These figures are much improved by using ETAAS; in this case, typical DLs are usually lower than $1\,\mu g\,l^{-1}$ (ppb). ICP-OES can be considered 'half way' as it offers DLs usually better than in FAAS but poorer than in ETAAS. Considering flame-based techniques, flame OES is advantageous just for alkali metals and, occasionally, calcium. Flame AFS might provide sensitivity similar to or better than FAAS (fluorescence signals have a low background). The limited use of AFS in routine laboratories is not due to technical disadvantages; rather it can be attributed to the fact that it does not offer additional advantages to those from the well-established AAS and OES techniques.

Concerning the linear ranges of calibration, more limited linear (straight-line) ranges are obtained with AAS than with OES and AFS, which span about three orders of magnitude above the corresponding limit of quantification for AAS and five to six orders of magnitude for OES and AFS methods.

In FAAS, relative standard deviations (RSDs) observed for measurements within the linear range are always better than $\pm 1\%$. This value is usually better than the RSDs observed using ETAAS, OES or AFS measurements.

1.2.3.3 Versatility and Sample Throughput

About 70 elements of the Periodic Table can be determined by optical techniques of atomic spectrometry. AAS techniques are basically considered as single element (particularly so for ETAAS, where the lamp and the atomisation conditions have, as a rule, to be selected individually for each element). This feature determines that the sample throughput in AAS (especially with ETAAS) is comparatively low.

In contrast, OES techniques (particularly those using a hot spectrochemical source such as an ICP) are intrinsically multielemental and this offers the possibility of a very high sample throughput in routine analysis. To counterbalance such a clear advantage, AAS techniques are simpler and cheaper than ICP-OES.

1.2.3.4 Comparative Analytical Assessment of the Most Common Analytical Techniques Based on Optical Spectrometry

FAAS, ETAAS and ICP-OES are probably today's 'workhorses' for the routine analysis of dissolved samples. Instrumental development and analytical applications have grown extensively throughout the years. A general comparative assessment is given in Table 1.1. Assuming that any generalisation is risky, it can be stated that FAAS dominates elemental inorganic analysis carried out in rather small laboratories when only a few analytes (probably at $mg\,l^{-1}$ levels) have to be determined. Such long-lasting use can be attributed to several main facts, including its robustness and comparatively low cost, well-established and validated analytical methods and fast analyses (*i.e.* low turnaround time). When sensitivity at the $\mu g\,l^{-1}$ level is required, the technique of choice is ETAAS, even though its simplicity, robustness and speed are worse than those of FAAS. Finally, ICP-OES seems to be the most popular routine

Table 1.1 Comparative advantages and limitations of the most common atomic "workhorses" of dissolved samples analysis by optical spectrometry.

	FAAS	ETAAS	ICP-OES
General advantages	Simple and reliable Most widespread Moderate interferences Ideal for unielemental monitoring in small labs High sample throughput	Sub-ppm ($mg\,l^{-1}$) DLs Microsamples (< 1 ml)	Multielemental High temperature Relatively low matrix interferences High dynamic range
Cost	Low cost	Higher cost	High instrument cost
Limitations	Usually unielemental	Unielemental	Serious spectral interferences
	Sub-ppm ($mg\,l^{-1}$) DLs	Time consuming	Sub-ppm-ppb DLs
	Low temperature (interferences by refractory compounds)	Not so easy to optimize	Expensive to run
	For metal & metalloids	Problems with background For metal & metalloids	Also for some non-metals

technique for inorganic multielemental analysis of dissolved samples, even though the initial investment and subsequent running expenses are much higher than those needed for AAS.

1.3 Atomic Mass Spectrometry

One of the more recent branches of atomic spectrometry, although perhaps the most exciting one, is *atomic mass spectrometry*, which has had a very important impact on science and technology. At present, atomic mass spectrometry is ordinarily performed using inductively coupled plasma ion sources and either a quadrupole or a scanning sector-field mass spectrometer as an analyser. The remarkable attributes of such a combination, being an indispensable tool for elemental analysis, include:

- very low detection limits for many elements;
- availability of isotopic information in a relatively simple spectrum;
- acceptable levels of accuracy and precision.

The success of inductively coupled plasma mass spectrometry (ICP-MS) has resulted in a broad availability of sophisticated instrumentation packages with user-friendly software and sample-analysis 'cookbooks' at reasonable cost [10].

Despite these strengths, ICP-MS has also some important drawbacks, many of them related to the spectral isotopic and/or chemical interferences, which affect analyte signal intensities and, therefore, the applicability of the technique. The complexity of the optimisation of the methodological and operating conditions, the differences in the ionisation rates of the various elements, the sequential isotopic measurements and the limited speed of signal acquisition (a serious drawback in multielemental analysis of fast transient signals) are some other problems to be considered.

In order to overcome, or at least minimise, such drawbacks we can resort to the use of chemometric techniques (which will be presented in the following chapters of this book), such as multivariate experimental design and optimisation and multivariate regression methods, that offer great possibilities for simplifying the sometimes complex calibrations, enhancing the precision and accuracy of isotope ratio measurements and/or reducing problems due to spectral overlaps.

1.3.1 Fundamentals and Basic Instrumentation of Inductively Coupled Plasma Mass Spectrometry

Since the launch of the first commercial quadrupole ICP-MS instrument in 1983, the technology has evolved from large, floor-standing, manually operated systems, with limited functionality and relatively poor detection limit capabilities, to compact, sensitive and highly automated routine analytical instruments. In principle, all ICP-MS systems consist of similar components: a sample introduction system, the ICP ion source, an interface system, the mass analyser, the detector and a vacuum system [8,11].

1.3.1.1 The Sample Introduction System

This is needed to bring the sample into the ICP plasma, where the ions are generated. Solution-based samples are currently introduced via a nebuliser (employed to produce small and narrow aerosol particle size distributions). This system is known to be a critical factor in achieving low detection limits because different designs (pneumatic nebulisers, ultrasonic nebulisers, high-pressure hydraulic design, thermospray or direct injection nebuliser systems, among others) differ in the efficiency of the mass transport (mass of analyte transported to the plasma per unit mass of analyte introduced from the sample). Also, an important body of work in the ICP-MS literature deals with the development of alternative sample introduction systems for both solution and solid sample forms. For example, hydride generation methods, flow injection analysis and slurry nebulisation methods are commonly used.

1.3.1.2 The ICP Ion Source

The ICP system is, in principle, identical with that used for ICP-OES and ICP-AFS, as it described earlier in this chapter. In the ICP, the major mechanism by which ionisation occurs is thermal ionisation. When a system is in thermal equilibrium, the degree of ionisation of an atom is given by the Saha equation:

$$\frac{n_i n_e}{n_a} = 2\frac{Z_i}{Z_a}\left[2\pi m k_B \frac{T}{h^2}\right]^{\frac{3}{2}} \exp(-E_i/k_B T) \tag{1.11}$$

where n_i, n_e and n_a are the number density of ions, free electrons and atoms, respectively, Z_i and Z_a are the ionic and atomic partition functions, respectively, m is the electron mass, k_B is the Boltzmann's constant, T is the absolute temperature, h is the Plank's constant and E_i is the first ionisation energy.

From the Saha equation, we can see that the degree of ionisation is dependent on the electron number density, the temperature and the ionisation energy of the element of interest. The typical ICP electron densities and temperatures result in all elements with first ionisation potentials below 8 eV being completely ionised and even elements with first ionisation potentials between 8 and 12 eV are ionised by more than 10%. However, ICP is not an efficient ionisation source for elements with ionisation energies above approximately 10 eV. Moreover, both the temperatures and the electron number densities are dependent on the operating conditions of the plasma, in particular the forward power and the carrier gas flow rate, on the solvent loading, on the torch design, on the inner diameter of the injector and on the generation characteristics (frequency). Hence apparently small changes in the operating conditions can lead to a totally different plasma in terms of ionisation efficiencies, therefore affecting the tolerance to solvent loading, sensitivity and susceptibility to matrix effects.

1.3.1.3 The Interface System

An interface is needed for pressure reduction, so that the ions generated in the atmospheric pressure ICP become efficiently transported to the ion lens system of the mass spectrometer where they are focused into the mass analyser. It should be noted that an enormous pressure change is required (reduction by a factor of 10^8–10^{12}) and most of it has to be accomplished over a very short distance (< 10 cm) using relatively small vacuum pumps. All modern ICP-MS instruments employ very similar sampling interfaces in which there are a number of common components. The basic layout is shown in Figure 1.9. The plasma expands from the atmosphere through a first orifice on the tip of a metal cone (sampler) to form a 'free jet' into a region whose pressure is a few Torr. Behind the sample cone is the expansion chamber, which is pumped by a single- or double-stage rotary pump producing an operating pressure in the 2–5 mbar range. The cloud of gas atoms, molecules, ions and electrons entering the expansion chamber quickly increases in speed and expands outwards under the influence of the reduced pressure in this region, resulting in the formation of a free jet. Spaced behind the sample cone is a second cone with a smaller orifice in its tip, called the skimmer cone. The centreline flow passes through a molecular beam skimmer into the downstream vacuum system pumped to a pressure of

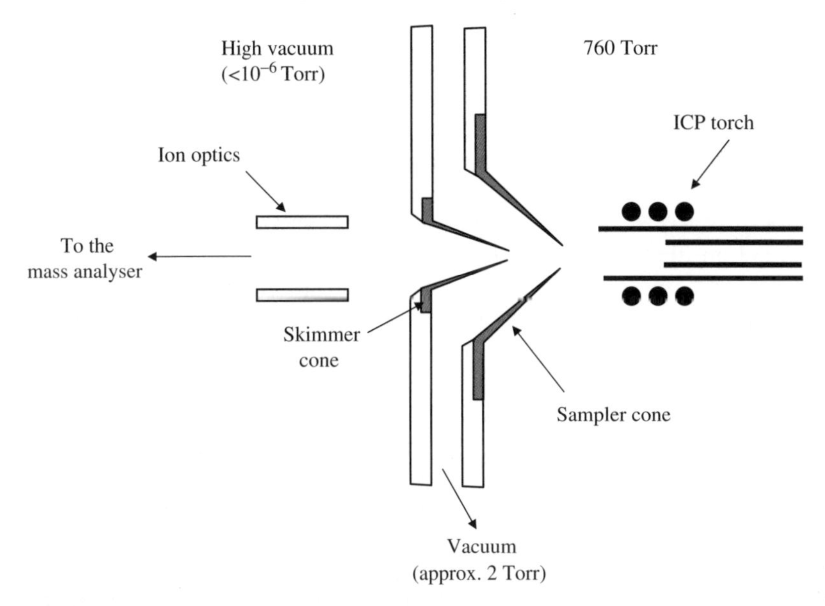

Figure 1.9 Basic layout of an ICP-MS interface.

10^{-3}–10^{-4} Torr. Operating pressures and cone shapes within the sampling interface strongly influence analyte sensitivity, mass response, matrix tolerance and levels of molecular interference. Ion sampling interfaces for ICP-MS have been designed using theory derived from molecular beam studies.

1.3.1.4 Ion Optics

Once the ions have entered the skimmer tip, it is necessary to extract and focus them into the analyser by subjecting the charged ions to constant electric fields. In order to construct an effective ion optical array, it is necessary to calculate the path followed by the ions in the electrostatic fields. We can resort to a number of mathematical models, such as SIMION, for a better understanding and optimisation of the ion-optical design for ICP-MS and the processes involved [12].

1.3.1.5 The Mass Analyser, the Detector and the Vacuum System

The mass analyser sorts the ions extracted from the ICP source according to their mass-to-charge ratio (m/z). The mass spectrum is a record of the relative numbers of ions of different m/z, which is characteristic of the analyte compound (see Figure 1.1). Functionally, all mass analysers perform two basic tasks:

1. separation of ions according to their mass-to-charge ratio;
2. measurement of the relative abundance of ions at each mass.

Most work in ICP-MS has been conducted using quadrupole-based mass spectrometers. The quadrupole mass filter is comprised of four parallel electrically conducting rods. Opposite rods are connected and parallel rods are supplied with a dc voltage, one pair being held at $+U$ V and the other set at $-U$ V. The first set of rods is supplied with an rf voltage and the second set of rods is supplied with an rf voltage out of phase by 180°. This combination results in the formation of an oscillating hyperbolic field in the area between the rods. Ions, when inside the quadrupole, are forced to follow certain trajectories that are dependent on the geometry of the field, the amplitude and the angular frequency of the alternating potential, the magnitude of the dc bias applied to the rods, the m/z of the ions, the initial conditions (position and velocity) and the phase angle with which the ions enter the field. Hence the quadrupole mass analyser has the ability to transmit certain ions and reject others depending on the stability of their path. Stable ions are transmitted through the length of the quadrupole whereas unstable ions hit the rods. The mass spectrum is scanned by varying the amplitude of the potential and the magnitude of the dc bias applied to the rods, while maintaining a constant ratio between those parameters. Quadrupole mass analysers are essentially sequential instruments, although they can be used to scan rapidly over 200 m/z in around 1 ms. Such a system typically provides a resolution of 0.5–1 atomic mass unit (amu).

However, nowadays some other different mass spectrometers are used for ICP-MS: 'time-of-flight' (TOF) systems for multielemental analysis of transient signals, ion trap analysers for ion storage, multicollector instruments for precise isotope ratio measurements and double-focusing sector field mass spectrometers for high mass resolution, but still the majority of instruments are equipped with quadrupole filters, which are simpler and cheaper.

We should not forget that an appropriate detector, a Faraday cup or a secondary electron multiplier equipped with a conversion dynode, is needed for ion detection. Most commercial instruments are equipped with a secondary electron multiplier, which can be operated in a low amplification mode, the analogue mode, and with a high gain, the counting mode, where each ion is counted. With this dual mode, a linear dynamic range of up to nine orders of magnitude can be achieved, so that major and minor components of the sample can be measured in one run.

Finally, successful operation of the mass analyser requires a collision-free path for ions. To achieve this, the lens system, mass analyser and detector are operated in a high-vacuum environment (below 10^{-6} Torr).

1.3.2 Analytical Performance Characteristics and Interferences in ICP-MS

The mass spectrum generated with an ICP-MS system is extremely simple (see Figure 1.1). Each elemental isotope appears at a different mass (*e.g.* ^{27}Al would appear at 27 amu) with a peak intensity directly proportional to the initial concentration of that isotope in the sample solution. A large number of elements ranging from lithium at low mass to uranium at high mass are simultaneously analysed typically within 1–3 min. With ICP-MS, a wide range of elements at concentration levels from ppt to ppm can be measured in a single analysis.

Detection limits reported for the analysis of solutions by ICP-MS depend strongly on different variables, particularly the mass analyser, the sample matrix and the analyte under investigation. Moreover, small variations in many different instrumental parameters (including those affecting the plasma generation and the efficiency of ion sampling or transmission) can govern the sensitivity and detection limits when working with ICP-MS. However, for most elements, typical DLs in solution are in the range 0.1–10, 1–100 and 0.01–0.1 $ng\,l^{-1}$ for the quadrupole filter, the time-of-flight and the sector-field-based MS systems, respectively. In most cases, these DLs are 100–1000 times superior to those achieved routinely by plasma emission (ICP-OES) and fluorescence (ICP-AFS) spectrometry [13].

1.3.2.1 Quantification Procedures

Quantitative analysis in ICP-MS is typically achieved by several univariate calibration strategies: external calibration, standard addition calibration or internal standardisation. Nevertheless multivariate calibration has also been applied, as will be presented in Chapters 3 and 4.

Conventional external calibration uses pure standard solutions (single- or multielement) and is therefore unable to compensate for matrix effects, fluctuations or drifts in sensitivity. Matrix effects can be compensated for by using matrix-matched calibration solutions. In this case, the degree of compensation depends on the proper matrix adjustment.

Different mathematical approaches have been applied to enhance the performance of external calibrations in ICP-MS. To some extent, the observed drifts can be compensated for by regularly repeating the calibration or by repeated measurement of one standard, which allows for a mathematical drift correction [14]. Also, external calibration by weighted linear regression offers significant advantages over simple regression, especially for the determination of analytes at low concentrations. Confidence intervals are equivalent to those obtained by simple regression for analyte concentrations around 10 times the limit of quantification or lower. On the other hand, accurate results can be obtained even though calibration ranges are not adapted to the analyte concentration, which implies working with sufficiently wide calibration intervals where the main contribution to total uncertainty arises from the calibration itself.

Standard addition calibration is more robust and reliable than conventional external calibration, but is more time consuming and costly if it is applied separately for each sample. A major advantage of standard addition is the correction of multiplicative matrix effects such as alteration of nebulisation efficiency. The intensities of all samples (and spiked samples) change by the same factor, which leads to an altered calibration slope. However, for additive effects, such as interferences caused by the matrix, the calibration line is shifted parallel and the intercept changes, which results in biased analyte concentrations. In some cases, this bias can be avoided (or indeed identified) by choosing another isotope and comparing the results for each. Standard addition has no inherent compensation for instrumental drifts in the ICP-MS system. However, a reduction of the drifts, which limit the applicability of standard addition for ICP-MS, was achieved by applying a chemometric method (a bracket approach, where the spiked sample is measured between two different measurements of the sample) [15].

Internal standard (IS) calibration requires ratioing of an analytical signal to an IS which has very similar characteristics to that of the analyte of interest (an element which is similar to the analyte either in mass, ionisation potential or chemical behaviour). Quantitative analysis applying internal standardisation is the most popular calibration strategy in ICP-MS, as improvements in precision are obtained when the technique is appropriately used. Of course, the validity of this calibration method requires that one ensures a good selection of the correct internal standard. For this purpose it is possible to resort to chemometric methods [16].

An alternative to quantitative analysis by ICP-MS is **semiquantitative analysis**, which is generally considered as a rapid multielement survey tool with accuracies in the range 30–50%. Semiquantitative analysis is based on the use of a predefined response table for all the elements and a computer program that can interpret the mass spectrum and correct spectral interferences. This approach has been successfully applied to different types of samples. The software developed to perform semiquantitative analysis has evolved in parallel with the instrumentation and, today, accuracy values better than 10% have been reported by several authors, even competing with typical ones obtained by quantitative analysis. The development of a semiquantitative procedure for multielemental analysis with ICP-MS requires the evaluation of the molar response curve in the ICP-MS system (variation of sensitivity as a function of the mass of the measured isotope) [17]. Additionally, in the development of a reliable semiquantitative method, some mathematical approaches should be employed in order to estimate the ionisation conditions in the plasma, its use to correct for ionisation degrees and the correction of mass-dependent matrix interferences.

1.3.2.2 Interferences in ICP-MS

Spectroscopic interferences have been recognised as one of the main limitations of the most often used quadrupole-type ICP-MS since its initial development. Such interferences, which appear when an interfering species has the same nominal m/z as the analyte, may be subdivided as follows.

1. *Isobaric overlaps* appear when isotopes of different elements have the same nominal mass. Many of them can be overcome by choosing an alternative less interfered isotope of the element of interest, although a sacrifice in sensitivity may result.

2. Some elements in the low mass range, such as Ce and Ba, have second ionisation potentials low enough to yield significant quantities of *doubly charged* ions. In general, mass resolutions between 2000 and 10 000 are required to separate them.
3. *Molecular (polyatomic) ions* are the main source of spectral overlaps in ICP-MS. Ions from the plasma gas (such as Ar^+ and Ar^{2+}) can result in spectral overlaps with $^{40}Ca^+$ and $^{80}Se^+$, respectively. Acids used to digest and preserve samples can also result in intense signals from molecular ions that can overlap with major isotopes of some elements (*e.g.* $^{15}N^{16}O^+$ from HNO_3 has the same nominal mass as $^{31}P^+$). Molecular oxides formed from elements present in the sample are also common. Polyatomic interferences are more difficult to correct for, as they are less predictable because they depend on the abundance of at least two isotopes and also depend strongly on the sample (analytes and matrix) and the operational parameters of the ICP-MS system.

Some of the most prominent spectral interferences can be resolved with a resolution from 4000 up to 10 000, depending on the analytical problem. It can be tempting to calculate the resolution necessary to resolve two masses based only on their exact masses and the specified resolving power of the instrument. However, the resolution required will depend on the relative magnitude of the spectral overlap and analyte ion signals. For example, to resolve the overlap of $^{37}Cl^+$ and $^1H^{36}Ar^+$, a resolution of 3900 would be sufficient when considering the exact masses alone. However, as can be seen in Figure 1.10, a resolution of 10 000 is needed to provide baseline resolution of the two peaks (because the $^1H^{36}Ar^+$ ion is much more intense).

ICP-MS instruments based on a quadrupole mass analyser typically provide a mass resolution not better than 0.6 mass units, clearly insufficient for many applications. The development of high-resolution ICP-MS, in the late 1980s, made it possible to resolve analytes from interferences using a double-focusing instrument on the basis of a magnetic and an electric sector field. Additional time is required to acquire these data, but the chances of overlooking a potential overlap are greatly diminished. High mass resolution is one of the most important features of double-focusing instruments, but not the only one. When working at 'low' resolution, they also show a higher sensitivity, as compared with quadrupole devices, and have much lower background noise. However, the trade-off for increasing resolution in sector-based mass spectrometers is a decrease in sensitivity. Typically, the sensitivity decreases by at least a factor of 6–8 when the resolution is increased from 4000 to 10 000.

Although this technology is effective in resolving a wide range of polyatomic interferences, the increased cost associated with this type of instrumentation (more than twice the price of a quadrupole instrument) limits its use in most routine laboratories, hence alternative methods of interference reduction have been sought for. The use of chemical extraction and chromatography (in order to separate the analyte from the matrix prior to analysis) or the operation of the ICP-MS under so-called 'cool plasma' conditions, allows the elimination of

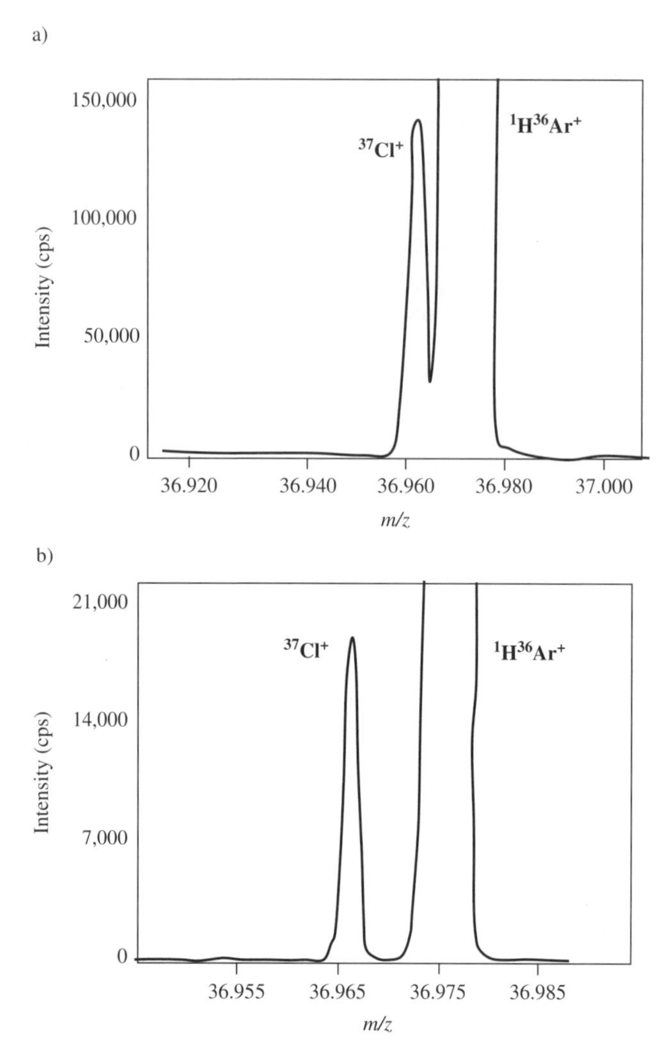

Figure 1.10 Mass spectra near $^{37}Cl^+$. (a) Resolution of 4000 (signal of peak $^1H^{36}Ar^+$ about 20 000 000 cps). (b) Resolution of 10 000 provided by sector-based mass spectrometer (signal of peak $^1H^{36}Ar^+$ about 3 000 000 cps).

many polyatomic ion interferences. However, several drawbacks still remain with these modes of operation: they are sometimes difficult to optimise and are suitable for only a few interferences.

More recently, the advent of the collision/reaction cell technology has revolutionised commercial quadrupole ICP-MS systems. A gas, such as hydrogen, helium or ammonia, is introduced into the reaction cell (placed inside the mass spectrometer and preceding the analyser quadrupole), where it reacts and dissociates or neutralises the polyatomic species or precursors. Through collision and reaction with appropriate gases in a cell, interferences such as $^{40}Ar^{+}$ can be almost completely eliminated while leaving analyte ions (*e.g.* ^{80}Se) relatively unaffected.

However, often these possibilities are not available or not applicable, due to either a lack of appropriate technology or financial or personnel constraints. Alternatively, we could also apply a wide variety of multivariate methods and chemometric approaches to correct or minimise for ICP-MS spectral interferences (some of them are referred to in Chapter 2, when discussing experimental design and optimisation), particularly when the ratio of overlap ion signal to analyte ion signal is not too large. The general mathematical strategy to correct for isobaric interferences in atomic mass spectrometry is subtraction on the contribution of interfering isotopes from measurements of non interfering isotopes. In the case of polyatomic interferences from two or more species on a particular mass, the chemometric approaches require consideration of correction factors calculated using the natural isotope abundances of the atoms from which the interference is formed. In such a situation, the complexity of the mathematical approach is high. In the following chapters different tools will be introduced that can be applied in order to solve this problem of ICP-MS (*e.g.* multivariate methods, which have been applied by several authors for the correction of spectral and non-spectral interferences [18]).

1.3.3 Isotope Ratio Measurements and Their Applications

The scope of isotopic analysis is extremely wide nowadays and natural or induced variations in the isotopic composition of target elements are being investigated for several purposes, including bioavailability studies, nuclear chemistry, age determinations and environmental, geological and clinical applications. Precise and accurate isotope ratio measurements have traditionally been carried out by thermal ionisation mass spectrometry (TIMS). However, the capability of ICP-MS (a technique that is easier to handle, with a higher sample throughput and widespread availability) to provide isotopic elemental information permits not only the determination of isotopic ratios but also the use of isotope dilution and its corresponding improvement in accuracy (of special interest for quantification purposes). Also, isotopic patterns are extremely useful in ICP-MS to confirm the identity of sought-for elements.

As the precision of the ICP-MS isotope ratio is poor compared with the precision using TIMS, the range of applications for ICP-MS have traditionally been limited to measuring induced changes in the isotopic composition of a target element (for example, to calibrate by means of isotope dilution). However, the introduction of multicollector ICP-MS systems to enhance precision and accuracy in isotopic analysis opened up novel applications.

1.3.3.1 Precision and Accuracy in Isotope Ratio Measurements by ICP-MS

When performing accurate isotope ratio measurements with ICP-MS, the following issues should, at least, be considered.

Required precision. This will lead you to the instrument you need. Quadrupole ICP-MS is easy to use, robust and relatively inexpensive. In general, these instruments permit good precision of isotopes ratio measurements ranging from 0.1 to 0.5%. Applying high-resolution ICP-MS precision in isotope ratio measurements can generally be improved by a factor of 5–10 (mainly because of the flat-topped peak shapes and fewer spectral overlaps obtained with these high-resolution instruments). Multicollector ICP-MS systems increase precision due to the collection of all isotopes of interest simultaneously in a multicollector array and so they provide an opportunity to measure the isotopic composition of many elements more accurately than other ICP-MS instruments.

Spectroscopic interferences. These can affect the ion intensity of an isotope and, therefore, the isotope ratio measurement. One would try to minimise such interferences by resorting to any of the techniques and technologies, and also the mathematical correction strategies, described in the previous section.

Mass discrimination. In ICP-MS devices, ions of different mass-to-charge ratios are transmitted with different efficiencies and therefore the instrument produces different responses for ions of different masses. This systematic error is called mass discrimination. Typically, the mass discrimination for ICP-MS instruments is about 1% per mass unit (at mass 100). As the ion kinetic energy is dependent on the mass, any energy-dependent process in the instrumentation (*e.g.* sampling of ions from the ICP, transfer of ions, mass separation and ion detection) will contribute to the mass discrimination. Accurate isotope ratio measurements require mathematical corrections for mass discrimination. External calibration is frequently used, offering sufficient accuracy. For this, isotopic reference materials with known isotopic composition or samples in which the element has a known natural isotopic composition are used. The correction factor for mass discrimination (the so-called K-factor) can be easily calculated based on the 'true' value (given by the certified isotope ratio) and the 'observed' value (that is, the measured isotope ratio including the bias caused by mass discrimination). However, with reduced uncertainty (*e.g.* using a multicollector ICP-MS instrument), complex mathematical models should be used for appropriate mass discrimination correction.

Detector dead time. This is the time required for the detection and electronic handling of an ion pulse. If another ion strikes the detector surface within the time required for handling the first ion pulse, the second ion will not be detected and, hence, the observed count rate will be lower than the actual value. If this is not corrected for, inaccurate isotope ratio results will be reported. In ICP-MS, several mathematical methods should be applied for its evaluation and correction.

Data acquisition parameters. Precision and accuracy in the measurement of isotope ratios can be improved if the number of measurements is increased (*e.g.* if the measurement time is increased). Various measurement protocols can be applied and those whereby the time actually spent on measuring the isotope ratios of interest is maximised are preferable. The data acquisition parameters of an ICP-MS device that can be changed to improve the isotope ratio precision are the integration time (dwell time) per acquisition point, the number of acquisition points per spectral peak and the number of sweeps, among others. Many of these parameters have been optimised using chemometric approaches. Many examples will be given in Chapter 2. The measurement time can sometimes be used more efficiently by increasing the acquisition time for the less abundant isotope(s) relative to that of the most abundant isotope(s). Finally, simultaneous monitoring of all the isotopes is performed in multicollector ICP-MS instrumentation, which results in a superior isotope ratio precision, similar to that offered by state-of-the-art TIMS. A multicollector ICP-MS system is operated in a static mode during the measurements, which means that neither the accelerating field nor the strength of the magnetic field is changed during data acquisition.

1.3.3.2 Isotope Dilution Analysis

Inductively coupled plasma isotope dilution mass spectrometry (ICP-IDMS) is a well-known analytical technique based on the measurement of isotope ratios in samples where their isotopic composition was altered by the addition of a known amount of an isotopically enriched element.

ICP-IDMS has high potential for the routine analysis of trace elements if accuracy is of predominant analytical importance [19]. In contrast to other calibration approaches, IDMS does not directly suffer from long-term changes or drifts in instrument sensitivity. Moreover, provided that isotopic exchange between the sample and spike is ensured, losses of analyte do not affect the analytical results. Additionally, IDMS can also be used to prevent the final analytical result being affected by analyte losses during sample pretreatment.

In the last few years, we have seen the application of isotope dilution methodologies to some new analytical fields. One of these is 'elemental speciation', where the aim is to determine individual chemical species in which an element is distributed in a given sample. IDMS has also proved its usefulness in element speciation, in which either species-specific or species-unspecific spikes can be used. For example, species-specific IDMS is nowadays used in several laboratories as an effective tool to validate analytical procedures for speciation and to investigate and document eventual interconversion between species. In addition, the study of induced variations in the isotopic composition of a target element can also provide insight into various (bio)chemical and physical processes; isotopic analysis is, therefore, also of increasing importance in biological studies.

The principle of isotope dilution analysis is surprisingly simple. It relies on the intentional alteration of the isotope abundance of an endogenous element in a given sample by the addition of a known amount of an enriched isotope of the same element (spike). Therefore, the element to be analysed must have, at least, two stable isotopes that can be measured free of spectral interferences in a mass spectrometer. This principle is illustrated in Figure 1.11 for an element containing two different isotopes, a and b. As can be observed, the a isotope is the most abundant one in the sample whereas the spike is isotopically enriched in the b isotope. It is clear that the abundances of the two isotopes

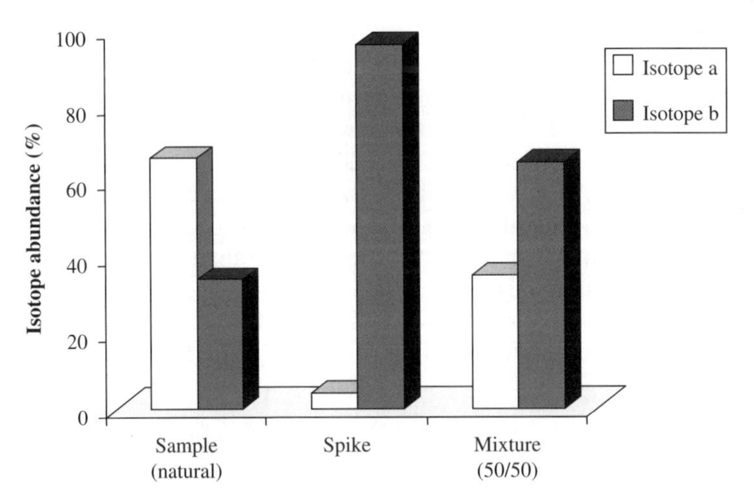

Figure 1.11 Illustration of the principle of isotope dilution analysis for an element containing two different isotopes (a and b).

and, hence, the isotope ratio in the mixture will be intermediate between those in the sample and the spike and it will depend both on the amount of spike added and on the initial amount of the element in the sample. These relationships can be expressed mathematically using the isotope dilution equation, which can be written in different forms depending on the complexity of the application [20].

IDMS is based on measurements of masses and isotope ratios only. Some important advantages, compared with other calibration strategies, such as external calibration and standard additions, are that instrumental instabilities such as signal drift and matrix effects will have no influence in the final concentration in the sample, high accuracy and small measurement uncertainties are enabled, possible loss of substance of the isotope-diluted sample will have no influence on the final result and there is no need to resort to an external instrumental calibration or standard additions to the sample.

Isotope dilution analysis is now internationally regarded as a reference or highly qualified primary method thanks to all these advantages.

1.4 Flow Systems with Atomic Spectrometric Detection

The coupling between flow injection (FI) systems and atomic spectrometric detectors is a strategy now well established for inorganic elemental analysis. Flow injection analysis (FIA) is a powerful, convenient analytical tool highly suitable for automating sample pretreatment, which is often required before measuring by atomic spectrometry (including sampling dissolution and/or dilution, matrix removal, preconcentration, *etc.*). Additionally, flow manifolds can simplify the well-known problem of sample introduction to atomisers or even reduce/minimise interferences (*e.g.* by coupling on-line separation techniques before the detector). All these facts explain the great importance of FI systems as practical troubleshooters when coupled to atomic spectrometric detectors.

1.4.1 Flow Injection Analysis and Atomic Spectrometry

The basics of an FI experiment are very simple. A discrete volume of a sample is injected into a continuous liquid stream. The sample becomes mixed with the carrier stream by a number of processes collectively known as 'dispersion' and is conducted to the detector, where the analyte is monitored on-line. Some reagents can also be added to the FI system and merged with the sample (before arriving at the detector) at a confluence point (they perform sample conditioning, analyte derivatisation, *etc.*). As a result of the dispersion processes involved in an FI manifold, the concentration–time profile of the analyte at the detector results in a peak. Depending on the experimental conditions, the peak may be skewed or may be close to a symmetrical Gaussian shape. The design of an FI system requires consideration and optimisation of different operational parameters, many of them interrelated, by using appropriate chemometric tools (*e.g.* the simplex method). Examples will be given in Chapter 2.

Flow systems are developed mainly for liquid samples and their complexity can range from simple to very complex manifolds to deal with ultratrace amounts of the target analyte in complex matrices, which often require on-line separation/preconcentration steps. As a wide variety of chemical manipulations can be carried out in an FI manifold, the scope of the FI applications is enormous. Not only liquid samples, but also both gas and solid samples, can be also introduced into the liquid flow manifold if special adaptations are made. Gas samples simply require impermeable tubing. Solids can be either introduced into the system and leached with the help of auxiliary energy (*e.g.* ultrasound) or introduced as slurries.

Concerning the requirements of the detector, it is important to stress that interfacing a detector with an FIA system yields transient signals. Therefore, desirable detector characteristics include fast response, small dead volume and low memory effects. FI methods have been developed for UV and visible absorption spectrophotometry, molecular luminescence and a variety of electrochemical techniques and also for the most used atomic spectrometric techniques.

A large part of the success of the combination of FI and atomic spectrometry is due to its ability to overcome interference effects. The implementation of some pretreatment chemistry in the FI format makes it possible to separate the species of the analyte from the unwanted matrix species (*e.g.* by converting each sample into a mixture of analyte(s) and a standard background matrix, designed not to interfere in the atom formation process and/or subsequent interaction with radiation in the atom cell). Often such separation procedures result also in an increased analyte mass flux into the atom source with subsequent improvements in sensitivity and detection limits.

In general, FI procedures are used in conjunction with atomic spectrometry for any of the following purposes:

1. The use of discrete sample volumes to provide improved tolerance of the detector to dissolved solids, organic solvents and sample viscosity. FI provides on-line dilution and a suitable means of handling slurried samples.
2. Retention of the analyte on a solid-phase extractant, followed by dissolution in a clean matrix (such as dilute nitric acid) to remove interferences and preconcentrate the analyte.
3. To implement an easy and automated means for chemical vapour generation procedures (hydride generation for arsenic, selenium, *etc.*, and cold vapour mercury), which allows for a reduction on the interferences caused by first-row transition metals (such as copper and nickel). FI methods may be readily coupled with almost all the atomic-based spectroscopic techniques (including graphite furnace atomisers).
4. Manifolds have been described for the addition of internal standards in order to automate the standard additions method.
5. In addition to solid-phase extraction and chemical vapour generation, other sample pretreatment procedures (including liquid–liquid extraction, precipitation, dialysis and even distillation) can be automated and coupled to the spectrometer.

In principle, all these capabilities will enhance the performance of any type of atomic spectrometry, independently of the nature of the spectroscopic technique used (*e.g.* a procedure that separates trace elements from a large volume of a highly saline medium and releases them into a smaller volume of dilute nitric acid can be used in conjunction with any type of spectrometer).

Nowadays, the outstanding advantages of using flow manifolds as sample preparation systems for atomic detectors have been demonstrated for a variety of techniques. Characteristic examples of the instrumentation required and typical applications are presented in a comprehensive monograph dedicated to FI and atomic spectrometric detectors [21].

1.4.1.1 Coupling FI Systems to Atomic Detectors

The on-line interface of flow manifolds to continuous atomic spectrometric detectors for direct analysis of samples in liquid form typically requires a nebuliser and a spray chamber to produce a well-defined reproducible aerosol, whose small droplets are sent to the atomisation/ionisation system. A variety of nebulisers have been described for FAAS or ICP experiments, including conventional cross-flow, microconcentric or Babington-type pneumatic nebulisers, direct injection nebuliser and ultrasonic nebulisers. As expected, limits of detection have been reported to be generally poorer for the FIA mode than for the continuous mode.

Alternatively, one can resort to the introduction of analytes as gaseous derivatives, which offers special advantages in atomic detection (*e.g.* improved detection limits, reduction of matrix effects and simplicity of the coupling to the atomic detector). Flow injection is a particularly useful procedure for the implementation of chemical vapour generation (CVG) methods. These procedures are based on the generation of a volatile chemical derivative of the species of the analyte (*e.g.* conversion of an analyte to its hydride, often by means of a tetrahydroborate reduction), removal of the generated volatile species from solution to the gaseous phase (by a gas–liquid separation device) and transport of the released compound by a carrier gas flow to the atomiser/detector. In terms of the number of analyses performed by CVG, the determination of mercury by the generation of the monatomic elemental vapour is by far the most widely used procedure. However, considerable attention has been also paid to the generation of the volatile species of arsenic, selenium and cadmium. For CVG in a typical flow system, a constant flow of sample solution is mixed with a constant flow of reducing solution (*e.g.* tetrahydroborate) and of the purge gas. Liquid and volatile generated gaseous species are then separated in a gas–liquid separator yielding two outlet flows. The gaseous analyte with hydrogen and purge gas flows to the atomiser/detector, while the liquid effluent is drained. Clearly, many different experimental parameters should be optimised in this design (*e.g.* the nature and concentration of the reducing reagents, the liquid flows of sample and reagents, the flow of the carrier gas or the nature of the gas–liquid separator), in order to ensure efficient generation of the volatile analyte and its transport to the spectrometer. Again, as multiple variables should be optimised (many of them interdependent), multivariate optimisation methods are critical. An important aspect of the FI–CVG procedure is that after appropriate optimisation of the experimental setup, certain interferences can be minimised.

Flow systems for volatile analyte generation have also been coupled to other atomic excitation sources for optical emission, atomic fluorescence and mass spectrometric detection (*e.g.* ICP-OES, ICP-MS, AFS, MIPs and AAS). Using this approach, it is possible to trap the volatile analyte hydride in the interior of a graphite furnace, thereby allowing a preconcentration prior to atomisation. Hydride generation is also a powerful tool for the elimination of matrix interferences in ICP-QMS. As a typical example, by using on-line flow hydride generation coupled to ICP-QMS, the sensitive determination of Se in biological or environmental matrices can easily be carried out, avoiding most of the isobaric interferences typically present on several of its most abundant isotopes.

Finally, we should consider that the intrinsic discontinuous nature of the ETAAS technique has limited the interest in interfacing basic continuous flow manifolds to this detector. However, several flow approaches offer special attraction for their combination with ETAAS, particularly:

- separation and preconcentration by on-line column sorption and solvent extraction;

- formation of volatile derivatives of the analyte and their preconcentration on a graphite tube;
- slurry sampling.

1.4.1.2 FIA Strategies for Calibration and Standardisation in Atomic Spectrometry

The contribution of flow analysis to improving the performance of atomic spectrometry is especially interesting in the field of standardisation. FIA can provide a faster and reliable method to relate the absorbance, emission or counts (at a specific mass number) to the concentration of the elements to be determined. In fact, flow analysis presents specific advantages to solving problems related to the sometimes short dynamic concentration ranges in atomic absorption spectrometry, by means of on-line dilution. The coupling of FI techniques to atomic spectrometric detectors also offers tremendous possibilities to carry out standard additions or internal standardisation.

However, it is worth noting that the basic advantage provided by FIA to atomic spectrometry is the ability to provide data on different wavelengths or mass numbers as a function of time for the same sample or standard injected, comparable to hyphenated gas chromatography–mass spectrometry (GC–MS) or two-dimensional nuclear magnetic resonance techniques. In this respect, the systems are unaffected by multiple spectral interferences which have no real detrimental effect on calibration apart from decreased signal averaging. Standardisation is theoretically still possible by a second-order standard additions method when matrix effects are present in the samples, provided that different instrumental responses are obtained for a standard solution of an analyte and a solution of the same analyte in the presence of other matrix compounds. Appropriate mathematical methods should be employed for such second-order calibration, simply by treating the signal as a function of time or introducing simple non-chromatographic separation methods based on FI principles.

1.4.2 Chromatographic Separations Coupled On-line to Atomic Spectrometry

In the past, most analytical problems related to environmental or biological systems were addressed by measuring the total concentrations of the elements. However, at present, there is an increasing awareness of the importance of the chemical form in which an element is present (*e.g.* the oxidation state, the nature of the ligands or even the molecular structure) since its chemical, biological and toxicological properties critically depend on it. Hence there is a clear need for rapid and robust analytical tools to perform chemical speciation, and atomic spectroscopy is undoubtedly one of the most important tools for such studies.

Many analytical strategies and methods have been described for elemental speciation. However, the so-called hyphenated (coupled or hybrid) techniques,

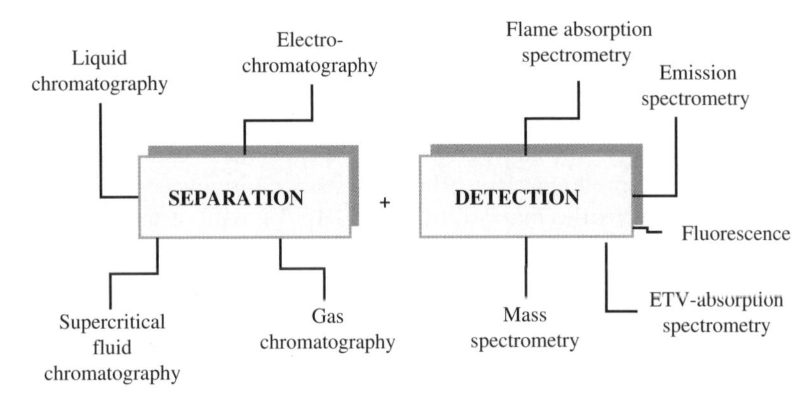

Figure 1.12 Species-selective hyphenated techniques for speciation analysis.

consisting in the on-line coupling of an efficient separation technique [such as GC, high-performance liquid chromatography (HPLC) or, more recently, electrophoresis] with a sensitive and element-specific atomic detector (usually an atomic absorption, emission or mass spectrometer) have become a fundamental tool for speciation analysis [22]. Some of the hyphenated techniques available for species-selective analysis in biological and environmental materials are summarised schematically in Figure 1.12. The choice of the hyphenated technique depends primarily on the objective of the research. In addition, speciation analysis in environmental and/or biological samples faces two main challenges because of the usually low concentrations of the analytes (below 1 $\mu g\, g^{-1}$) and the complexity of the matrix itself.

The success of an analytical speciation approach depends critically on the achievement of a good separation between the different species of the element. The chromatographic technique should guarantee that each signal corresponds to only a particular species. Despite the need for thermally stable and volatile analytes, GC remains the preferred sample introduction technique for the time-resolved introduction of analytes into an atomic spectrometer thanks to its high resolution and sensitivity, caused by the quasi-100% sample introduction efficiency and virtually no energy losses for the vaporisation and desolvation of the mobile phase. Although not so advantageous as GC in terms of detection limits and resolution, HPLC and capillary electrophoresis (CE) are the common choice for sample introduction into a plasma to determine species that either are non-volatile by themselves or that cannot be volatilised by a derivatisation reaction. The wide variety of separation mechanisms and mobile phases that preserve the species identity and also the coupling simplicity (in particular, the compatibility of the mobile phase composition and flow rate) to accepted standard nebulisers made HPLC–ICP-MS coupling an established procedure.

The choice of the detector becomes crucial when the concentration of analyte species in the sample is very low and low limits of detection are required. For element-specific detection, the major atomic spectrometric techniques, flame AAS, OES, AFS and ICP-MS, are specially suited as chromatographic detectors. An important problem when coupling the separation technique to the atomic detector consists on interfacing the chromatographic system and the atomic spectrometer, as the separation conditions may not be compatible with those required by the detector in terms of flow rate and the mobile phase composition. Usually, chromatography and spectrometry can be coupled on-line. However, the preference for a highly sensitive discrete atomisation technique such as electrothermal vaporisation (ETV) with atomic absorption spectrometric detection (ETAAS) or with ICP-MS detection may justify off-line coupling.

On-line coupling between a gas chromatograph and an atomic spectrometry detector is fairly simple. Typically, the output of the CG capillary column is connected to the entrance of the atomisation–ionisation system simply via a heated transfer line. When separation is performed by liquid chromatography (LC), the basic interface is straightforward: a piece of narrow-bore tubing connects the outlet of the LC column with the liquid flow inlet of the nebuliser. Typical LC flow rates of 0.5–2 ml min^{-1} are within the range usually required for conventional pneumatic nebulisation.

HPLC–AAS was the first hyphenated technique employed to determine metal–protein complexes. Although AAS is not a truly multielement technique, some instruments can measure up to four elements simultaneously, which is sufficient for a number of practical speciation applications. FAAS can be coupled with HPLC directly. This technique is compatible both with the flow rates and with the mobile phase composition (including organic solvents) commonly used in HPLC. Main applications include AAS detection of complexes with metals that yield intense responses in AAS (Cd, Zn, Cu) and species that can be converted on-line into volatile hydrides (*e.g.* As, Se, Cd). ETAAS is more sensitive than FAAS but its coupling is not so straightforward. The off-line ETAAS analysis of metallothionein-bound metals by fraction collection after HPLC separation has been a common approach. Use of an autosampler and flow-through cells allows for a high degree of automation, leading to a quasi-on-line coupling.

Microwave-induced plasma optical emission spectrometry (MIP-OES) is very sensitive for volatile species containing metals. Hence its use has been also proposed as a detector in the development of hyphenated techniques for speciation. GC–MIP-OES has been successfully applied for the speciation of alkylmetal species of low molecular weight (Hg, Sn and Pb compounds) in many different environmental applications [23].

Because MIPs are formed at low temperatures, liquid samples cannot be introduced because they extinguish the plasma, even small amounts of organic vapour. However, the on-line coupling of HPLC to MIP-OES has been described for the speciation of mercury and arsenic compounds. Continuous cold vapour (CV) or hydride generation (HG) techniques were used as interfaces between the exit of the HPLC column and the MIP, held in a surfatron at reduced pressure [24].

When ultrasensitive detection is required, ICP-MS is virtually the only technique capable of coping, in an on-line mode, with such trace element concentrations. Due to the multielement capability and high sensitivity of ICP-MS, along with the possibility of measuring different isotopes of a given element, its coupling to high-resolution separation techniques is well recognised as one of the most powerful tools for elemental speciation. HPLC and GC couplings are especially simple since the gas or liquid flows can be directly introduced into the ICP torch with only slight disturbances of the plasma, without any splitting or dilution process. In contrast, CE requires sophisticated interfaces and nebuliser systems in order to coupling it to an ICP-MS.

In addition, the isotope specificity of ICP-MS offers a still underexploited potential for improved accuracy when quantifying via the use of isotope dilution techniques. The application of isotope dilution analysis (IDA) in elemental speciation allowed for the development of highly accurate and precise quantification approaches for the determination of a wide range of elemental species even when analysing complicated matrices [25]. When several species of the same element need to be analysed, each compound can be enriched in a different isotope of the element, opening up a unique capability for quantification: multiple spiking species-specific IDA. Resorting to this powerful approach requires rather complex mathematical treatments. In fact, depending on the availability of the isotopically enriched species and the complexity of the speciation problem, a more or less sophisticated and specific mathematical approach must be developed to quantify the processes and, finally, the concentrations [26].

In all these hyphenated techniques, many different experimental parameters affecting the chromatographic separation, the interface and the detector should be carefully optimised. The use of mathematical approaches for adequate optimisation and development of the hyphenated systems is unavoidable (see Chapter 2 for many examples).

1.4.3 Detection of Fast Transient Signals

When discrete amounts of analyte (sample) are carried into a stream of a flowing fluid (a gas or a liquid carrier), analyte signals do not reach a steady state and transient time-dependent signals are obtained. Data acquisition is a discontinuous process which can be characterised by the number of data collected per unit time (acquisition frequency) and the time spent sampling each datum (sampling period). The acquisition frequency must be adapted to the signal to be sampled; too low a frequency can result in a loss of information, too high a frequency can overload the system without improving the measurements. The adjustment of such parameters is of particular importance when detecting transient signals.

Transient signals are typically obtained in atomic spectrometry when samples are introduced by flow injection techniques or when the spectrometer is used as an element-specific detector in hyphenated techniques. Inductively coupled plasma mass spectrometry has nowadays become the detection technique of choice for multielement-specific detection in speciation as it allows multielemental

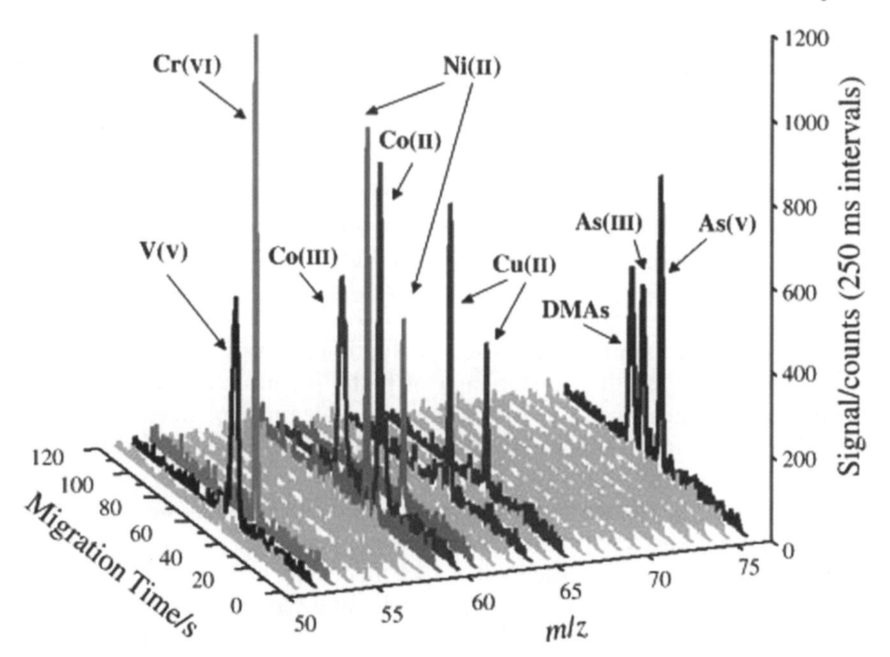

Figure 1.13 CE-ICP-TOFMS separation and multielemental detection of V(V), Cr(VI), Co(II), Co(III), Ni(II), Cu(II), As(V), As(III) and DMAs cyanide complexes. Reproduced from J. M. Costa, N. Bings, A. Leach and G. M. Hieftje, *J. Anal. At. Spectrom.*, 2000, **15**, 1063–1067, with permission.

analysis in a great variety of sample types and is virtually the only technique capable of achieving the stringent detection limits required in most practical cases.

When using ICP-MS as an elemental detector in hyphenated techniques, simultaneous detection of more than one analyte in the short transient signals generated by the chromatographic separations is typically required (*e.g.* see Figure 1.13). In such a situation, data acquisition parameters play a crucial role to enhance the detection limits and also the precision and accuracy. The same applies to isotope and isotope ratio measurements made using transient signals. Data acquisition in such systems is controlled by the number of points or channels per spectral peak, the number of sweeps (the number of scans along the mass spectrum to obtain a single reading) and the dwell time (the time spent counting ions per channel). It is essential to optimise those parameters carefully, which depend on the nature of the ICP-MS system, the nature of the transient signal (duration and intensity) and the number or isotopes to be monitored simultaneously. To obtain a reliable profile of the transient signals, it is common practice to work in the peak hopping mode using a single point per mass, one sweep per reading and a dwell time shorter than 100 ms.

Hyphenation between a chromatographic separation system and an ICP mass spectrometer often leads to transient signals of very short duration (from less than 5 s for GC up to about 60 s for LC). However, the best precision for the measurement of isotope ratios by ICP-MS is obtained using steady-state signals of several minutes or even longer, instead of short transient signals. Thus, the simplification of the sample preparation procedure achieved by the on-line coupling of a separation technique may be offset by a reduction in precision resulting from measurements made on short transient signals.

The common mass spectrometers used in ICP-MS today are scanning-based analysers, such as the quadrupole mass filter (ICP-QMS). Unfortunately, they suffer from important performance limitations when used as detectors of short transient signals (*e.g.* those generated in speciation analysis) derived from their inability to perform true simultaneous multielemental analysis. With scanning-based instruments, individual mass-to-charge ratios are measured in a sequential mode, from one isotope to the next. As a result, some difficulties (in terms of precision, sensitivity or accuracy of isotopic and isotope ratio measurements) are expected when fast transient or time-dependent signals (such as those produced by electrothermal vaporisation, laser ablation, chromatography or capillary electrophoresis) are used to determine a large number of analytes in a single peak.

The precision attainable in ICP-MS is limited by counting statistics and can be improved by increasing the integration time, which implies increasing the dwell time and/or the number of sweeps. For a fixed number of sweeps, increasing the dwell time is beneficial in reducing noise and hence improving limits of detection, although the number of points per peak is also reduced and the peak profile may not be described properly. Furthermore, the multielemental nature of ICP-MS adds another level of complexity to the monitoring of transient signals, because the number of readings has to be distributed between the different isotopes measured. This means that the mass range ratio of an analysis can be increased only by sacrificing sensitivity and precision. Furthermore, the measurement of sequential m/z values at different points within the time-dependent concentration profile of a transient signal can result in peak distortions and quantitation errors commonly referred to as 'spectral skew'. Finally, non-simultaneous ion extraction in scanning mass analysers hampers the use of ratioing techniques to reduce multiplicative noise associated with sample introduction and plasma fluctuation.

To minimise these shortcomings of scanning ICP-MS instruments, important efforts have been made to investigate other types of mass analysers. Particularly, time-of-flight mass spectrometry (TOFMS) should be well suited for the measurement of time-dependent transient signals [27]. In TOFMS spectrometers, all m/z values are extracted simultaneously (as a packet of ions) for mass analysis so that the 'spectral skew' usually associated with the measurement of transient signals is eliminated (see Figure 1.14). It has been pointed out that simultaneous ion extraction compensates for the effects of drift and multiplicative (flicker) noise components in the source by using ratioing techniques. Not surprisingly, ICP-TOFMS has been used already for multielemental detection in transient signals derived from several hybrid techniques, including CE and GC, for speciation purposes. A comparison between the performance

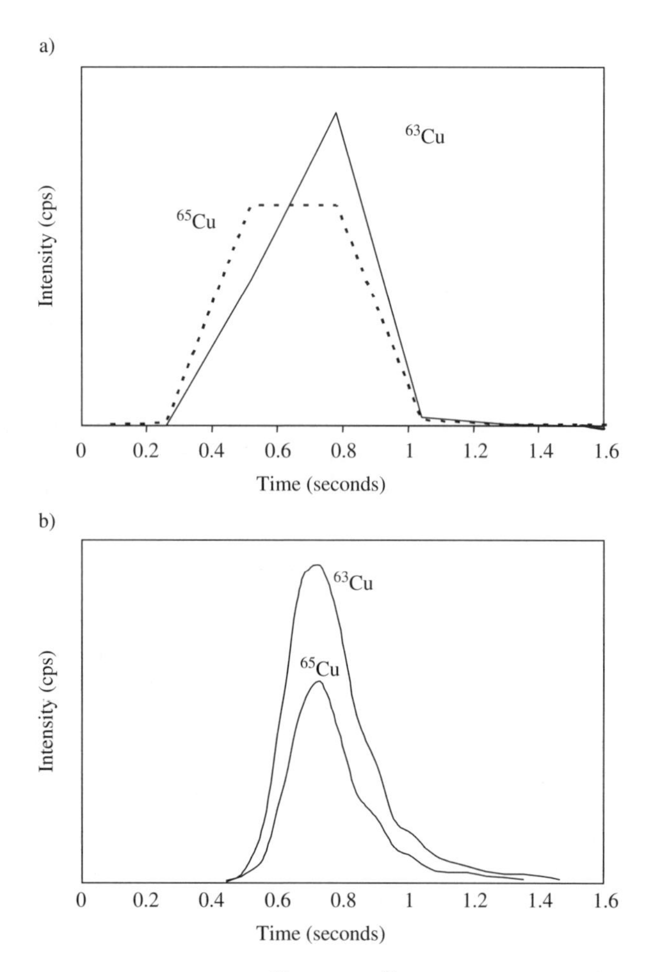

Figure 1.14 Transient signals for ^{63}Cu and ^{65}Cu produced by an electrothermal vaporization device and detected with (a) a quadrupole-based ICP-MS instrument and (b) a ICP-TOFMS instrument, when five isotopes were measured simultaneously.

of different ICP-MS systems (based on quadrupole and TOF mass analysers) [28] demonstrated that the best precision in isotope ratio measurements was obtained using ICP-TOFMS when transient signals faster than 8 s FWHM were monitored or when the number of isotopes to be measured was above 15. Conversely, the quadrupole mass analyser offered similar precision to ICP-TOFMS for transient signals of about 10 s or more, even when 25 isotopes were measured. In such measurements, the accuracy using a TOF mass analyser was slightly superior to that observed using an ICP-QMS instrument, probably as a result of the sequential nature of the Q-mass analyser.

However, the main limitation of ICP-TOFMS at present is, perhaps, its comparatively low sensitivity: detection limits for ICP-TOFMS are roughly one order of magnitude poorer than those reported for ICP-QMS, monitoring a single *m/z* under similar conditions [28]. On the other hand, as speciation of too many elements in a single chromatographic (or electrophoretic) peak is seldom needed, it appears that the ICP-QMS stands up in a general comparison for speciation analysis purposes.

As discussed above, ICP-IDMS is becoming a highly valuable method for trace element and element-speciation analysis. For ICP-IDMS the highly precise and accurate isotope ratio measurements that are currently required can be made by resorting to multicollector-ICP-MS.

The main limitation of ICP-IDMS is that isotope ratio measurements in transient signals generated by hyphenated techniques suffer from significant drifts. The main source of bias in the measured isotope ratios is in the ICP–MS (and not in the chromatographic separation) [29]. When the precision of the hyphenated chromatography–multicollector-ICP-MS method is compared with those of other quadrupole ICP-MS-based techniques, it can be clearly stated that this approach is more powerful than any quadrupole-based method which detects the ions sequentially. However, even using multicollector-ICP-MS as an element-specific detector, the precision of isotope ratio measurements in such transient signals, compared with the results from continuous sample introduction, is reduced by about one order of magnitude [29].

All isotope ratio measurements have to be corrected for instrumental mass bias by normalising to an invariant isotope of the same element (internal correction) or, whenever the internal approach cannot be applied, to a well-characterised isotope standard material (external correction). However, the external correction method requires the mass discrimination of an element being identical for the sample and the standard, which is not always the case. A large benefit of the hyphenated chromatography–ICP-MS system is that all measurements of standards and real samples can be carried out with exactly the same matrix – the eluent of the HPLC system.

1.5 Direct Analysis of Solids by Spectrometric Techniques

Advantages brought about by the direct analysis of solid samples as compared with the analysis of dissolved samples include a shorter total analysis time (prior dissolution steps are not required), low cost (chemical reagents are not used), less risk of contamination and less destruction of the sample. In addition, some techniques can extract information about chemical speciation (*e.g.* XPS provides information about oxidation states and chemical bonds) and spatial composition, *i.e.* information with lateral resolution allowing mapping of the surface and analysis with depth resolution, of particular interest for thin-film analysis.

A few representative and widely used techniques based on optical and mass spectrometry for direct solid analysis have been selected for further explanation here. As was stated in the introductory section (see Section 1.1), analytical techniques based on electron spectroscopy, although offering high interest for surface identification, will not be described in this chapter.

1.5.1 Elemental Analysis by Optical Spectrometry

Very often, techniques for direct solid analysis are classified into two groups according to whether they provide bulk information (of interest for homogeneous samples) or analytical information with lateral and/or depth resolution.

1.5.1.1 Bulk Analysis Techniques

Spark-source optical emission spectrometry (SS-OES) and XRF are well-established routine techniques, providing bulk information, which play an important role nowadays in industrial process monitoring (raw materials and final products). Whereas SS-OES is clearly an atomic technique, where atoms are formed directly from the solid sample by virtue of the high energy of an electrical spark, the inclusion of XRF among the atomic techniques deserves some explanation. In this latter technique, the sample is investigated for its composition at room temperature and so atoms are not formed during the analysis. In XRF, a primary beam of X-rays is used to excite and eject electrons from inner shells of the atoms (*e.g.* K or L shells) of a solid. After such excitation, electrons of the outer shells fall spontaneously to fill in the 'holes' originated and the difference in the energy between the two levels is released in the form of electromagnetic radiation (fluorescent or secondary X-rays), providing elemental information.

SS-OES is used mainly for the analysis of electrical conductors. The surface of the sample is first ground flat and placed against the spark stand, where it is flooded with argon. The spark includes two phases: the first consists on a low-energy discharge produced by a primary circuit which applies a potential in excess of 10 kV for a few microseconds to ionise the argon and create a conducting plasma. As soon as the plasma is formed, the second phase starts, melting the sample and evaporating it at the spark's point of impact. The elements present in the plasma are excited and emit their characteristic spectra. The total duration of both phases of the spark is only a few milliseconds. Unfortunately, matrix effects, self-absorption and even self-reversal problems are usually observed in SS-OES.

In general, analytical techniques based on the use of electromagnetic radiation for excitation purposes allow for the direct analysis of any solid sample independently of its conductivity; however, thin conductive coatings and/or other charge-balancing techniques are usually required when using charged particles for excitation. Therefore, direct analysis of conducting and insulating samples with XRF is feasible. Regrettably, XRF suffers from severe matrix effects and this constitutes its most serious drawback. Matrix correction models are being developed to compensate for the absorption of light from other elements in the solid sample and also for secondary fluorescence coming from atoms of the analyte excited by light emitted by excited atoms of other elements in the sample.

1.5.1.2 Techniques Providing Lateral and/or Depth Information

Three techniques with spatially resolved information capabilities have been selected here for some further explanation: EPXMA, laser-induced breakdown spectroscopy (LIBS) and glow discharge optical emission spectrometry (GD-OES). Figure 1.15 summarises the lateral and depth resolution provided by the techniques described in this section. It is worth noting that the closer to the bottom left corner the technique is located, the higher (and so better) is the depth resolution.

As for XRF, gas atoms are not formed during EPXMA. In EPXMA, an electron probe is used to excite and eject electrons from the solid, yielding excited ions which relax and emit X-radiation. Electron guns can be focused easily on small areas of the solid surface, although exciting electrons cannot go too deep into the solid. Hence this technique obtains analytical information with some spatial resolution. The technique is prone to serious matrix interferences, like XRF.

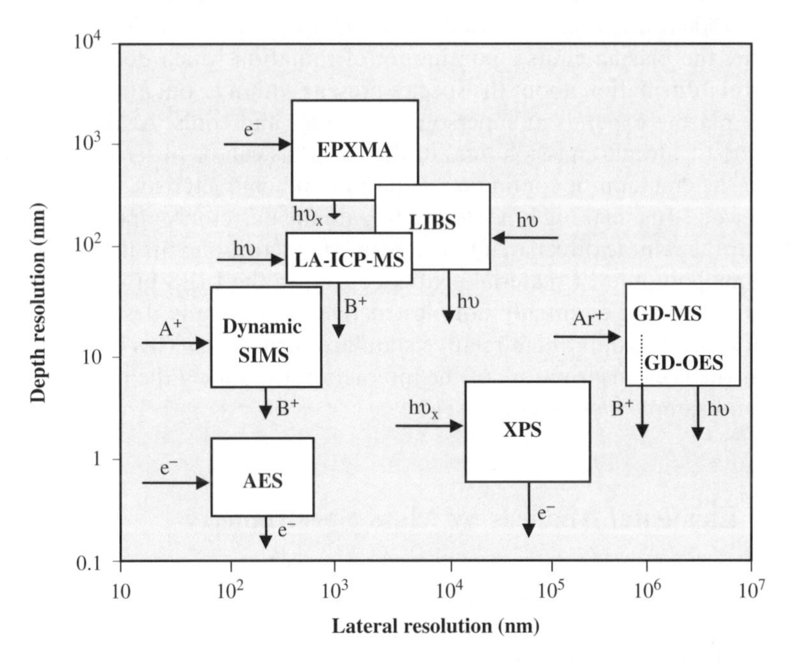

Figure 1.15 Comparison of the lateral and depth resolution allowed by different optical and mass spectrometric techniques used for direct solid analysis (A^+, B^+, incident and emitted ions, respectively; e^-, electrons; *hυ*, photons). XPS and AES are included in the graph for comparison.

The generation of a GD takes place typically in a low-pressure chamber through which argon flows continuously. The device consists of a grounded anode and a cathode (the sample). An electric current ionises the discharge gas, forming a plasma and yielding argon ions which are attracted towards the sample surface producing the sputtering (removal) of atoms, electrons and ions. The atoms of the analyte are excited and ionised through collisions in the plasma and, therefore, measurement by OES and MS is feasible. The uniform formation of craters on the sample surface leads to a good depth resolution [30] (see Figure 1.15). GD-OES offers fairly good detection limits (mg kg^{-1}) and high sample throughput. Further, the sputtering and excitation processes are rather separated, giving rise to minimal matrix effects as compared with other direct solids analysis techniques and this simplifies the quantitation of depth profiles. However, problems in depth quantitation still remain; for example for the analysis of multilayered samples containing layers of very different composition (from one layer to another, the composition of a given element can change from a very low concentration to almost 100%; therefore, if a highly sensitive emission line is chosen, self-absorption can occur). Moreover, algorithms need to be developed to correct for the effect of some light elements (such as hydrogen, nitrogen and oxygen) which can produce serious effects in the calibration curves.

LIBS operates by focusing a laser on to a small area at the surface of the specimen. The laser ablates a tiny amount of material, in the ng–pg range, which instantaneously generates a plasma plume with temperatures of about 10 000–20 000 K and breaks down the material into excited and ionic species. At that time, the plasma emits a continuum of radiation which does not contain any useful information about the species present within it, but after a very short time the plasma expands at supersonic velocities and cools. At this point, the characteristic atomic emission lines of the elements can be observed. The delay between the emission of continuum radiation and characteristic radiation is of the order of 10 μs, making it necessary to gate the detector temporally. LIBS is useful for the remote direct analysis of both conductors and insulators. Because a very small amount of material is ablated during the LIBS process, the technique is considered essentially non-destructive or minimally destructive. Typical data in LIBS exhibit high relative standard deviations. This is due, in part, to either the non-linear nature of the interaction producing the plasma, matrix effects and sample heterogeneity [31].

1.5.2 Elemental Analysis by Mass Spectrometry

Three direct solid analysis mass spectrometric techniques allowing for lateral and/or depth resolution have been selected in this section: laser ablation (LA) coupled to ICP-MS, secondary ion mass spectrometry (SIMS) and GD-MS.

LA–ICP-MS is suitable for the direct analysis of materials such as metals, semiconductors, ceramics and insulators at trace and ultratrace levels (detection limits ~1 ng g^{-1}) without sample preparation. The MS detection mode makes it possible isotope analysis and also isotope dilution methods using stable tracers to improve the accuracy and reproducibility of the analytical results. Unlike LIBS, LA–ICP-MS separates the ionisation step from the sampling step. In LA–ICP-MS, a short-pulsed, high-power laser beam is focused into the sample surface in an inert gas atmosphere (*e.g.* Ar) under normal pressure. The ablated material is transferred by an Ar gas stream into the ICP ion source of an ICP-MS instrument for atomisation and ionisation purposes. Therefore, both steps can be independently controlled and optimised. Further, in LA–IPC-MS a dry sample is introduced into the plasma, which results in a lack of polyatomic interfering species caused by the interaction of water and acid species with the argon plasma. However, limitations still occur and most of them come from the laser-based sampling process [32], for example, the occurrence of non-stoichiometric effects in the transient signals, defined as elemental fractionation. Moreover, matrix effects and non-linear calibrations are other frequent limitations of LA–ICP-MS.

In dynamic SIMS, a primary ion beam of energy, ranging from 0.5 to 20 keV, is used to sputter-remove successive layers of the sample in a well-defined area ranging in size from, typically, 1×1 mm to $10 \times 10\,\mu m$. This yields elemental information on the surface region from a few nanometres to several hundred micrometres in depth. The detection limits of the technique are in the ppm–ppb range. Unfortunately, quantification by SIMS is bedevilled by matrix effects. They arise because the particle emission and ionisation processes take place 'simultaneously'. If we were able to decouple sputtering from ionisation (ionisation occurring after the neutrals were moved away from the surface), the ion yield would be independent of the matrix and quantification would be easier.

Instruments based on GD-MS coupling have been employed most commonly for the quantitative analysis of trace and ultratrace amounts in high-purity materials. However, it has been demonstrated that, as in GD-OES, quantitative depth profile analysis by GD-MS is possible [33]. At present, a GD-MS prototype which allows the depth quantification of thin layers on conducting or insulating materials is being developed for commercial purposes [34].

Figure 1.15 shows the lateral and depth resolution achievable with the three mass spectrometric techniques described in this section. As can be seen, the depth resolution obtained with the GD techniques is similar to that with dynamic SIMS (with the additional advantage of less matrix effects in the GD sources). However, the lateral resolution obtained with SIMS is much better because the primary ion beam in SIMS is highly focused whereas in a GD the limitations in the source design make it necessary to sputter a sample area with a diameter of 1–4 mm. On the other hand, the depth resolution obtained with techniques based on lasers is not yet as good as with SIMS or GDs.

References

1. J. Koch, A. Zybin and K. Niemax, Narrow and broad band diode laser absorption spectrometry – concepts, limitations and applications, *Spectrochim. Acta, Part B*, 57, 2002, 1547–1561.

2. U. Heitmann, M. Schutz, H. Becker-Ross and S. Florek, Measurements on the Zeeman-splitting of analytical lines by means of a continuum source graphite furnace atomic absorption spectrometer with a linear charge coupled device array, *Spectrochim. Acta Part B*, 51, 1996, 1095–1105.
3. A. Sanz-Medel and R. Pereiro, *Atomic Absorption Spectrometry: an Introduction*, Coxmoor Publishing, Oxford, 2008.
4. M. G. R. Vale, N. Oleszczuk and W. N. L. dos Santos, Current status of direct solid sampling for electrothermal atomic absorption spectrometry – a critical review of the development between 1995 and 2005, *Appl. Spectrosc. Rev.*, 41, 2006, 377–400.
5. M. J. Cal-Prieto, M. Felipe-Sotelo, A. Carlosena, J. M. Andrade, P. López-Mahía, S. Muniategui and D. Prada, Slurry sampling for direct analysis of solid materials by electrothermal atomic absorption spectrometry (ETAAS). A literature review from 1990 to 2000, *Talanta*, 56, 2002, 1–51.
6. J. M. Mermet, Trends in instrumentation and data processing in ICP-AES, *J. Anal. At. Spectrom.*, 17, 2002, 1065–1071.
7. R. E. Sturgeon, Furnace atomisation plasma emission/ionisation: review of an underutilized source for atomic and molecular spectrometry, *Can. J. Anal. Sci. Spectrosc.*, 49, 2004, 385–397.
8. S. J. Hill, *Inductively Coupled Plasma Spectrometry and its Applications*, Blackwell Publishing, Oxford, 2007.
9. A. Montaser and D. W. Golightly, *Inductively Coupled Plasmas in Analytical Atomic Spectrometry*, VCH, Weinheim, 1992.
10. G. M. Hiefjte, Emergence and impact of alternative sources and mass analysers in plasma source mass spectrometry, *J. Anal. At. Spectrom.*, 23, 2008, 661–672.
11. S. M. Nelms, *ICP Mass Spectrometry Handbook*, Blackwell Publishing, Oxford, 2005.
12. S. D. Tanner, Ion optics for ICP-MS: modelling intuition or blind luck, in *Plasma Source Mass Spectrometry Development and Applications*, ed. G. Holland and S. D. Tanner, Royal Society of Chemistry, Cambridge, 1997, pp. 13.
13. G. Horlick and Y. Shao, Inductively coupled plasma-mass spectrometry for elemental analysis, in *Inductively Coupled Plasmas in Analytical Atomic Spectrometry*, ed. A. Montaser and D. W. Golighly, VCH, Weinheim, 1992, Chapter 12, pp. 551–612.
14. J. Vogl, Calibration strategies and quality assurance, in *ICP Mass Spectrometry Handbook*, ed. S. M Nelms, Blackwell Publishing, Oxford, 2005, pp. 147–181.
15. P. Abbyad, J. Tromp, J. Lam and E. Salin, Optimisation of the technique of standard additions for inductively coupled plasma mass spectrometry, *J. Anal. At. Spectrom.*, 16, 2001, 464–469.
16. I. E. Vasilyeva, E. V. Shabanova, Y. V. Sokolnikova, O. A. Proydakova and V. I. Lozhkin, Selection of internal standard for determination of boron and phosphorus by ICP-MS in silicon photovoltaic materials, *J. Anal. At. Spectrom.*, 14, 1999, 1519–1521.

17. J. I. García-Alonso, M. Montes-Bayón and A. Sanz-Medel, Environmental applications using ICP-MS: semiquantitative analysis, in *Plasma Source Mass Spectrometry New Development and Applications*, ed. G. Holland and S. D. Tanner, Royal Society of Chemistry, Cambridge, 1999, pp. 95–107.
18. M. Rupprecht and T. Probst, Development of a method for the systematic use of bilinear multivariate calibration methods for the correction of interferences in inductively coupled plasma-mass spectrometry, *Anal. Chim. Acta*, 358, 1998, 205–225.
19. K. G. Heumann, Isotope-dilution mass-spectrometry (IDMS) of the elements, *Mass Spectrom. Rev.*, 11, 1992, 41–67.
20. P. Rodríguez-González, J. M. Marchante-Gayón, J. I. García Alonso and A. Sanz-Medel, Isotope dilution analysis for elemental speciation: a tutorial review, *Spectrochim. Acta, Part B*, 60, 2005, 151–207.
21. A. Sanz-Medel (ed.), *Flow Analysis with Atomic Spectrometric Detectors*, Elsevier, Amsterdam, 1999.
22. J. Szpunar, R. Lobinski and A. Prange, Hyphenated techniques for elemental speciation in biological systems, *Appl. Spectrosc.*, 57, 2003, 102A–112A.
23. I. Rodríguez Pereiro and A. Carro Díaz, Speciation of mercury, tin and lead compounds by gas chromatography with microwave-induced plasma and atomic-emission detection (GC–MIP–AED), *Anal. Bioanal. Chem.*, 372, 2002, 74–90.
24. J. M. Costa-Fernández, F. Lunzer, R. Pereiro, N. Bordel and A. Sanz-Medel, Direct coupling of high-performance liquid chromatography to microwave-induced plasma atomic emission spectrometry via volatile-species generation and its application to mercury and arsenic speciation, *J. Anal. At. Spectrom.*, 10, 1995, 1019–1025.
25. K. G. Heumann, Isotope-dilution ICP-MS for trace element determination and speciation: from a reference method to a routine method?, *Anal. Bioanal. Chem.*, 378, 2004, 318–329.
26. P. Rodríguez-González, M. Monperrus, J. I. García Alonso, D. Amouroux and O. F. X. Donard, Comparison of different numerical approaches for multiple spiking species-specific isotope dilution analysis exemplified by the determination of butyltin species in sediments, *J. Anal. At. Spectrom.*, 22, 2007, 1373–1382.
27. N. H. Bings, J. M. Costa-Fernández, J. P. Guzowski Jr, A. M. Leach and G. M. Hieftje, Time-of-flight mass spectrometry as a tool for speciation analysis, *Spectrochim. Acta, Part B*, 55, 2000, 767–778.
28. M. Vázquez Peláez, J. M. Costa-Fernández and A. Sanz-Medel, Critical comparison between quadrupole and time-of-flight inductively coupled plasma mass spectrometers for isotope ratio measurements in elemental speciation, *J. Anal. At. Spectrom.*, 17, 2002, 950–957.
29. I. Gunther-Leopold, B. Wernli, Z. Kopajtic and D. Gunther, Measurement of isotope ratios on transient signals by MC-ICP-MS, *Anal. Bioanal. Chem.*, 378, 2004, 241–249.

30. J. Pisonero, B. Fernández, R. Pereiro, N. Bordel and A. Sanz-Medel, Glow-discharge spectrometry for direct analysis of thin and ultra-thin solid films, *Trends Anal. Chem.*, 25, 2006, 11–18.
31. D. L. Death, A. P. Cunningham and L. J. Pollard, Multielement analysis of iron ore pellets by Laser-induced breakdown spectroscopy and principal components regression, *Spectrochim. Acta, Part B*, 63, 2008, 763–769.
32. B. Fernández, F. Claverie, C. Pécheyran and O. F. X. Donard, Direct analysis of solid samples by fs-LA-ICP-MS, *Trends Anal. Chem.*, 26, 2007, 951–966.
33. M. Vázquez Peláez, J. M. Costa-Fernández, R. Pereiro, N. Bordel and A. Sanz-Medel, Quantitative depth profile analysis by direct current glow discharge time of flight mass spectrometry, *J. Anal. At. Spectrom.*, 18, 2003, 864–871.
34. A. C. Muñiz, J. Pisonero, L. Lobo, C. González, N. Bordel, R. Pereiro, A. Tempez, P. Chapon, N. Tuccitto, A. Licciardello and A. Sanz-Medel, Pulsed radiofrequency glow discharge time of flight mass spectrometer for the direct analysis of bulk and thin coated glasses, *J. Anal. At. Spectrom.*, 23, 2008, 1239–1246.

CHAPTER 2

Implementing A Robust Methodology: Experimental Design and Optimization

XAVIER TOMÁS MORER,[a] LUCINIO GONZÁLEZ-SABATÉ,[a] LAURA FERNÁNDEZ-RUANO[a] AND MARÍA PAZ GÓMEZ-CARRACEDO[b]

[a] Department of Applied Statistics, Institut Químic de Sarrià, Universitat Ramon Llull, Barcelona, Spain; [b] Department of Analytical Chemistry, University of A Coruña, A Coruña, Spain

2.1 Basics of Experimental Design

In this section, the basics of experimental designs are introduced. Many definitions are given to introduce the most usual nomenclature and to avoid some common misconceptions. A practical and easy-to-follow approach was preferred instead of a more formal one, and literature references are given for more advanced or interested readers.

2.1.1 Objectives and Strategies

An '*experiment*' is just a test or series of tests. Experiments are performed in all scientific disciplines and are an important part of the way we learn about how systems and processes work. The validity of the conclusions that are drawn from an experiment depends to a large extent on how the experiment was conducted. Therefore, the design of experiments plays a major role in the eventual solution of the problem that initially motivated the experiment and, as

RSC Analytical Spectroscopy Monographs No. 10
Basic Chemometric Techniques in Atomic Spectroscopy
Edited by Jose Manuel Andrade-Garda

Published by the Royal Society of Chemistry, www.rsc.org

a consequence, what it is currently understood by '*Experimental Design*' is an important tool for analytical chemists who are interested in improving the performance of an analytical procedure. The term '*performance*' might be rather ambiguous although, in general, we relate it to an analytical property such as recovery, signal-to-noise ratio, uncertainty and robustness, and we suppose that such a property is influenced by an undefined number of experimental conditions such as pH, reagent concentrations, analytical method, instrumental conditions, analysts and so on.

A critical question is how to perform the experiments in order to evaluate the influence of the experimental conditions on the property that measures the performance of the procedure.

Common practice consists in investigating the influence of one experimental variable (hereafter we will refer to it as a '*factor*') while keeping other factors at a fixed value. Then, another factor is selected and modified to perform the next set of experiments, and so forth. This one-factor-at-a-time strategy has been shown to be inefficient and expensive; it lacks the ability to detect the joint influence of two or more factors (*i.e.* it cannot address interactions) and often needs many experiments. An increase in efficiency can be achieved by studying several factors simultaneously and systematically by means of an appropriate type of experimental design. In such a way, the experiments will be able to detect the influence of each factor and also the influence of two or more factors because every observation gives information about all factors.

2.1.2 Variables and Responses: Factors, Levels, Effects and Interactions

The concepts related to 'experimental design' [1–3] were developed in the framework of applied statistics and so they use a particular language. Let us state some relevant technical words related to the main concepts of experimental design from a more chemical viewpoint.

1. The process we are studying (analytical procedure, sample pretreatment, measuring stage, *etc.*) is referred to as '*a system*'.
2. We will define '*an input to the system*' as a quantitative or qualitative variable (experimental condition) that might have an influence upon the system. We can measure such an influence as a change in one or more output variables named '*responses*'.
3. Variables can be quantitative, *e.g.* temperature, pH, concentration, or qualitative, such as type of solvent, presence or absence of a catalyst.
4. Responses can include such quantities and qualities as yield, recovery, colour, cost and accuracy, and normally are related to the objectives of the research.
5. A '*factor*' is defined as one of the variables contributing to a particular result or situation. It is an input variable that has (or may have) an influence upon the system. Factors can be controlled or uncontrolled by the experimenter. Controlled factors are identified experimental conditions

that can be fixed (*e.g.* temperature 25 °C) or settled at will (pH 5/pH 7). Each of these different values is called '*a level of the factor*'. Uncontrolled factors are unidentified variables or conditions that the experimenter voluntarily decides not to control (room temperature, atmospheric pressure, ionic strength, *etc.*). Uncontrolled factors are undesirable in experimental situations because their influence cannot always be easily or unambiguously detected or evaluated. Experimental error can be considered as an uncontrolled factor.

6. The '*effect of a factor*' is defined as the measurable change in a response that can be assigned to a change in the level of the factor. We must consider its magnitude (large, small) and also the sign: positive (increase in response) or negative (decrease in response).
7. Two factors are said to '*interact*' when the effect of one of them is different at different levels of the other. In general, when factors operate independently of each other, they do not exhibit interaction. As an example, consider that in a hypothetical reaction intended to improve the extraction of some metal from a sample matrix addition of 10 mg of a catalyst at a temperature of 25 °C increases the yield. We can try to predict what happens at 45 °C. If the yield decreases, it is a proof that an interaction between the factors catalyst and temperature exists. Why? Well, answering this question is not an experimental design task; rather, it is simply a chemical problem and it may be a challenge for chemists to explain it.
8. Of course, more than two factors can be involved in an interaction. The number of factors involved in an interaction is called '*the order of the interaction*'. Interactions are always a source of information for the experimenter. If an unexpected interaction exists, the total effect is not the simple sum of the effects of the factors but an interaction term is included. For instance, for two factors we can establish the following equation:

$$\text{total effect} = \text{effect of factor A} + \\ + \text{effect of the AB interaction} \quad (2.1)$$

2.2 Analysis of Experimental Designs

2.2.1 Factorial Designs

Factorial designs are a popular class of experimental designs that are often used to investigate multifactor response surfaces. The word '*factorial*' does not have its usual mathematical meaning of an integer multiplied by all integers smaller than itself (*e.g.* $5! = 5 \times 4 \times 3 \times 2 \times 1$); instead, it simply indicates that many factors are varied simultaneously in a systematic way. Major advantages of factorial designs are that they can be used to reveal the existence of factor interactions and to fit an empirical response surface.

Historically, factorial designs were introduced by Sir Ronald A. Fisher to refute the, then (1935), prevalent idea that if one were to discover the effect of a factor, all other factors must be held constant and only the factor of interest should be varied. Fisher showed that all factors of interest could be varied simultaneously and the individual factor effects and also the effects of interactions could be estimated by a proper mathematical treatment.

The '*treatments*' or '*experimental conditions*' for each run consist of all possible combinations of levels from the different factors. For example, in a complete factorial design (CFD) with two factors A and B, if they have levels a_1, a_2, a_3 and b_1, b_2, respectively, then the treatment combinations are (a_1, b_1), (a_1, b_2), (a_2, b_1), (a_2, b_2), (a_3, b_1) and (a_3, b_2). The number of experimental runs required by a CFD is the product of the number of levels of each factor and the number of times each combination is replicated.

If a complete factorial design has three levels (low, middle and high) for each of three factors (A = temperature; B = pH and C = reaction time), it is said to be a $3 \times 3 \times 3$ or 3^3 CFD and 27 runs are required, each one corresponding to a particular combination of the factor levels. Furthermore, if we replicate each run k times, then the number of experiments needed is $k \times 3^3$.

By means of an analysis of variance (ANOVA), we can detect the influence of each factor (A, B and C), each interaction (AB, AC, BC, ABC) and, if replicates are available, estimate the purely experimental error. If no replicates were performed, according to the ANOVA rationale, we can consider the high-order interaction as an estimation of the pure experimental error.

Knowing which factor and interaction have an influence on the response (y), we can try to fit an empirical model by the common least-squares criterion:

$$y = b_0 + b_1\mathrm{A} + b_2\mathrm{B} + b_3\mathrm{C} + b_{11}\mathrm{A}^2 + b_{22}\mathrm{B}^2 + b_{33}\mathrm{C}^2 + b_{12}\mathrm{AB} + b_{13}\mathrm{AC} + \ldots + b_{123}\mathrm{ABC} \quad (2.2)$$

This model allows us to estimate a response inside the experimental domain defined by the levels of the factors and so we can search for a maximum, a minimum or a zone of interest of the response. There are two main disadvantages of the complete factorial designs. First, when many factors were defined or when each factor has many levels, a large number of experiments is required. Remember the expression: number of experiments = replicates $\times$ (levels)^{factors} (*e.g.* with 2 replicates, 3 levels for each factor and 3 factors we would need $2 \times 3^3 = 54$ experiments). The second disadvantage is the need to use ANOVA and the least-squares method to analyse the responses, two techniques involving no simple calculi. Of course, this is not a problem if proper statistical software is available, but it may be cumbersome otherwise.

2.2.2 2^f Factorial Designs

The most common types of factorial designs are those that have all factors at two levels, let us call them 'high' and 'low'. The number of different treatment combinations or experiments is 2^f and if k replicates are performed at each treatment combination, the total number of runs is $N = k \times 2^f$. To study two-level factorial designs, the levels can be coded -1 (low level) and $+1$ (high level), so each treatment combination is then represented by a series of plus and minus signs [*e.g.* (–, +, –) means (low, high, low level)]. For a continuous factor such as temperature with levels 25 and 45 °C, the levels can be coded by the following equations:

$$\text{low level} = 2(\text{low value} - \text{mean})/\text{range} = 2(25 - 35)/(45 - 25) = -1 \quad (2.3)$$

$$\text{high level} = 2(\text{high value} - \text{mean})/\text{range} = 2(45 - 35)/(45 - 35) = +1 \quad (2.4)$$

For a discontinuous factor such as 'solvent' with levels 'water' and 'ethanol', the levels can be coded at your convenience, *e.g.* 'water' (–1) and 'ethanol' (+1).

The '*design matrix*' of a factorial design is a list detailing the total number of treatments, combinations or experiments. Columns represent each of the factors being studied, denoted by capital letters, and each row corresponds to an experiment. Hence a 2^2 factorial design would have 4 runs with the following design matrix:

	Factor	
Run	*A*	*B*
1	–	–
2	+	–
3	–	+
4	+	+

Similarly, a 2^3 factorial design has 8 runs and its design matrix is:

	Factor		
Run	*A*	*B*	*C*
1	–	–	–
2	+	–	–
3	–	+	–
4	+	+	–
5	–	–	+
6	+	–	+
7	–	+	+
8	+	+	+

If we analyse the columns, we can easily discover a general rule to construct any design matrix. First column alternates 1 minus and 1 plus signs; second column alternates 2 minus and 2 plus signs; third column alternates 4 minus and 4 plus signs; fourth column alternates 8 minus and 8 plus signs, and so on. Now one is able to establish the matrix design of any 2^f factorial design.

A '*matrix design*' is a suitable way of obtaining all the treatment combinations implicated in a 2^f factorial design, but it is not a handy system to notate them. We are obliged to specify the complete sequence of signs of the corresponding row in the matrix design (*e.g.* in a 2^5 design: $-+-++$). Standard notation, developed by Frank Yates, makes the identification of each run easier. It uses the lower-case letter of the factor to indicate that this factor is in its high level in the run. Hence the previous run will be expressed as *bde* in its standard notation and means that factors B, D and E are in their high level and factors A and C in their low level. If all the factors are in their low level, the symbol used to indicate this is (1). As an example, the standard notation for the matrix design of a 2^2 factorial design is (1), *a*, *b*, *ab* and that of a 2^3 factorial design is (1), *a*, *b*, *ab*, *c*, *ac*, *bc*, *abc* (the order is also standardised and is termed '*standard order*').

2.2.3 Algorithms: BH^2 and Yates

Once you have selected the factors to study, defined their low and high levels, constructed the matrix design and carried out the 2^f runs in the laboratory, you should have obtained the corresponding 2^f responses. Now, the interest is in evaluating the different effects.

The net effect of each factor can be calculated as the mean of the responses obtained when the factor is in its high level (+) minus the mean of the responses obtained when the factor is in its low level (−). Thus, denoting as Yi the response of run i, the effect of factor A in a 2^2 factorial design is given by

$$\text{effect A} = (Y2 + Y4)/2 - (Y1 + Y3)/2 = (-Y1 + Y2 - Y3 + Y4)/2 \quad (2.5)$$

We can observe that the responses are in the standard order and the sign of each response is the same as that of the column corresponding to factor A, and the divisor is just half of the total number of experiments.

Two algorithms are available to perform all the calculations in a very simple way, namely the Box, Hunter and Hunter (BH^2) algorithm and Yates's algorithm. Both are considered below for a typical and simple example of a 2^3 factorial design. Assume we are studying the influence of pH (A), temperature (B) and time (C) over the yield (response in %) of the extraction of a metal from a complex analytical matrix, just before conducting the extracts to an ICP device. The levels of each factor, fixed by the analyst, are: pH (A), 3 (−), 5(+); temperature (B), 40 (−), 60 °C (+) and time (C), 1 (−), 2 h. (+). The matrix design and the experimental data are as follows:

Run	*A*	*B*	*C*	*Yield*
1	−	−	−	82
2	+	−	−	88
3	−	+	−	82
4	+	+	−	86
5	−	−	+	81
6	+	−	+	84

(*Continued*).

Run	*A*	*B*	*C*	*Yield*
7	–	+	+	85
8	+	+	+	89

How can we evaluate the effect of factors A, B and C and also those of interactions AB, AC, BC and ABC?

2.2.3.1 *BH*^{2} *algorithm*

The algorithm proposed by Box, Hunter and Hunter (BH^2) starts with the matrix design and the concept of effect of a factor. As shown in Table 2.1, the columns of signs corresponding to interactions are obtained by multiplying the elements of their respective factors (*e.g*. the column of signs for the interaction AB is obtained by multiplying together the signs for A and B).

Three additional rows labelled 'Total', 'Divisor' and 'Effect', are added. Each element of the 'Total' row is obtained by adding the response elements taking into account the signs of the column under consideration (*e.g*. for A: Total = $-82+88-82+86-81+84-85+89=17$). 'Divisor' elements are half of the number of runs, except for column I, which is exactly the number of runs. The last row, 'Effect', is just the ratio of the elements in the respective 'Total' and 'Divisor' rows (A effect = $17/4=4.25$).

As can be seen, all the elements of the extra column I are '+', so that the respective effect that they evaluate is just the arithmetic mean of the responses, named in experimental design '*Mean effect*'. Although the BH^2 algorithm has many uses, the most rapid way to calculate effects (particularly when we consider many factors) is by means of an algorithm developed by Yates.

Table 2.1 Signs to calculate the effects on a 2^3 factorial design according to the BH^2 algorithm.

Run	*I*	*A*	*B*	*C*	*AB*	*AC*	*BC*	*ABC*	*Response*
(1)	+	–	–	–	+	+	+	–	82
a	+	+	–	–	–	–	+	+	88
b	+	–	+	–	–	+	–	+	82
ab	+	+	+	–	+	–	–	–	86
c	+	–	–	+	+	–	–	+	81
ac	+	+	–	+	–	+	–	–	84
bc	+	–	+	+	–	–	+	–	85
abc	+	+	+	+	+	+	+	+	89
Total	677	17	7	1	–1	–3	11	3	
Divisor	8	4	4	4	4	4	4	4	
Effect	84.625	4.25	1.75	0.25	–0.25	–0.75	2.75	0.75	

Table 2.2 Yates's algorithm for the 2^3 factorial design.

Run	*Response*	*E1*	*E2*	*E3*	*Divisor*	*Effect*
(1)	82	170	338	677	8	84.625
a	88	168	339	17	4	4.25
b	82	165	10	7	4	1.75
ab	86	174	7	−1	4	−0.25
c	81	6	−2	1	4	0.25
ac	84	4	9	−3	4	−0.75
bc	85	3	−2	11	4	2.75
abc	89	4	1	3	4	0.75

2.2.3.2 Yates's algorithm

Yates's algorithm (named after Frank Yates, a co-worker of Ronald Fisher, 1902–94) is applied to the observations after they have been arranged in the standard order. As shown in Table 2.2, the Yates calculations start by evaluating as many auxiliary columns as factors are considered, in our example three columns E1, E2 and E3 for a 2^3 design.

The column 'Response' contains the yields for each run, which are now considered in successive pairs. The first four elements in column E1 are obtained by adding the pairs together ($88+82=170$, $86+82=168$, and so on). The last four elements in column E1 are obtained by subtracting the top response from de bottom response of each pair, thus $88-82=6$, $86-82=4$, and so on. In the same way that column E1 is obtained from the 'Response' column, column E2 is obtained from column E1 and column E3 is obtained from column E2. The elements in column E3 are precisely the values that are obtained in the row 'Total' by the BH^2 algorithm. To obtain the net effects of the factors, we have only to divide those values by the appropriate divisor, as before, which is 8 (the number of runs) for the first element and 4 (half of the runs) for the others. The effects are identified by locating the plus signs in the design matrix or directly from the standard notation. Thus in the second row we identify the effect of factor A and in the sixth row the effect of the interaction AC.

2.2.4 Graphical and Statistical Analysis

According to the definition of effect, we can interpret it as the change induced in the response due to the change of a factor from its low level to the high level. A positive effect suggests an increase in the response value whereas a negative effect means a decrease in the response. The '*magnitude of the effect*' is the extent of this change measured in the response units (remember that the levels are coded).

The simplest way to analyse the effects is, perhaps, by means of proper graphics, among which the Pareto chart and the 'main effects' or 'interactions effects' plots are used widely. In a Pareto chart we represent the different effects ordered by magnitude (absolute value, on the vertical axis) and the magnitude

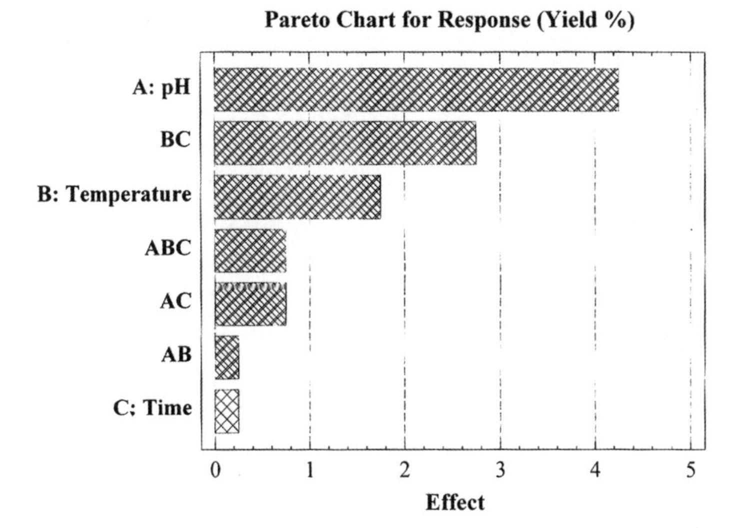

Figure 2.1 Pareto chart for a 2^3 design.

of each effect (on the horizontal axis), as shown in Figure 2.1. The Pareto chart allows us to visualise which effects are really important. According to Pareto's principle, only some effects will be relevant (the name was assigned after Italian economist Vilfredo Pareto, who observed that 80% of income in Italy went to 20% of the population; this was generalised to a common rule of thumb in business and quality control, *e.g.* '80% of your sales and/or quality faults come from 20% of your clients/technical problems), and so we might discard the others. As can be seen in our example, the main effect is due to factor A (pH), followed by the effect of the interaction BC (temperature × time) and factor B (temperature). The other effects are small, even compared with the effect of the highest interaction (ABC), which is currently considered as the experimental error in statistics whenever no replications of the experiments are available.

The 'main effects plot' represents the response (vertical axis) *versus* the levels of the factor under consideration (horizontal axis, see Figure 2.2). Each point represents the mean response obtained at each level. For our example (see Figure 2.2), the mean responses (yield) for factor A at the low and high levels are 82.5% and 86.75%, respectively. That is to say, the increase in yield (effect of A, pH) is 4.25% when we change the pH from 3 to 5.

Since there is a non-negligible interaction between temperature (B) and time (C), the effects of these factors must be considered jointly. An 'interaction effect plot' is a simple way to carry out this analysis. In Figure 2.3, the vertical axis represents, again, the response whereas the horizontal axis shows the levels of one of the factors involved in the interaction. For each of these levels we represent the mean response obtained at each combination of levels of the other factor. As presented in Figure 2.3, when time (C) is set at its low level, a change from the low to the high level of temperature (B) decreases the response (yield)

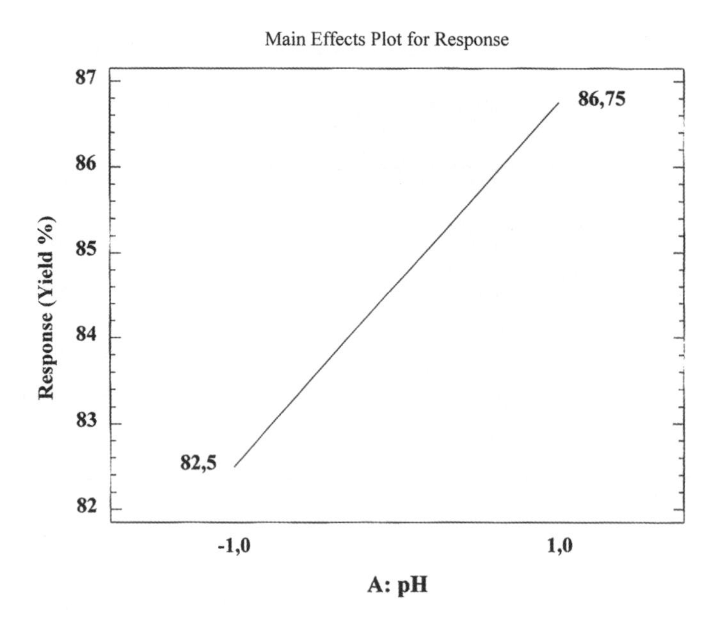

Figure 2.2 Main effects plot for a 2^3 design.

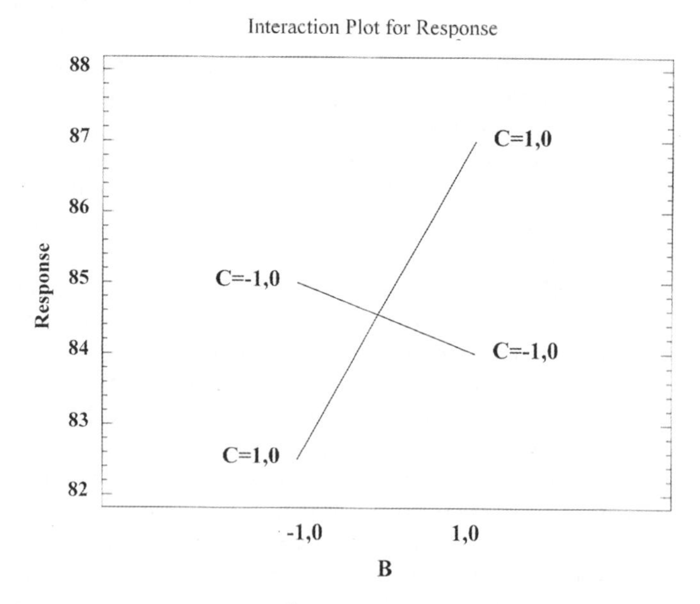

Figure 2.3 Interaction plot for a 2^3 design (the response is given as yield(%), see text for details).

by 1%, whereas when time is at its high level, the same change in temperature increases the yield by 4.5%.

Experimental design does not explain the nature of any interaction, as this is a chemical task, but it provides helpful suggestions to explain it. Fairly often, graphical analyses of the response provide us with enough information to state convenient conclusions. Going back to our example, if we are looking for a high extraction yield we can conclude that the best experimental conditions are those with all experimental parameters at their highest levels: pH = 5, temperature = 60 °C and time = 2 h. In doing so, we expect an average yield of about 87%.

In addition to a graphical analysis, a statistical study can also be performed to analyse the experimental results. As any factorial design matrix is a structure that allows for the analysis of the responses by means of an ANOVA, this is a common way to detect which effects have statistical significance. ANOVA needs an estimation of the mean square error that can be obtained by performing replicates of the runs or assuming that the effect of the higher order interaction would measure differences arising principally from experimental error, and then the mean square error is the mean of the sum of squares of the n interactions considered and has n degrees of freedom (df) (if $n = 1$, then $\mathrm{df} = \infty$).

The sum of squares of each factor and interaction can be easily evaluated if we have used the BH^2 or Yates's algorithm. In the BH^2 algorithm, each sum of squares is equal to the square of the corresponding element in row 'Total' divided by the total number of runs (2^f). In Yates's algorithm, each sum of squares is equal to the square of the corresponding element in the last auxiliary column (E3 in the example) also divided by the total number of runs. Table 2.3 shows the ANOVA table of our example. Note that all elements in the degrees of freedom column are equal to 1 as our design is a two-levels design. Column 'p-Value' shows that only factors A (pH), B (temperature) and the interaction BC (temperature × time) have statistical significance. Of course, calculations can be performed quickly using a statistical software package if available.

Fitting the data to an empirical surface response can also be used to estimate the response within the range of experimental conditions. The simplest model that we can consider is the following:

$$\text{response} = b_0 + b_1\mathrm{A} + b_2\mathrm{B} + \ldots + b_{12}\mathrm{AB} + b_{13}\mathrm{AC} + b_{123}\mathrm{ABC} + \ldots \quad (2.6)$$

Table 2.3 ANOVA table to study a 2^3 factorial design.

Source	*Sum of squares*	*df*	*Mean square*	*F-ratio*	*p-Value*
Total	59.875	7	59.875		
A	36.125	1	36.125	32.11	0.0000
B	6.125	1	6.125	5.44	0.0196
C	0.125	1	0.125	0.11	0.7389
AB	0.125	1	0.125	0.11	0.7389
AC	1.125	1	1.125	1.00	0.3173
BC	15.125	1	15.125	13.44	0.0002
ABC	1.125	1	1.125	1.00	

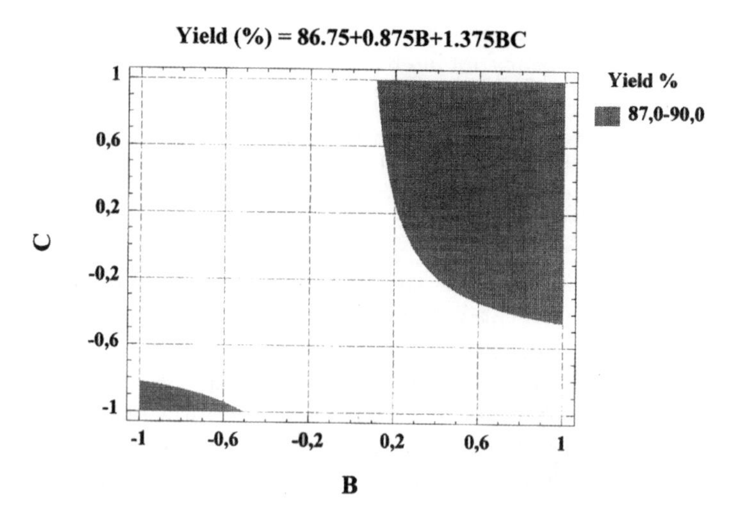

Figure 2.4 Surface response plot for a 2^3 design.

We can remove from this model all terms related to a non-significant factor or interaction. Therefore, in our example, the model is simplified to

$$\text{yield} = b_0 + b_1\text{A} + b_2\text{B} + b_{23}\text{BC} \tag{2.7}$$

Regression coefficients can either be calculated by the least-squares method or set directly if we calculated the effects in advance. Hence b_0 is the mean effect and the remaining coefficients are equal to half the corresponding effect. Recall that we are working with coded factors that range from -1 to $+1$. Then, our model for the example is yield = 84.625 + 2.125A + 0.875B + 1.325BC.

Figure 2.4 displays a contour plot of this model (representing the response surface) for different levels of the response at pH = 5 (A = 1). It is expected that any point inside the shaded zones will have a yield between 87 and 90%. Thus, we can easily define the working levels for B (temperature) and C (time) to achieve a yield equal to or higher than 87%.

2.2.5 Blocking Experiments

In many experimental design problems, it is necessary to design the trials or runs in such a way that the variability arising from some nuisance factors can be controlled. These nuisance factors (or '*blocking variables*') may affect the response but are neither factors nor fixed variables. They can be analysts, instruments, reagents, sample materials, working day, *etc.* Let us consider the simple case of a 2^2 factorial design to improve an automated extraction step in a continuous flow injection system where factor A is pH and factor B is temperature. We must do four runs, but let us consider that a whole sample unit (*e.g.* a container) is consumed every two runs. We need two sample units (containers) but we cannot fully guarantee that they are exactly equal. If we assign sample 1 to the two runs with factor A in its low level and sample 2 to the runs with A in its high level, we cannot identify the effect of factor A (pH) and the effect of the blocking variable (sample). We can only estimate the joint effect and we would say that factor A and the blocking variable are confounded (this is expressed as A ≡ blocking variable). Note that we have grouped the runs in two blocks with coincident levels. Perhaps a better blocked design would be obtained if the levels of the blocking variable are coincident with the levels of the interaction AB (AB ≡ sample) and we can estimate the net effects of both factors. In this case, the two blocks of runs are sample 1 [(1), *ab*] and sample 2 [*a*, *b*]. A complete study of blocked designs is outwith the scope of this introductory chapter and we strongly recommend interested experimenters to consult any text devoted to experimental design for details (*e.g.* [1–3]).

2.2.6 Confounding: Fractional Factorial Designs

As the number of factors in a 2^f factorial design increases, the number of runs required increases rapidly. For example, a 2^5 design requires 32 runs. Using this design, we can estimate the effect of 5 main factors, 10 effects correspond to two-factor interactions and 16 effects to high-order interactions, that is, three-factor and higher order interactions. Often, there is little interest in these high-order interactions, particularly when we first begin to study a system. If we can assume that certain high-order interactions are negligible, a fractional factorial design involving less than 2^f runs can be used to obtain information on the main effects and low-order interactions.

As an example, consider a 2^{3-1} design that is a half fraction of the 2^3 design. That design needs only four runs, in contrast to the full factorial that would require eight runs. The matrix design of the 2^{3-1} design, needing four runs, can be generated starting from a matrix design of a 2^2 factorial design (for factors A and B) and adding a third column for factor C. The signs of this column are those of the interaction AB, so factor C is confounded with the interaction AB (*i.e.* C ≡ AB). If we do so, C ≡ AB is called '*the generator*' of this particular design; see Table 2.4.

Note that as the 2^3 full factorial designs needs 8 runs, it is also possible to select the other half fraction of the experiments, with the complementary signs in column C. In this case the generator would be denoted C ≡ −AB. Furthermore, the product ABC renders a column, denoted I, with all elements showing positive signs. So we call I ≡ ABC the '*defining relation*' for this design.

Table 2.4 Matrix design of a 2^{3-1} fractional design. Generator: C≡AB.

Run	*A*	*B*	*C≡AB*
1	−	−	+
2	+	−	−
3	−	+	−
4	+	+	+

Using the BH^2 algorithm to estimate the effects, we note that column AB has the same signs as column C, column AC has those of column B, column BC those of column A and column ABC those of I. Hence the linear combination of observations in column A, l_A, can be used to estimate not only the main effect of A but also the BC interaction ($l_A = A + BC$; $l'_A = A - BC$ if we select $C \equiv -AB$ as generator). Two (or more) effects that share this type of relationship are termed '*aliases*'. As a consequence, '*aliasing*' is a direct result of fractional replication. In many practical situations, it will be possible to select a fraction of the experiments so that the main effect and the low-order interactions that are of interest become aliased (confounded) only with high-order interactions, which are probably negligible.

To assess the '*alias structure*' for a design, we can use the defining relation. In our example above, $I \equiv ABC$. Multiplying any effect by the defining relation yields the aliases for that effect. For example, the alias of A is $A = A \times ABC = A^2BC = I \times BC = BC$. The full alias structure is named a '*confusion matrix*' and in this example it is $A \equiv BC$; $B \equiv AC$; $C \equiv AB$; $I \equiv ABC$. It is worth noting that since we can define the generator of a design, we can know the corresponding alias structure before any experiment is performed. That is, we know the 'quality' of the design (and future conclusions) in advance.

Sometimes sequences of fractional factorial designs can be used to estimate effects. For example, suppose we have run the principal fraction of experiments of the 2^{3-1} design with generator $C \equiv AB$. At this point we are willing to assume that the two factor interactions are negligible. However, after running the principal fraction of experiments we may still be uncertain about the interactions. It is possible to estimate them by running the alternate fraction, that is, the four runs of the complementary (folded) design that has $C \equiv -AB$ as generator. Of course this is a simple example. In a 2^{4-1} fractional factorial design, the additional factor D can be confounded with interaction AB, AC, BC or ABC, yielding four possible designs with different alias structures.

On the other hand, we might be interested in introducing more than one additional factor in a fractional factorial design. It is as simple as selecting as many interactions as factors to be confounded, *e.g.* $D \equiv AB$ and $E \equiv ABC$. The alias structure is without doubt more complex and these designs are called 2^{f-p} fractional factorial designs. For a complete study of fractional factorial designs we also recommend experimenters to consult any text devoted to experimental designs for additional details and discussions.

2.2.7 Saturated Designs: Plackett–Burman Designs. Use in Screening and Robustness Studies

As we might expect as a generalisation of the previous paragraphs, in a 2^3 factorial design we can accommodate up to seven factors confounding each additional factor with one interaction, that is, $D \equiv AB$, $E \equiv AC$, $F \equiv BC$ and $G \equiv ABC$. This is a '*nodal (saturated) design*' because for eight runs it includes the largest number of factors and all interaction effects are confounded. In the early days of the bombing of London in World War II, two scientists, Robin Plackett and Peter Burman, derived a new class of two-level designs applied to the development of a new weapon. It was later realised that these arrangements were in fact Hadamard matrices. These type of arrangements are called '*orthogonal arrays*' and are denoted L_n. These saturated designs allow us to study the effects of $(n-1)$ factors in only n runs, but only when n is a multiple of 4 (which it is not a serious constraint in most practical situations). Main applications of saturated or Plackett–Burman designs are the screening of factors in preliminary studies and the assessment of the robustness of an analytical procedure. Therefore, we will be capable of assessing the effects of 3 factors in 4 runs, 7 factors in 8 runs, 11 factors in 12 runs and so on. Table 2.5 shows the first row of the matrix design for 3, 7, 11, 15 and 19 factors.

For illustration, the 8-run Plackett–Burman matrix design may be written starting from its first row of signs. The next six rows are obtained by successively shifting all the signs one step to the right (or left) and moving the sign that falls off the end to the beginning of the next row. The last row consists solely of minus signs; see Table 2.6. Once the experimental responses have been measured, the effects of each factor can be evaluated by applying the BH^2 algorithm.

As a practical application of this design, consider that we are implementing a two-step microwave digestion of a complex lyophilised vegetable matrix for the determination of lead by graphite furnace atomic absorption spectrometry. The final objective, from an analytical point of view is to replace the classical and time-consuming digestion procedure by a microwave digestion, carry out a screening of

Table 2.5 Plackett–Burman designs. First row of the matrix design.

Factors	*Runs*	*First row*
3	4	+ + −
7	8	+ + + − + − −
11	12	+ − + − − − + + + − +
15	16	+ + + + − + − + + − − + − − −
19	20	+ + − − + + + + − + − + − − − − + + −

Table 2.6 The eight runs of a Plackett–Burman matrix design.

Run	*A*	*B*	*C*	*D*	*E*	*F*	*G*
1	+	+	+	−	+	−	−
2	−	+	+	+	−	+	−
3	−	−	+	+	+	−	+
4	+	−	−	+	+	+	−
5	−	+	−	−	+	+	+
6	+	−	+	−	−	+	+
7	+	+	−	+	−	−	+
8	−	−	−	−	−	−	−

the most influencing factors on the recovery of the procedure and estimate which experimental variables (factors) have no influence. Seven variables were selected by the analysts as possible factors influencing the recovery of lead by this procedure. Selected factors and the correspondent levels are as follows:

A: Number of digestion vessels (4; 8)
Digestion Step 1:
B: Power 1 (40%; 60%)
C: Pressure 1 (60 psi; 80 psi)
D: Digestion time 1 (5 min; 10 min)
Digestion Step 2:
E: Power 2 (40%; 60%)
F: Pressure 2 (100 psi; 120 psi)
G: Digestion time 2 (10 min; 15 min)
Response: Lead recovery (%)

From preliminary assays, the experimental error was estimated as ±2.50%, expressed as percentage recovery. Note that the complete factorial design is a 2^7, requiring 128 runs, whereas the Plackett–Burman design needs only 8 runs to estimate the effects. The responses to the 8 runs corresponding to the design matrix in Table 2.6 were as follows:

Run	1	2	3	4	5	6	7	8
Recovery (%)	90.2	81.2	87.6	91.4	95.3	73.2	83.0	71.4

and the effects of the seven factors obtained by the BH^2 algorithm were

Factor	A	B	C	D	E	F	G
Effect	0.58	6.53	–2.23	3.28	13.93	2.23	1.23

As the estimated experimental error is ±2.50%, we can conclude that the three most influencing factors are the power of the microwave oven in both steps (factors E and B) and the digestion time of the first step (D). These three have positive signs, so working at their high levels should lead to a high recovery. The remaining factors (A, number of digestion vessels; C, pressure step 1; F, pressure step 2; and G, digestion time step 2) have no influence so we can say that the procedure is robust in relation to them and the experimenter can select the most favourable level for each one.

2.3 Taguchi's Approach to Experimental Design

2.3.1 Strategies for Robust Designs

We have discussed in the previous sections how to define and analyse a factorial design in order to estimate the effect of a series of variables on the response of a process. Compared with factorial designs, both complete and fractional, Taguchi's approach has a series of distinct differences but also very important similarities. Of course, experimental designs developed by Taguchi (a Japanese engineer and statistician, who worked with R. A. Fisher and W. Shewhart and developed some novel approaches to experimental designs around 1950) are, indeed, factorial designs, so what is different about them?

In general, the experimenter starts the use of any factorial design by defining the number of factors or variables to be considered in his/her study. At that moment, a range of possibilities are available: a complete factorial design, several fractional factorial designs or a saturated design. Usually, the selection of a specific design is based on the information in which we are interested. If we need maximum information, then we should choose a complete design (to evaluate the main effects and all the interaction effects), otherwise we can resort to a saturated design (to evaluate only main effects).

In the strategy proposed by Taguchi, some initial information is needed, not only the number of factors but also which specific interactions we want to estimate. Only then can we select a specific experimental design. In any case, the logical sequence to select a proper design is as follows: select number of factors → define information needed → set number of runs. However, whereas in factorial designs the interactions become fixed by the selected design, the Taguchi approach selects the interactions to define the design. Furthermore, the Taguchi approach has another objective: to achieve a robust system.

It was discussed above that two types of factors can influence our process or system: controlled factors ('*design factors*' in Taguchi's terminology) and uncontrolled factors ('*noise factors*'). The latter are inherent to the experimentation and can only be estimated by replication of runs. If the variability between replications is too large, any conclusion drawn from our study may have no meaning at all. Further, their variability can be as important as the mean of the replicates. Noise factors must be identified properly and, if possible, simulated in our experimentation. Sometimes noise factors are actually uncontrolled and, in such a case, we must be able to simulate them by means of some alternative parameter controlled during the experiments. For instance, temperature inside an oven can be a noise factor, but we can measure it at different locations, and accordingly, the effect of 'temperature inside' is simulated by a 'location' factor.

Any experimenter is usually interested in obtaining an 'optimal' response (a nominal value, a maximum or a minimum response) which is related directly to changes in the levels of the factors, but he/she needs also to guarantee a constant quality and, as a consequence, a minimum variability in the response. The main objective is then to control, although partially, the noise factors and obtain a system which is insensitive ('*robust*') to them. In such a way, we can discard the noise factors in future studies. In short, a robustness design identifies the levels of the experimental factors that reduce the effect of the noise factors and consequently minimise the variability in the response in a simple and economical way.

2.3.2 Planning Experiments: Orthogonal Arrays

To improve quality and to study the relationships between the response and both the design and noise factors, Taguchi proposed the use of experimental designs. Nevertheless, it is clear that before drawing any conclusion, we must assure an important characteristic of any experimental design: the reproducibility of the results. In effect, recall that after all the objective is to compare factors under different conditions in an efficient way. A suitable way to achieve this is by the use of orthogonal designs or orthogonal arrays, which have been applied widely.

An orthogonal array is an $n \times m$ matrix (n = trials or experiments, m = factors) with s different levels (elements) in each factor so that any pair of columns has all the s^2 possible pairs of elements with the same frequency. Taguchi represented these matrices by the $L_n(s^m)$ nomenclature. For instance, an $L_8(2^7)$ matrix describes 8 runs and has 7 columns. It is orthogonal because for any pair of columns the four combinations (1,1), (1,2), (2,1) and (2,2) appear with the same frequency.

These designs are of general use because they are reasonably small and easily adaptable to different problems. Some of Taguchi's orthogonal matrices consider the case where not all factors have the same number of levels and they are denoted $L_n(s^m \times t^u)$, where s and t denote the number of levels decided for some of the factors. Hence we would say that 'm' factors have 's' different levels, whereas 'u' factors have 't' levels. Taguchi described 18 orthogonal matrices, 12 for those cases where all factors have the same number of levels and 6 for the opposite situation. These matrices are as follows:

- Two-level orthogonal arrays:
  $L_4(2^3)$ $L_8(2^7)$ $L_{12}(2^{11})$ $L_{16}(2^{15})$ $L_{32}(2^{31})$ $L_{64}(2^{63})$
- Three-level orthogonal arrays:
  $L_9(3^4)$ $L_{27}(3^{13})$ $L_{81}(3^{40})$
- Four-level orthogonal arrays:
  $L_{16}(4^5)$ $L_{64}(4^{21})$
- Five-level orthogonal array:
  $L_{25}(5^6)$
- Mixed orthogonal arrays:
  $L_{18}(2^1 \times 3^7)$ $L_{32}(2^1 \times 4^9)$ $L_{50}(2^1 \times 5^{11})$
  $L_{36}(2^{11} \times 3^{12})$ $L_{36}(2^3 \times 3^{13})$ $L_{54}(2^1 \times 3^{25})$

Table 2.7 shows the orthogonal matrix $L_8(2^7)$. In an orthogonal matrix, symbols used to identify the levels are arbitrary. For example, symbols (1,2) proposed by Taguchi can be replaced by (0,1), (−,+), (*a*,*b*) or (−1,1) as level codes.

If we analyse Table 2.7, we can observe that columns 1, 2 and 4 are those of a 2^3 factorial design (replace symbols [1,2] by [−1, + 1] and see columns A, B and C in Table 2.1) and the remaining columns are those of the respective interactions in the BH^2 algorithm. This is because the Taguchi $L_8(2^7)$ matrix is in fact a 2^3 factorial design. Likewise, the orthogonal matrix $L_{16}(2^{15})$ in Table 2.8 is a 2^4 factorial design.

Table 2.7 Taguchi's orthogonal array $L_8(2^7)$.

	Columns						
Run	*1*	*2*	*3*	*4*	*5*	*6*	*7*
1	1	1	1	1	1	1	1
2	1	1	1	2	2	2	2
3	1	2	2	1	1	2	2
4	1	2	2	2	2	1	1
5	2	1	2	1	2	1	2
6	2	1	2	2	1	2	1
7	2	2	1	1	2	2	1
8	2	2	1	2	1	1	2

Table 2.8 Taguchi's orthogonal array $L_{16}(2^{15})$.

	Columns														
Run	*1*	*2*	*3*	*4*	*5*	*6*	*7*	*8*	*9*	*10*	*11*	*12*	*13*	*14*	*15*
1	1	1	1	1	1	1	1	1	1	1	1	1	1	1	1
2	1	1	1	1	1	1	1	2	2	2	2	2	2	2	2
3	1	1	1	2	2	2	2	1	1	1	1	2	2	2	2
4	1	1	1	2	2	2	2	2	2	2	2	1	1	1	1
5	1	2	2	1	1	2	2	1	1	2	2	1	1	2	2
6	1	2	2	1	1	2	2	2	2	1	1	2	2	1	1
7	1	2	2	2	2	1	1	1	1	2	2	2	2	1	1
8	1	2	2	2	2	1	1	2	2	1	1	1	1	2	2
9	2	1	2	1	2	1	2	1	2	1	2	1	2	1	2
10	2	1	2	1	2	1	2	2	1	2	1	2	1	2	1
11	2	1	2	2	1	2	1	1	2	1	2	2	1	2	1
12	2	1	2	2	1	2	1	2	1	2	1	1	2	1	2
13	2	2	1	1	2	2	1	1	2	2	1	1	2	2	1
14	2	2	1	1	2	2	1	2	1	1	2	2	1	1	2
15	2	2	1	2	1	1	2	1	2	2	1	2	1	1	2
16	2	2	1	2	1	1	2	2	1	1	2	1	2	2	1

In addition to the $L_8(2^7)$ and $L_{16}(2^{15})$ orthogonal matrices, matrices $L_{27}(3^{13})$ and $L_{32}(2^{31})$ are often used successfully. The L_{32} matrix is applied when the factors have two levels whereas the L_9 and L_{27} matrices work with factors at three levels.

Table 2.9 presents the $L_9(3^4)$ orthogonal matrix, used several times in atomic spectrometry (*e.g.* [4,5]).

$L_{18}(2^1 \times 3^7)$ is also a useful orthogonal matrix. It allows for the study of seven factors at three levels and a factor at two levels, and also the interaction of the latter with the factor situated in column 2, as displayed in Table 2.10.

The methodology proposed by Taguchi to plan experiments has three main tools: orthogonal matrices (experimental designs), linear graphs and interaction tables [6]. An experimental design is constructed on the basis of an orthogonal matrix, although linear graphs and/or interaction tables are also used to assign

Table 2.9 Taguchi's orthogonal array $L_9(3^4)$.

Run	*Columns*			
	1	*2*	*3*	*4*
1	1	1	1	1
2	1	2	2	2
3	1	3	3	3
4	2	1	2	3
5	2	2	3	1
6	2	3	1	2
7	3	1	3	2
8	3	2	1	3
9	3	3	2	1

Table 2.10 Taguchi's orthogonal array $L_{18}(2^1 \times 3^7)$.

	Columns							
Run	*1*	*2*	*3*	*4*	*5*	*6*	*7*	*8*
1	1	1	1	1	1	1	1	1
2	1	1	2	2	2	2	2	2
3	1	1	3	3	3	3	3	3
4	1	2	1	1	2	2	3	3
5	1	2	2	2	3	3	1	1
6	1	2	3	3	1	1	2	2
7	1	3	1	2	1	3	2	3
8	1	3	2	3	2	1	3	1
9	1	3	3	1	3	2	1	2
10	2	1	1	3	3	2	2	1
11	2	1	2	1	1	3	3	2
12	2	1	3	2	2	1	1	3
13	2	2	1	2	3	1	3	2
14	2	2	2	3	1	2	1	3
15	2	2	3	1	2	3	2	1
16	2	3	1	3	2	3	1	2
17	2	3	2	1	3	1	2	3
18	2	3	3	2	1	2	3	1

each column of the orthogonal matrix to the appropriate factor and, maybe, interaction. The term 'graph' here may be misleading for most spectroscopists because it does not refer to 'graphical plots' but to simple representations of the interactions between the columns of the designs (*i.e.* the + and − signs). Therefore, '*linear graph*' is a display of some relationship between the columns of an orthogonal matrix. Each orthogonal matrix has an interaction table, that is, a matricial display of the different linear graphs containing all possible interactions between the matrix columns. Figure 2.5 displays the interaction table for the L_8 matrix. For example, the interaction between columns 1 and 6 is shown in column 7.

1	2	3	4	5	6	7
(1)	3	2	5	4	7	6
	(2)	1	6	7	4	5
		(3)	7	6	5	4
			(4)	1	2	3
				(5)	3	2
					(6)	1
						(7)

Figure 2.5 Triangular table for L_8.

The overall characteristics of a design generated from an orthogonal matrix are defined by how factors and interactions are assigned to the different columns of the matrix. Certainly, different assignations lead to different experimental designs, but in any case a main effect should be confounded with other main effects.

The Taguchi procedure to use the so-called linear graphs and to assign columns properly to factors follows several steps [7]:

1. Define the number of levels for each factor you want to consider in the problem at hand in addition to the interactions of interest.
2. Consider the overall number of effects (main effects and interaction effects) to be evaluated and select the smaller orthogonal matrix yielding a plan capable of estimating them. To select the orthogonal matrix we must evaluate the degrees of freedom that we would need and select the closest matrix, avoiding larger matrices. The required number of degrees of freedom is the sum of the degrees of freedom of each factor (levels of the factor minus 1) and each interaction (this is calculated as the product of the degrees of freedom of the factors involved). The number of degrees of freedom of an orthogonal matrix is the sum of the degrees of freedom of each column (levels in the column minus 1). That is, a column with two levels has one degree of freedom and a column with three levels has two degrees of freedom. For example, the $L_{16}(2^{15})$ orthogonal matrix allows for the study of 15 factors at two levels. It has 15 columns with one degree of freedom each, that is, the matrix has 15 degrees of freedom. The $L_{18}(2^1x3^7)$ orthogonal matrix has eight columns, one for a factor at two levels (1 df) and seven for factors at three levels (2 df), so the matrix also has 15 degrees of freedom.
3. To assign factors and interactions to the columns of the orthogonal matrix, we must start by placing the factor involved in more interactions in any column. The next assignment is for the following factor in importance, which is assigned to a free column in such a way that the interactions in which it is involved can be assigned also to free columns.

The process continues until all factors involved in interactions, and also all interactions, are assigned. Finally, factors not involved with interactions are assigned to free columns.

4. If the selected orthogonal matrix does not permit one to estimate all the effects, we can select a larger matrix or modify properly the requirements of the problem to be solved.

The next example illustrates this procedure. Suppose we want to perform an experiment with five factors at two levels each to estimate the main effects of factors A, B, C, D and E and also the effects of interactions AB and BE. We accept in advance that all remaining interactions have negligible effects.

The required degrees of freedom are, accordingly, seven, so we can start by selecting the L_8 orthogonal matrix (also with seven degrees of freedom) for the final experimental design. As factor B is involved in two interactions, we assign B first, later we assign A and E, involved in one interaction each, and finally we assign the remaining factors C and D. For instance, we assign factor B to column 1 and factor A to column 2. The interaction between both factors (columns) is placed in column 3 (see the L_8 interactions matrix in Figure 2.5). So far, we have occupied columns 1, 2 and 3. Now we must assign factor E to a free column so that its interaction BE (factor B is in column 1) will be also a free column. If factor E is assigned to column 4, interaction BE is placed in column 5, also a free column. The two remaining factors, C and D, can be placed in the two free columns, 6 and 7.

The final assignment is then 1(B), 2(A), 3(AB), 4(E), 5(BE), 6(C) and 7(D). The matrix is displayed in Table 2.11 and the matrix of experiments that have to be made in the laboratory is given in Table 2.12.

However, where are the remaining interactions? What is the confounding structure? To answer these questions, we will use the L_8 interactions table (see Figure 2.5). For instance, interaction AC corresponds to the interaction between columns 2 and 6 and, as the interaction table shows, is column 4. That is, interaction AC is confounded with factor E. Proceeding for all the second-order

Table 2.11 Assignment of columns for an orthogonal array with seven degrees of freedom.

	Effects						
Run	*B*	*A*	*AB*	*E*	*BE*	*C*	*D*
1	1	1	1	1	1	1	1
2	1	1	1	2	2	2	2
3	1	2	2	1	1	2	2
4	1	2	2	2	2	1	1
5	2	1	2	1	2	1	2
6	2	1	2	2	1	2	1
7	2	2	1	1	2	2	1
8	2	2	1	2	1	1	2

Table 2.12 Matrix of laboratory experiments derived from the assignment of columns in Table 2.11 (see text for details).

Run	*Factors*				
	A	*B*	*C*	*D*	*E*
1	1	1	1	1	1
2	1	1	2	2	2
3	2	1	2	2	1
4	2	1	1	1	2
5	1	2	1	2	1
6	1	2	2	1	2
7	2	2	2	1	1
8	2	2	1	2	2

interactions in the same manner, we obtain the following structure:

$$B \equiv CD \quad A \equiv CE \quad AB \equiv DE \quad E \equiv AC \quad BE \equiv AD \quad D \equiv BC \quad C \equiv AE \equiv BD$$

It sometimes happens that we need designs different from the 18 described by Taguchi. There exist techniques to modify the original designs in order to generate more levels (*i.e.* combine columns) or, simply, reduce the number of levels (modify only a column) [8]. The four main techniques are (i) the '*column-merging*' method, to originate multiple levels, (ii) the '*dummy-level*' method, (iii) the '*combination design*' method (also called '*combining columns*') and (iv) the *idle-column* method to use an empty column [9]. They are briefly reviewed here and more interested readers are encouraged to consult some other references given in this section.

It has to be stressed that only by originating multiple levels can the orthogonality of the design can be guaranteed. The use of this method and/or the dummy-level technique permits the combination of columns to obtain a column with more levels. Furthermore, the column-merging method allows for the introduction of factors at four and eight levels in orthogonal matrices of systems at two levels and also factors at nine levels in orthogonal matrices of systems at three levels.

In order to insert a two-level factor in an orthogonal matrix to obtain a three-level system, the factor is formally transformed into a three-level one. Just assign one of the two already defined levels as the third level. This is the dummy-level technique.

The combination of columns is also an effective technique to assign a two-level factor to an orthogonal matrix with three levels. In this case, an original column with three levels is split into two columns, each one being a factor at two levels.

The idle-column method is used to assign two or more factors with three levels to some of the following orthogonal matrices: $L_8(2^7)$, $L_{16}(2^{15})$, $L_{32}(2^{31})$ and $L_{64}(2^{63})$.

Although the designs generated by these techniques (except for multiple levels) are not orthogonal, they require fewer experiments that the orthogonal designs.

2.3.3 Robust Parameter Design: Reducing Variation

To study how the variability in the response due to noise factors changes with the design factors, Taguchi suggested analysing the response at systematic levels of noise in different design factors combinations. In this way, the mechanism that may cause the variability in the response is the same for any combination of the design factors. This is termed 'looking for a *parameter design*' and Taguchi proposed two experimental plans, one for each type of factors. The experimental plan for the design factors is named '*inner array*', that for noise factors is termed '*outer array*' and the global experiment is denoted the '*direct product design*'. In laboratory experiments, design factors are varied according to the combinations of the inner array; and for each row of the inner array, noise factors are varied according to the combinations of the outer array. Figure 2.6 displays a simple crossed matrix. The inner array for 6 design factors needs 8 runs and the outer array for 2 noise factors (U and V) has 4 runs, that is, the global matrix needs 32 runs.

The aim of the '*parameter design*' is to establish a combination of levels of the design factors which leads to a minimum in the variability around a mean, minimum or maximum value in the response. The idea behind the 'parameter design' is to find a set of analytical conditions which are both functional (*i.e.* they fulfil the objective) and robust (*i.e.* they are not sensitive to noise factors), as far as possible.

Taguchi classified robust designs according to the main objective when measuring the response in three basic types: '*nominal-is-best*' (minimal variability around a target value), '*larger-the-better*' (minimal variability for a maximum in the response) and '*smaller-the-better*' (minimal variability for a minimum in the response).

								Experiment number				Noise
								1	2	3	4	factors
	Control factors							1	1	2	2	U
Exp. No.	A	B	C	D	E	F	e	1	2	1	2	V
1	1	1	1	1	1	1	1					
2	1	1	1	2	2	2	2					
3	1	2	2	1	1	2	2					
4	1	2	2	2	2	1	1					
5	2	1	2	1	2	1	2					
6	2	1	2	2	1	2	1					
7	2	2	1	1	2	2	1					
8	2	2	1	2	1	1	2					

Figure 2.6 Structure of parameter design.

As was explained above, these designs are based on orthogonal matrices. They are composed of an inner array (for the design factors) and an outer matrix (for the noise factors). It can be proved that heterogeneity in the variability is a consequence of the interaction between both types of factors [10]. Taguchi suggested summarising the responses for each run of the design matrix as a signal-to-noise ratio, and then combining the mean response with its variability, leading to a quality measurement.

Thus, in the '*nominal-is-best*' case, the signal-to-noise (s/n) ratio is defined, for each run of the design matrix, as

$$\mathrm{s/n}_{\mathrm{NOMINAL}} = 10\log\left(\frac{\bar{y}_i^2}{s_i^2}\right) \tag{2.7}$$

For the '*larger-the-better*' and '*smaller-the-better*' cases, the respective expressions for the s/n ratios are given below. In all cases, y_i are the observed responses and n_i the number of replicates for each run.

$$\mathrm{s/n}_{\mathrm{BIGGER}} = -10\log\left(\frac{1}{n_i}\sum_{i=1}^{n_i}\frac{1}{y_i^2}\right) \tag{2.8}$$

$$\mathrm{s/n}_{\mathrm{SMALLER}} = -10\log\left(\frac{1}{n_i}\sum_{i=1}^{n_i}y_i^2\right) \tag{2.9}$$

According to these definitions, in the '*smaller-the-better*' case, the s/n ratio increases as the mean and variability decrease. Similarly, in the '*larger-the-better*' case, the s/n ratio increases as the mean increases and variability decreases.

On the basis of his experience in the engineering field, Taguchi recommended the use of the s/n ratio with a logarithmic transformation, as it is useful for most practical situations whenever the standard error of the response is a function of the response value. Although in general the above three equations are used widely, more than 70 ratios were defined by Taguchi [11].

Inner and outer arrays are selected according to the procedure described in the previous paragraphs, the experimental runs carried out and the mean response and s/n ratio for each run of the inner matrix are evaluated.

The analysis of the mean response and the s/n ratio can be performed employing the usual ANOVA and/or hypothesis tests to detect which factors or interactions have statistical significance. Taguchi proposed a conceptual approach based on the graphical display of the effects (they are called '*factor plots*' or '*marginal means*') followed by a qualitative evaluation. This provides objective information and a test for the significance of each design factor on the two observed responses: mean and s/n ratio.

Taguchi also suggested the use of Pareto's ANOVA [12]. This technique does not require any statistical assumption so a statistical analysis of the responses cannot be performed. Figure 2.7 shows a Pareto's ANOVA table.

Factors, interactions and error columns		A	B	C	D	e	e	Total
Sum at column level	1	A_1	B_1	C_1	D_1	e_1	e_1	
	2	A_2	B_2	C_2	D_2	e_2	e_2	
	3	--	--	C_3	--	--	e_3	
Sum of squares of differences		$S_A=(A_2-A_1)^2$	$S_B=(B_2-B_1)^2$	$S_C=(C_2-C_1)^2+ + (C_3-C_1)^2 + + (C_3-C_2)^2$	$S_D=(D_2-D_1)^2$	$S_{e1}=(e_2-e_1)^2$	$S_{e2}=(e_2-e_1)^2 + + (e_3-e_1)^2 + + (e_3-e_2)^2$	
Degrees of freedom		$d_A=1$	$d_B=1$	$d_C=2$	$d_D=1$	$d_{e1}=1$	$d_{e2}=2$	
Total sum of squares		S_A	S_B	S_C	S_D	$S_{e1}+S_{e2}$		
Total degrees of freedom		$d_A=1$	$d_B=1$	$d_C=2$	$d_D=1$	$d_e=3$		
Mean squares of differences		$M_A=S_A/d_A$	$M_B=S_B/d_B$	$M_C=S_C/d_C$	$M_D=S_D/d_D$	$M_e=S_e/d_e$		$M_T=M_A+...+M_e$
Contribution ratio (%)		$(M_A/M_T)\cdot 100$	$(M_B/M_T)\cdot 100$	$(M_C/M_T)\cdot 100$	$(M_D/M_T)\cdot 100$	$(M_e/M_T)\cdot 100$		100%

Figure 2.7 Pareto's ANOVA table to assess the statistical significance of design factors according to Taguchi.

We start by evaluating the mean response for each level and each factor. Then, we calculate the sum of squares of the differences of each factor for all possible differences between levels (for example, one for two levels, three for three levels, six for four levels and so on). Dividing the sum of squares by the degrees of freedom, we can obtain the mean square of the differences. Means are compared among themselves as percentages. Finally, the effects are ordered by magnitude. When empty columns are present, their sums of squares are added in order to obtain an estimate of the residual mean square and determine which factors are important and which are negligible.

Factors can, accordingly, be classified in three categories:

1. Factors controlling the s/n ratio. They have influence both on the variability of the process (s/n ratio) and on the mean response.
2. Signal factors. They have a significant effect on the mean of the response but not on the s/n ratio.
3. Factors without a significant effect neither on either the mean response or the s/n ratio.

In the '*nominal-is-best*' case, optimisation is performed in two steps:

1. Selection of factor control levels in order to maximise the s/n ratio.
2. Selection of signal factor levels to adjust the mean response to the target nominal value, keeping variability constant.

For the '*larger-the-better*' and '*smaller-the-better*' cases, Taguchi suggested identifying the optimal levels of the design factors or just maximising the corresponding s/n ratio.

2.3.4 Worked Example

To determine Sb in marine sediments by ETAAS, a direct method was developed based on quantitating the analyte in the liquid phase of the slurries (prepared directly in autosampler cups). The variables influencing the extraction of Sb into the liquid phase and the experimental setup were set after a literature search and a subsequent multivariate optimisation procedure. After the optimisation, a study was carried out to assess robustness. Six variables were considered at three levels each (see Table 2.13). In addition, two noise factors were set after observing that two ions, which are currently present into marine sediments, might interfere in the quantitations. In order to evaluate robustness, a certified reference material was used throughout, BCR-CRM 277 Estuarine Sediment (guide value for Sb $= 3.5 \pm 0.4\,\mu g\,g^{-1}$). Table 2.13 depicts the experimental setup.

All results presented here correspond to the concentrations of Sb found for each experimental run (referred to the original sample). The two noise factors correspond to copper and sulfate, as they may interfere in the measurements (mainly when they are at high concentrations, as previous studies had revealed).

Table 2.13 Experimental conditions for the worked example.

Factor	*Level*		
	1 (low)	*2 (medium)*	*3 (high)*
A: Triton X-100 (% m/v)	0	0.04	0.08
B: HF concentration (% v/v)	10	20	30
C: HNO_3 concentration (% v/v)	0	5	10
D: Sample mass (mg)	20	50	80
E: Volume of slurry (ml)	0.50	0.75	1.00
F: Ultrasonic bath agitation (min)	5	10	15

Table 2.14 Inner array for the worked example.

Experiment	*A*	*B*	*C*	*D*	*E*	*F*
1	1	1	1	1	1	1
2	1	2	2	2	2	2
3	1	3	3	3	3	3
4	2	1	1	2	2	3
5	2	2	2	3	3	1
6	2	3	3	1	1	2
7	3	1	2	1	3	3
8	3	2	3	2	1	1
9	3	3	1	3	2	2
10	1	1	3	3	2	1
11	1	2	1	1	3	2
12	1	3	2	2	1	3
13	2	1	2	3	1	2
14	2	2	3	1	2	3
15	2	3	1	2	3	1
16	3	1	3	2	3	2
17	3	2	1	3	1	3
18	3	3	2	1	2	1

Since their concentrations in real samples cannot be controlled at all, three levels were considered (low, medium, high). These were 5, 50 and 100 $\mu g\,ml^{-1}$ for SO_4^{2-} and 2, 10 and 20 $\mu g\,ml^{-1}$ for Cu^{2+} (concentrations are referred to the extract).

The design factors require 12 df so an L_{18} orthogonal array (with 15 df) was selected. Hence we can study a factor at two levels and seven factors at three levels each. The matrix is adapted to our needs by discarding column 1 (designed for a variable with two levels) and column 7 (not needed in this example). This yields 3 df to calculate the residuals. Hence the experimental matrix is as presented in Table 2.14.

Each run of this matrix (inner array) is repeated nine times (outer array), following a CFD 3^2, for both noise factors. After all experiments were made in the laboratory, calculations were done to refer the concentration of the slurry extracts to the original sample. They are shown in Table 2.15.

Table 2.15 Experimental responses for the worked example.

	Cu^{2+} ($\mu g\,ml^{-1}$)								
	2			*10*			*20*		
	SO_4^{2-} ($\mu g\,ml^{-1}$)								
Experiment	*5*	*50*	*100*	*5*	*50*	*100*	*5*	*50*	*100*
1	1.93	1.88	1.86	1.88	1.83	1.81	1.85	1.84	1.85
2	5.28	5.09	5.14	5.10	5.15	5.21	5.22	5.21	4.99
3	5.89	5.79	5.77	6.05	5.81	5.79	5.98	5.84	5.86
4	2.02	2.01	2.00	2.04	2.03	2.01	2.05	2.03	2.01
5	4.00	4.05	4.07	4.46	4.41	4.46	4.93	4.94	4.93
6	2.92	2.83	2.81	2.73	2.74	2.73	2.83	2.78	2.74
7	2.93	2.94	2.93	3.39	3.23	3.29	3.61	3.57	3.59
8	4.61	4.37	4.24	4.31	4.06	3.99	4.97	4.80	4.50
9	3.76	3.72	3.75	3.77	3.82	3.80	4.01	3.97	4.01
10	3.13	3.11	3.14	3.37	3.35	3.32	3.50	3.49	3.44
11	2.42	2.40	2.40	2.48	2.44	2.41	2.45	2.45	2.47
12	5.82	5.67	5.74	5.68	5.60	5.71	5.59	5.68	5.64
13	2.69	2.65	2.62	2.58	2.54	2.51	2.67	2.69	2.67
14	2.76	2.74	2.76	2.73	2.76	2.74	2.79	2.74	2.74
15	3.12	3.05	3.02	3.16	3.11	3.12	3.14	3.09	3.10
16	2.92	2.91	2.92	2.97	3.05	3.04	3.14	3.06	3.04
17	4.18	4.05	3.89	3.79	3.58	3.47	4.49	4.20	4.12
18	3.03	3.02	3.03	3.47	3.46	3.42	4.02	3.99	4.04

Since the objective is to obtain the guide value stated on the CRM certificate, *i.e.* 3.5, Taguchi's term '*nominal is best*' should be followed and we should proceed in two stages: first the selection of the control factor levels in order to minimise the effect of the noise factors and second the selection of the signal factor levels in order to obtain the Sb certified value.

We start the data analysis by calculating, for each experiment, the average of the nine replicates, its standard deviation and the s/n ratio '*nominal is best*'. These values are given in Table 2.16.

If we examine these values, we can obtain some interesting information. For example, the averages are in the range 1.86–5.87 (our target is 3.5) and the standard deviations are in the range 0.02–0.43. Results close to the target value (3.5) are those of experiments 7, 9, 10 and 18 and the best average is obtained in experiment 18. However, although the average is correct, its deviation is too large, in fact the highest among the 18 experiments. That is, we can obtain an average concentration equal to 3.5 but the variability in the concentration of Sb can be from 3.02 to 4.04. Experiment 10 leads to a lower average (3.32) but with a lower variability (3.11–3.50). Recall that only a reduced fraction (18 out of $3^6 = 729$, around 2.5%!) of the possible experiments were carried out, so the question is whether there exists a combination of levels with a best response. What are the levels of the design factors leading to a response equal to 3.5 and, in addition, to a minimal deviation?

The answers need an analysis of the s/n ratio. Table 2.17 summarises the Pareto analysis of variance (see Figure 2.7 for the calculations involved) and Figure 2.8 displays the factor plots associated with the calculations.

Table 2.16 Summary of responses for the worked example.[a]

Experiment	*A*	*B*	*C*	*D*	*E*	*F*	*1*	*7*	*Average*	*SD*	*s/n*
1	1	1	1	1	1	1	1	1	1.86	0.03	34.77
2	1	2	2	2	2	2	1	2	5.15	0.09	35.59
3	1	3	3	3	3	3	1	3	5.87	0.09	35.87
4	2	1	1	2	2	3	1	3	2.02	0.02	41.82
5	2	2	2	3	3	1	1	1	4.47	0.39	21.24
6	2	3	3	1	1	2	1	2	2.79	0.07	32.61
7	3	1	2	1	3	3	1	2	3.28	0.29	21.13
8	3	2	3	2	1	1	1	3	4.43	0.33	22.65
9	3	3	1	3	2	2	1	1	3.85	0.12	30.37
10	1	1	3	3	2	1	2	2	3.32	0.16	26.59
11	1	2	1	1	3	2	2	3	2.44	0.03	38.45
12	1	3	2	2	1	3	2	1	5.68	0.07	37.90
13	2	1	2	3	1	2	2	3	2.62	0.07	31.74
14	2	2	3	1	2	3	2	1	2.75	0.02	43.04
15	2	3	1	2	3	1	2	2	3.10	0.04	37.16
16	3	1	3	2	3	2	2	1	3.00	0.08	31.57
17	3	2	1	3	1	3	2	2	3.97	0.32	21.80
18	3	3	2	1	2	1	2	3	3.50	0.43	18.18

[a]Abbreviations: SD, standard deviation; s/n, signal-to-noise ratio.

Table 2.17 Pareto ANOVA table to assess the s/n ratio in the worked example.[a]

Column	*A*	*B*	*C*	*D*	*E*	*F*	*e*	*e'*
Level 1	34.86	31.27	34.06	31.36	30.24	26.77	30.67	33.15
Level 2	34.60	30.46	27.63	34.45	32.60	33.39	31.83	29.15
Level 3	24.28	32.02	32.06	27.94	30.91	33.60		31.45
SS	218.47	3.63	65.01	63.69	8.86	90.55	1.33	24.23
df	2	2	2	2	2	2	1	2
SS'	218.47	3.63	65.01	63.69	8.86	90.55	25.57	
df'	2	2	2	2	2	2	3	
MS	109.24	1.81	32.51	31.84	4.43	45.28	8.52	
Contribution	46.76	0.78	13.91	13.63	1.90	19.38	3.65	

[a]Abbreviations: SS, sum of squares; df, degrees of freedom; MS, mean square; *e* and *e'*, are two error estimates.

It can be seen that factors E and B (volume of slurry and concentration of HF, respectively) do not affect the s/n ratio (their contribution is lower than the experimental error, 3.65). The other factors can be considered as control factors, A (Triton X-100) being the most important. It is worth interpreting this plot from an atomic spectroscopist's point of view. Indeed, its interpretation cannot be separated from the tables and graphs, so both can be explained as follows.

Since Sb is determined in slurry extracts of sediments and we aim for a good analyte extraction, HF is, without doubt, the most satisfactory acid because of its power to solubilise silicon matrices. Noteworthily, the experiments where HF is at its lowest level tend to yield low concentrations (poor extraction). In

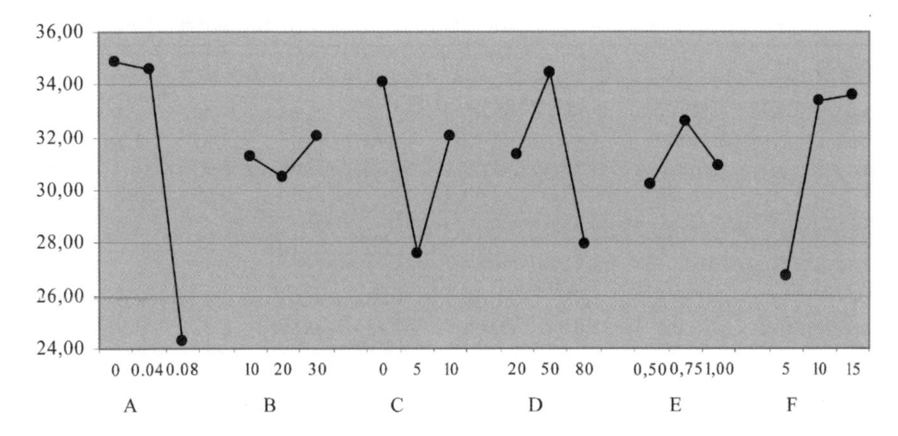

Figure 2.8 Factor plots for a worked example of the Taguchi methodology.

this particular example, nitric acid influenced the s/n ratio significantly. In fact, the oxidative character of nitric acid may affect the Sb extraction because it may cause the formation of insoluble Sb(V)–silicate complexes, yielding low extract recoveries and non-reproducible results (for more details, see [13]). Therefore, it is advisable not to use it in this application.

The remarkable influence of Triton X-100 on the s/n ratio is explained because of the high viscosity of the final solutions when the highest level is used (the higher the viscosity, the more defective is the pipetting). The effect of the ultrasonic bath (note that an ultrasonic probe cannot be used because HF degrades Ti) is relevant for both the s/n and overall effect because some agitation is needed to improve the contact between the solid particles and the extractant. Provided that this is achieved, the effect is not relevant (thus, 5 min seems sufficient). The sample mass shows its typical influence on slurry sampling measurements since too high amounts of solid sample may make the extraction difficult and increase the concomitants. While the total volume allows for good slurry formation, the variable is not relevant.

As our objective is to maximise the s/n ratio, we select factor A at level 1 or 2 indistinctly, factor F at level 2 or 3, factor C at level 1 and factor D at level 2. With this selection we are sure to obtain minimal variability. However, we also need an average equal to 3.5. Table 2.18 summarises the Pareto ANOVA for the average and Figure 2.9 displays the contribution of each term when the average response of the experiments in the experimental matrix is considered.

Those factors not affecting the s/n ratio (B or E) but only the averages are considered as signal factors. As E is not an important factor in relation to the average response, the only signal factor is, therefore, B. Then, keeping the control factors at those levels that minimise the variability (noise), we try to modify the level of B in order to reach the objective value (3.5).

According to the analysis performed for the s/n ratio, the levels of factors C and D are fixed, but the optimal levels for factors A and F are not defined (they

Table 2.18 Pareto ANOVA for the average values of the worked example.

Column	*A*	*B*	*C*	*D*	*E*	*F*	*e*	*e*
Level 1	4.05	2.68	2.87	2.77	3.56	3.44	3.75	3.60
Level 2	2.96	3.87	4.12	3.90	3.43	3.31	3.38	3.60
Level 3	3.67	4.13	3.69	4.02	3.69	3.93		3.48
SS	1.84	3.57	2.40	2.85	0.10	0.63	0.14	0.03
Df	2	2	2	2	2	2	1	2
SS′	1.84	3.57	2.40	2.85	0.10	0.63	0.17	
df′	2	2	2	2	2	2	3	
MS	0.92	1.78	1.20	1.43	0.05	0.32	0.06	
Contribution	16.00	31.01	20.86	24.76	0.89	5.52	0.97	

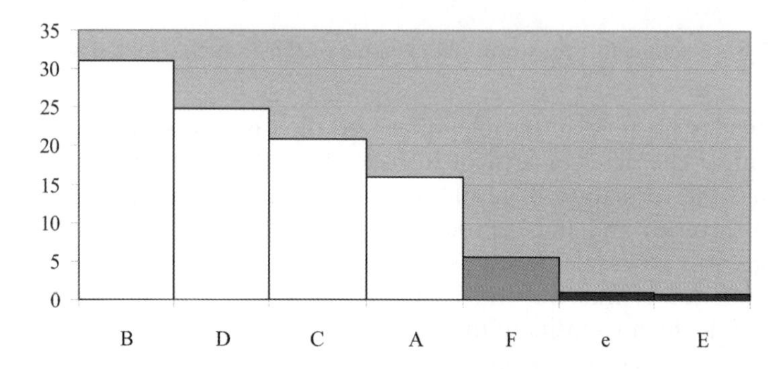

Figure 2.9 Contributions to the average value of the worked example.

Table 2.19 Predicted average values for the worked example.

		A(1) F(2)	*A(1) F(3)*	*A(2) F(2)*	*A(2) F(3)*
B	(1)	2.57	3.19	1.48	2.10
	(2)	3.76	4.37	2.66	3.28
	(3)	4.02	4.64	2.92	3.54

Note: Factor C at level 1 and factor D at level 2.

have a similar s/n ratio). Table 2.19 summarises the estimated average response for each possible combination of these levels.

Estimations are evaluated by adding the contribution of each level and then subtracting the average as many times as necessary to be included only once. For example, for the combination A(1) B(1) C(1) D(2) F(3):

$$\text{estimated response} = 4.05 + 2.68 + 2.87 + 3.90 + 3.93 - (4 \cdot 3.56) = 3.19$$

All the combinations shown in the table have minimal variability (high values of s/n ratio) but they lead to different average values. The closest average to the target is 3.54 for the combination A(2) B(3) C(1) D(2) F(3). The estimated s/n ratio for this combination of levels, evaluated by the same procedure, is 44 and, from the applied s/n expression, the standard deviation is 0.25.

The next experimental step is the validation of the conclusions and the best way is to carry out some experiments in order to confirm the expected results. To do this, design factors were fixed as follows: A(2) B(3) C(1) D(2) E(3) F(3); factor E is in its highest level to guarantee a good mass/volume ratio for the preparation of the slurry and to improve the Sb extraction to the liquid phase. Nine replicates in such conditions were carried out with the following responses: 3.52, 3.47, 3.39, 3.37, 3.56, 3.55, 3.52, 3.59, 3.58. The average response of these replicates is 3.51, the standard deviation is 0.08 and the s/n ratio is 33. This is, by far, the experiment leading to the best results, which confirms the suitability of Taguchi's methodology.

2.4 Optimisation

2.4.1 Experimental Optimisation

The experimental designs that we have seen in the previous sections allow us to establish the factors that influence the answer most and which levels have to be set in order to obtain a good result. However, if we were to choose other values for each of the levels, the answer would, perhaps, be better. The process by which the value of the level of each factor that provides the best answer obtained is termed '*optimisation*'.

The most intuitive optimisation method is called OFAT (one-factor-at-a-time). It consists of fixing each factor to one level, choosing a factor and varying its levels until the best answer is reached, fixing that value and varying the levels of a second factor until we find the level that provides the best answer, and so forth. This strategy, despite being very simple, has two important problems: (a) it is not able to find optimal answers if interactions between the factors occur, because a best answer can be found by varying all the factors simultaneously, something this approach cannot perform; and (b) it forces the spectroscopist to perform more experiments than when either experimental designs or optimisation methods (we will explain some common ones next) are used [14].

In order to profit from the advantages of the experimental designs and to solve the OFAT problems, Box and Wilson [15] published in 1951 a new method to obtain optimal conditions experimentally. The method was known as EVOP (evolutionary operations). There, a 2^k factorial design was moved in an iterative form until the optimal conditions were reached. In 1962, Spendley, Hext and Himsworth [16] published another method which potentiated the advantages of the basics of the EVOP techniques. A so-called '*simplex design*' replaced the 2^k design and defined an algorithm to move this design towards the optimum, it could be programmed in a computer and it allowed for the automation of the searching process. This method was known as the '*simplex sequential method*'. The experimental optimum obtained depended on the location and the size of the initial simplex because the form of the design is not modified during the process, the design can only be moved and redirected.

The 'rigidity' that prevented an accurate optimal point from being obtained was solved by Nelder and Mead [17] in 1965. They proposed a modification of the algorithm that allowed the size of the simplex to be varied to adapt it to the experimental response. It expanded when the experimental result was far of the optimum – to reach it with more rapidly – and it contracted when it approached a maximum value, so as to detect its position more accurately. This algorithm was termed the '*modified simplex method*'. Deming and it co-workers published the method in the journal *Analytical Chemistry* and in 1991 they published a book on this method and its applications.

A note is in order here to stress that a special form of experimental designs can be used for optimization. They are known as Doehlert designs, of which several applications are reviewed in Table 2.34. As this chapter is intended at an introductory level, no further details will be give here.

2.4.2 The Simplex Method

We will present the basics of the simplex method with the aid of a simulation and then describe the algorithm. As an example, Soylak *et al.* [18] optimised a procedure to preconcentrate lead (the studied response, Y) using a 2^3 factorial design in which the factors were:

RC: the reagent concentration (mol l^{-1}): low level = 5×10^{-6}; high level = 5×10^{-5}.
pH: low level=6.0; high level = 8.0.
ST: the shaking time (min): low level = 10, high level = 30.

Let us suppose that we are interested in implementing this procedure in our laboratory and we fix the time of agitation at 10 min. So, we want to look for the RC (X_1) and pH (X_2) values that provide the largest percentage lead recovery (Y), and we will use the simplex method defined by Spendley *et al* [16].

If we call f the number of factors, a simplex is a '*convex*' and '*closed*' figure formed by $f+1$ vertices in a space of f dimensions. This is a triangle for the case when we consider two factors ($f=2$) or a tetrahedron if $f=3$; for greater values of f the design cannot be drawn and we need to resort to matrix notation; see Table 2.20.

Initially, we establish the working intervals (boundaries) of each factor: X_1, between 0 and 60×10^{-6} (mol l^{-1}); and X_2, between 5.0 and 9.0. Then, we define

Table 2.20 Defining a regular simplex.

		Factors			
Simplex		X_1	X_2	...	X_f
Vertex number	**1**	$X_{1,1}$	$X_{2,1}$	...	$X_{f,1}$
	2	$X_{1,2}$	$X_{2,2}$	...	$X_{f,2}$
	...	...	...	...	...
	f	$X_{1,f}$	$X_{2,f}$	...	$X_{f,f}$
	$f+1$	$X_{1,(f+1)}$	$X_{2,(f+1)}$	...	$X_{f,(f+1)}$

Table 2.21 Initial simplex for the worked example.

	X_1	X_2
Boundaries	(0; 60)	(5.0; 9.0)
Step size	$s_1 = 9.0$	$s_2 = 0.6$
Starting vertex	$X_{1,1} = 6.0$	$X_{2,1} = 5.4$

Table 2.22 How to set the experimental conditions of the 'initial corner simplex'.

		Factors			
Initial corner simplex		X_1	X_2	...	X_f
Vertex number	1	$X_{1,1}$	$X_{2,1}$	...	$X_{f,1}$
	2	$X_{1,2} = X_{1,1} + s_1$	$X_{2,2} = X_{2,1}$	...	$X_{f,2} = X_{f,1}$
	3	$X_{1,3} = X_{1,1}$	$X_{2,3} = X_{2,1} + s_2$		$X_{f,3} = X_{f,1}$
	...	...	...	...	...
	f	$X_{1,f} = X_{1,1}$	$X_{2,f} = X_{2,1}$	...	$X_{f,f} = X_{f,1}$
	$f + 1$	$X_{1,(f+1)} = X_{1,1}$	$X_{2,(f+1)} = X_{2,1}$	...	$X_{f,(f+1)} = X_{f,1} + s_f$

Table 2.23 Initial simplex of the worked example.

			Factors	
Corner initial simplex			X_1	X_2
Vertex number	(Starting vertex)	1	$X_{1,1} = 6.0$	$X_{2,1} = 5.4$
		2	$X_{1,2} = 6.0 + 9.0$	$X_{2,2} = 5.4$
		3	$X_{1,3} = 6.0$	$X_{2,3} = 5.4 + 0.6$

the initial simplex, *i.e.* we fix the initial levels of each factor: X_1 at 6.0×10^{-6} (mol l^{-1}) and X_2 at 5.4. These values establish the initial vertex of the simplex (starting vertex). Next we define the variations that each factor may have from one experiment to another (*step size*); this difference must be large enough that the answer changes clearly between the experiments (usually about 10 or 20% working intervals). Here, we establish $s_1 = 9.0 \times 10^{-6}$ (mol l^{-1}) for X_1 and $s_2 = 0.6$ for X_2; see Table 2.21.

Now we calculate the coordinates of the two vertices that define the first simplex. Several ways have been proposed to do this, the simplest being the so-called '*initial corner simplex*', which in general can be formulated as shown in Table 2.22.

Table 2.23 contains the experimental conditions for the initial simplex defined for our example and Figure 2.10 displays it in a two-dimensional space.

We perform the three experiments in the laboratory and measure the response (percentage lead recovery, Y) in each vertex of the simplex. The results are displayed in Table 2.24.

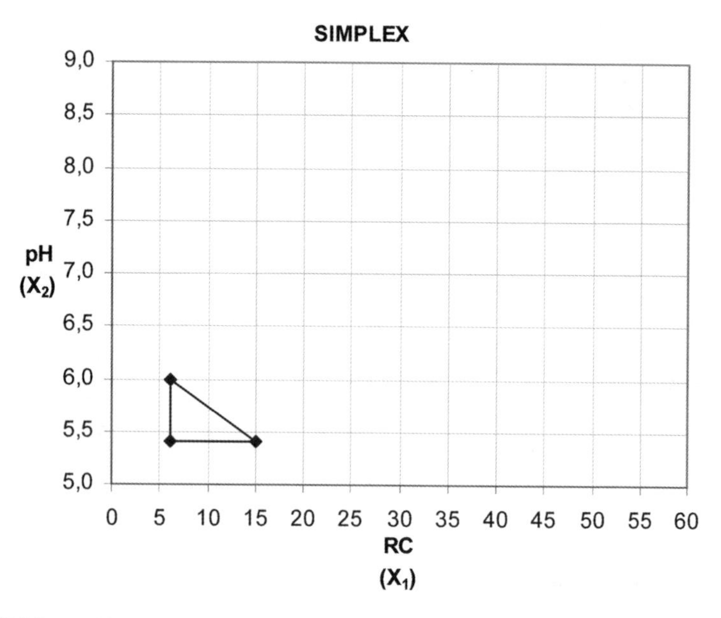

Figure 2.10 Initial simplex.

Table 2.24 Responses for the initial simplex of the worked example.

Simplex number	*Vertex number*	X_1	X_2	Y
1	1	6.0	5.4	50
	2	15.0	5.4	59
	3	6.0	6.0	63

Table 2.25 Ranked responses for the initial simplex of the worked example.

Simplex number	*Vertex number*	X_1	X_2	Y	*Order*	*Times retained*
1	1	6.0	5.4	50	W	
	2	15.0	5.4	59	N	
	3	6.0	6.0	63	B	1

Now, we have to continue and obtain simplex number 2 (and also the subsequent ones) following the five steps depicted below:

- *Step 1:* arrange the responses, assigning 'B' to the best, 'W' to the worst and 'N' to the next to the worst.
- *Step 2:* annotate the number of times that present vertex 'B' has been nominated as such (in this case, only 1 so far); when point B has been retained $2(f+1)$ times as the best vertex, finish the searching process, choosing that point as the optimal one (Table 2.25).

Table 2.26 Evaluating the vertex R of the worked example.

Simplex number	*Vertex number*	X_1	X_2	*Y*	*Order*	*Times retained*
1	1	6.0	5.4	50	W	
	2	15.0	5.4	59	N	
	3	6.0	6.0	63	B	1
G = (B + N)/2		10.5	5.7			
R = G*(G – W)	4	15.0	6.0	71		

Table 2.27 Second simplex for the worked example.

Simplex number	*Vertex number*	X_1	X_2	*Y*
2	4	15.0	6.0	71
	2	15.0	5.4	59
	3	6.0	6.0	63

- *Step 3:* to improve the response, the basic rule of the simplex method is: reflect the rejected vertex (usually W) through the centroid (G) of the other f vertices (they are said to be '*retained*'), obtaining the reflected vertex (R). Its experimental conditions are given by the following two expressions (apply them to each experimental factor):

  G = (sum all the coordinates of the factor except for W)/f

  R = G + (G – W)
- *Step 4:* carry out the experiment in the new conditions (reflected vertex). It sometimes happens that the proposed coordinates for the reflected point are outside boundaries, and in that case it is not possible to make the experiment and point R gets a response which is worse than that of the worst vertex. In our example, the new coordinates for R allow us to perform the experiment, so we obtain experimentally the response in the reflected R; in the example this is a 71% lead recovery (Y = 71). Now we have vertex number 4 (Table 2.26).
- *Step 5:* in order to obtain the following simplex (simplex number 2), we replace the W of the previous simplex by the reflected vertex R (Table 2.27). Figure 2.11 displays the two simplexes.

Repeating the previous five steps successively, the simplex moves towards an optimal point. In this case it will be the point that provides the highest percentage lead recovery (Y). Table 2.28 summarises the evolution of the simplex until the optimum is reached.

Figure 2.12 displays the evolution of the 18 simplexes; the maximum is at (X_1 = 42.0; X_2 = 7.2). Figure 2.13 displays the evolution of the response in the best vertex of each simplex; it is observed that we have obtained the optimum (Y = 93%) in simplex number 13 but we have decided to continue until simplex number 18 to confirm it. In total, we have carried out 20 experiments.

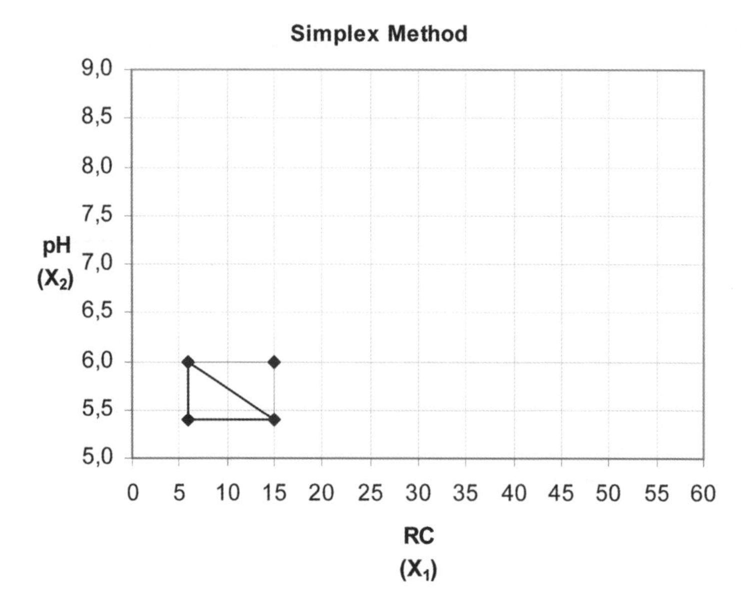

Figure 2.11 Second simplex.

2.4.2.1 *Summary of the Simplex Algorithm*

Below, a pseudo-code is presented to perform a simplex optimisation (using the original algorithm).

START
 Define response
 For each of the *f* factors:
  Boundaries
  Step size
 Construct the initial simplex:
  Initial vertex
  Calculate the other *f* vertices (ensure that all vertexes are within boundaries)
  Obtain the experimental responses for each vertex
LOOP WHILE times retained as the best is less than $2(f+1)$
 Step 1:
 Rank the responses: 'B' for the best, 'W' for the worst and 'N' for the next to the worst
 Step 2:
 Record the number of times a vertex was retained as B
 Stop Criterion: IF times retained as the best is less than $2(f+1)$ times continue with . . .
  Step 3:
  Choose the vertex to be rejected:
  IF W was R in the previous simplex

THEN reject N, so for each factor use

$$G = (\text{sum all except N})/f$$
$$R = G + (G - N)$$

ELSE: reject W, so for each factor use

$$G = (\text{sum all except W})/f$$
$$R = G + (G - W)$$

END IF

Now you have the R coordinates

Step 4: Get the response at R. If the coordinates proposed for the reflected point are outside the boundaries, the experiment cannot be performed and a response worse than that of the W vertex is assigned to the reflected point R

Step 5: To construct the next simplex, replace the rejected vertex (W or N) of the previous simplex by the reflected vertex R

ELSE:

The optimum is the best of the last simplex

END IF

END LOOP

END

2.4.3 The Modified Simplex Method

To compare this approach with the original simplex method, we will consider again the first initial simplex that we defined initially (see Table 2.29).

We see that between R and W, the lead recovery (response) has improved to 71–50% = 21%, so R should be the new B of simplex number 2. This suggests us that, perhaps, in the (W – R) direction we might obtain better responses. To assess this, a new point E (expanded) can be defined, whose coordinates are calculated as

$$E = G + 2(G - W)$$

As Table 2.30 demonstrates, the experimental response obtained after the experiment in those conditions ($Y = 78$) indicates that indeed we have improved more.

Now, to define the following simplex (simplex number 2), we replace vertex W of the previous simplex (simplex number 1) by the expanded vertex E. We obtain simplex number 2 (Table 2.31), the size of which is larger than the previous one and, more importantly, it is oriented in a good direction of improvement. Figure 2.14 displays the effect of the expansion movement.

It will not always be possible to make expansion movements because as we move closer to the optimum we must reduce the size of the simplex in order to locate the optimum accurately. This basic idea of adapting the size of the simplex to each movement is the one that sustains the modified simplex method proposed by Nelder and Mead [17]. Figure 2.15 displays the four possibilities to modify the size of the simplex and Table 2.32 gives their respective expressions for each factor.

Table 2.28 Evolution of the simplex for the worked example.

Simplex number	*Vertex number*	X_1	X_2	*Y*	*Order*	*Times retained*	*Rejected*
1	1	6.0	5.4	50	W		r
	2	15.0	5.4	59	N		
	3	6.0	6.0	63	B	1	
G = (B + N)/2		10.5	5.7				
R = G + (G – W)	4	15.0	6.0	71			
Simplex number	**Vertex number**	$\mathbf{X_1}$	$\mathbf{X_2}$	**Y**	**Order**	**Times retained**	**Rejected**
2	4	15.0	6.0	71	B	1	
	2	15.0	5.4	59	W		r
	3	6.0	6.0	63	N		
G = (B + N)/2		10.5	6.0				
R = G + (G – W)	5	6.0	6.6	71			
Simplex number	**Vertex number**	$\mathbf{X_1}$	$\mathbf{X_2}$	**Y**	**Order**	**Times retained**	**Rejected**
3	4	15.0	6.0	71	B	2	
	5	6.0	6.6	71	B	1	
	3	6.0	6.0	63	W		r
G = (B + N)/2		10.5	6.3				
R = G + (G – W)	6	15.0	6.6	78			
Vertices 4 and 5 have the same response, which is 'best', so they both increase the 'number of times retained'.							
Simplex number	**Vertex number**	$\mathbf{X_1}$	$\mathbf{X_2}$	**Y**	**Order**	**Times retained**	**Rejected**
4	4	15.0	6.0	71	W		r
	5	6.0	6.6	71	W as N		
	6	15.0	6.6	78	B	1	
G = (B + N)/2		10.5	6.6				
R = G + (G – W)	7	6.0	7.2	76			
Vertices 4 and 5 have the same response, which now is 'worst'; in this case we reject the oldest (vertex number 4) and the other one is considered as 'next to the worst'.							
Simplex number	**Vertex number**	$\mathbf{X_1}$	$\mathbf{X_2}$	**Y**	**Order**	**Times retained**	**Rejected**
5	7	6.0	7.2	76	N		
	5	6.0	6.6	71	W		r
	6	15.0	6.6	79	B	2	
G = (B + N)/2		10.5	6.9				
R = G + (G – W)	8	15.0	7.2	84			

Simplex number	Vertex number	X_1	X_2	Y	Order	Times retained	Rejected
6	7	6.0	7.2	76	W		r
	8	15.0	7.2	84	B	1	
	6	15.0	6.6	79	N		
G = (B + N)/2		15.0	6.9				
R = G + (G – W)	9	24.0	6.6	84			
Simplex number	**Vertex number**	**X_1**	**X_2**	**Y**	**Order**	**Times retained**	**Rejected**
7	9	24.0	6.6	84	B	1	
	8	15.0	7.2	84	B	2	
	6	15.0	6.6	79	W		r
G = (B + N)/2		19.5	6.9				
R = G + (G – W)	10	24.0	7.2	89			
Simplex number	**Vertex number**	**X_1**	**X_2**	**Y**	**Order**	**Times retained**	**Rejected**
8	9	24.0	6.6	84	W		
	8	15.0	7.2	84	W		r
	10	24.0	7.2	89	B	1	
G = (B + N)/2		24.0	6.9				
R = G + (G – W)	11	33.0	6.6	88			
Simplex number	**Vertex number**	**X_1**	**X_2**	**Y**	**Order**	**Times retained**	**Rejected**
9	9	24.0	6.6	84	W		r
	11	33.0	6.6	88	N		
	10	24.0	7.2	89	B	2	
G = (B + N)/2		28.5	6.9				
R = G + (G – W)	12	33.0	7.2	92			
Simplex number	**Vertex number**	**X_1**	**X_2**	**Y**	**Order**	**Times retained**	**Rejected**
10	12	33.0	7.2	92	B	1	
	11	33.0	6.6	88	W		r
	10	24.0	7.2	89	N		
G = (B + N)/2		28.5	7.2				
R = G + (G – W)	13	24.0	7.8	89			

Table 2.28 (*continued*).

Simplex number	*Vertex number*	X_1	X_2	*Y*	*Order*	*Times retained*	*Rejected*
Simplex number	**Vertex number**	$\mathbf{X_1}$	$\mathbf{X_2}$	**Y**	**Order**	**Times retained**	**Rejected**
11	12	33.0	7.2	92	B	2	
	13	24.0	7.8	89	W as N		
	10	24.0	7.2	89	W		r
G = (B + N)/2		28.5	7.5				
R = G + (G – W)	14	33.0	7.8	92			
As in simplex number 4, vertices 10 and 13 have the same answer, which is 'worst'; we reject the oldest (vertex number 10) and the other one is considered as 'next to the worst'.							
Simplex number	**Vertex number**	$\mathbf{X_1}$	$\mathbf{X_2}$	**Y**	**Order**	**Times retained**	**Rejected**
12	12	33.0	7.2	92	B	3	
	13	24.0	7.8	89	W		r
	14	33.0	7.8	92	B	1	
G = (B + N)/2		33.0	7.5				
R = G + (G – W)	15	42.0	7.2	93			
Before calculating G and R, it is advisable to review the experimental value of B to avoid the influence of experimental errors. This is done whenever the number of times it was retained as 'best' is $f + 1$ (this is called the '$f + 1$ rule'). Let us suppose that the response is correct.							
Simplex number	**Vertex number**	$\mathbf{X_1}$	$\mathbf{X_2}$	**Y**	**Order**	**Times retained**	**Rejected**
13	12	33.0	7.2	92	W		r
	15	42.0	7.2	93	B	1	
	14	33.0	7.8	92	W		
G = (B + N)/2		37.5	7.5				
R = G + (G – W)	16	42.0	7.8	92			
Simplex number	**Vertex number**	$\mathbf{X_1}$	$\mathbf{X_2}$	**Y**	**Order**	**Times retained**	**Rejected**
14	16	42.0	7.8	92	W as N		
	15	42.0	7.2	93	B	2	
	14	33.0	7.8	92	W		r
G = (B + N)/2		42.0	7.5				
R = G + (G – W)	17	51.0	7.2	91			
Remember the note in simplex number 11.							

Simplex number	Vertex number	X_1	X_2	Y	Order	Times retained	Rejected
15	16	42.0	7.8	92	N		r
	15	42.0	7.2	93	B	3	
	17	51.0	7.2	91	W		[r]
G = (B + W)/2		46.5	7.2				
R = G + (G – N)	18	51.0	6.6	88			

Remember the note in simplex number 12.

Vertex number 17 is W in the present simplex (simplex 15), but it was R in the previous simplex (simplex 14); therefore, if we rejected 17 we will obtain vertex 14 again, which will be W of the new simplex, and the process will oscillate between these two designs without being able to advance. In order to avoid this, vertex N (vertex 16) is projected according to:

G = (B + W)/2

R = G + (G – N).

Simplex number	Vertex number	X_1	X_2	Y	Order	Times retained	Rejected
16	18	51.0	6.6	88	W		[r]
	15	42.0	7.2	93	B	4	
	17	51.0	7.2	91	N		r
G = (B + W)/2		46.5	6.9				
R = G + (G – N)	19	42.0	6.6	89			

See previous note (simplex number 15).

Simplex number	Vertex number	X_1	X_2	Y	Order	Times retained	Rejected
17	18	51.0	6.6	88	W		r
	15	42.0	7.2	93	B	5	
	19	42.0	6.6	89	N		
G = (B + N)/2		42.0	6.9				
R = G + (G – W)	20	33.0	7.2	92			

Simplex number	Vertex number	X_1	X_2	Y	Order	Times retained	Rejected
18	20	33.0	7.2	92	N		
	15	42.0	7.2	93	B	6	
	19	42.0	6.6	89	W		

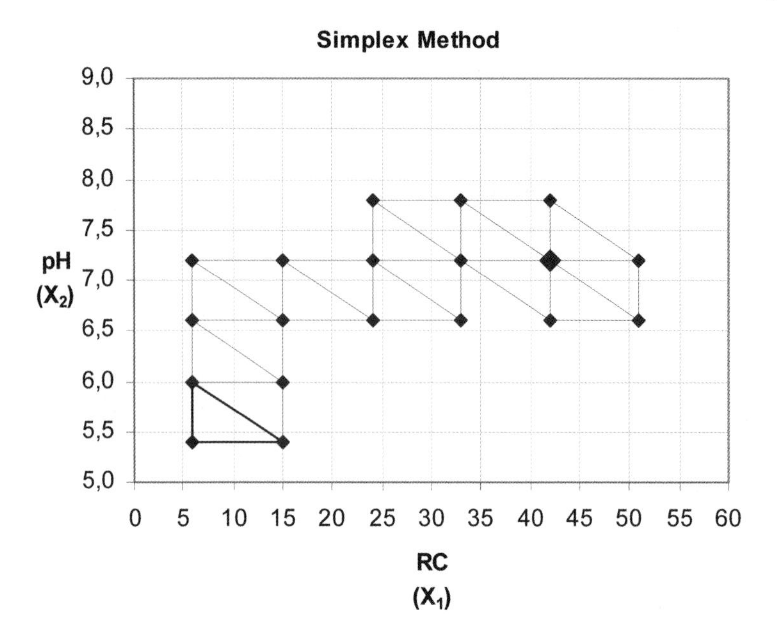

Figure 2.12 Evolution of the simplex.

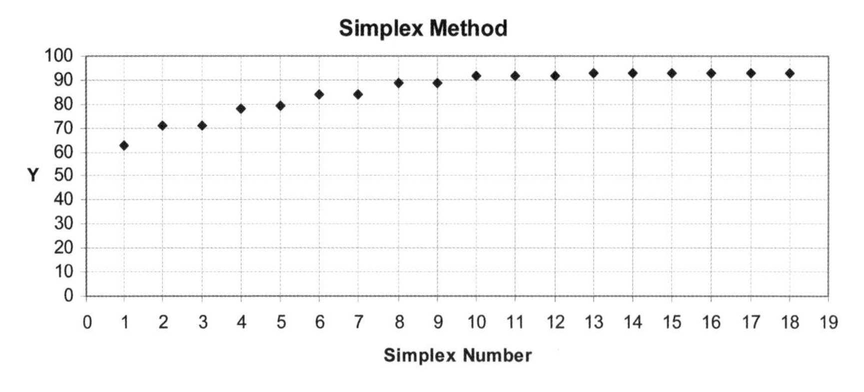

Figure 2.13 Evolution of the response in the best vertex.

2.4.3.1 Summary of the Modified Simplex Algorithm

A pseudo-code is presented below to perform a modified simplex optimisation.

START

    Define response
    For each of the *f* factors:
      Boundaries
      Step size

Table 2.29 Evaluating reflected vertex R for the worked example.

Simplex number	*Vertex number*	X_1	X_2	*Y*	*Order*	*Times retained*
1	1	6.0	5.4	50	W	
	2	15.0	5.4	59	N	
	3	6.0	6.0	63	B	1
G = (B + N)/2		10.5	5.7			
R = G* (G – W)	4	15.0	6.0	71		

Table 2.30 Expanded vertex for the worked example.

Simplex number	*Vertex number*	X_1	X_2	*Y*	*Order*
1	1	6.0	5.4	50	W
	2	15.0	5.4	59	N
	3	6.0	6.0	63	B
G = (B + N)/2		10.5	5.7		
R = G + (G – W)	4	15.0	6.0	71	R > B
E = G + 2(G – W)	5	19.5	6.3	78	

Table 2.31 Expanded simplex for the worked example.

Simplex number	*Vertex number*	X_1	X_2	*Y*
2	5	19.5	6.3	78
	2	15.0	5.4	59
	3	6.0	6.0	63

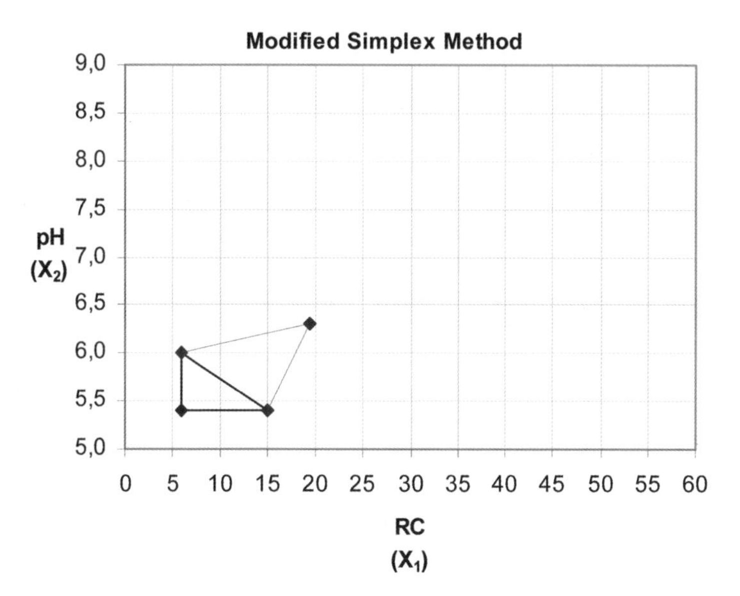

Figure 2.14 Modified simplex.

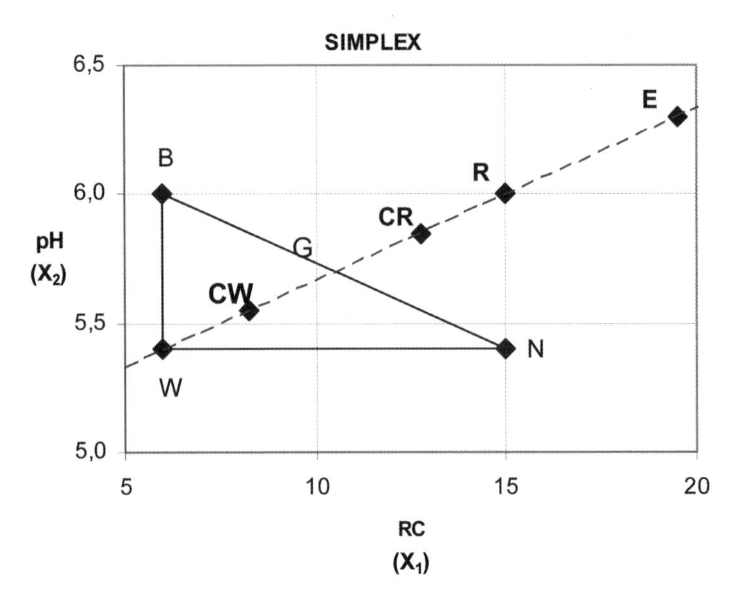

Figure 2.15 Size modification of the simplex design.

Table 2.32 General rules to expand a simplex; they yield four possible movements.

Equation for each factor	*Movement*
R = G + (G – W)	Reflection
E = G + 2(G – W)	Expansion
CR = G + 0.5(G – W)	Contraction on the R side
CW = G – 0.5(G – W)	Contraction on the W side

Construct the initial simplex:
Initial vertex
Calculate the other *f* vertices (ensure that all vertexes are within boundaries)
Obtain the experimental responses for each vertex
LOOP WHILE stop criterion is not achieved
Step 1:
Rank the responses: 'B' for the best, 'W' for the worst and 'N' for the next to the worst
Step 2: Have you achieved your stop criterion?
IF your answer in NO
THEN
Step 3:
Chose the vertex to be rejected
IF W was R in the previous simplex

```
            THEN reject N, so for each factor use
                     G = (sum all except N)/f
                     R = G + (G - N)
            ELSE: reject W, so for each factor use
                     G = (sum all except W)/f
                     R = G + (G - W)
            END IF
            Now you have the R coordinates
            Get experimental response at R
            Step 4:
            SELECT
            CASE: R is better than B:
                     Calculate E using
                              E = G + 2(G - W)
                     Get response at E
                     IF E is worse than B:
                     THEN R will become the new vertex of the next simplex
                     ELSE E will become the new vertex of the next simplex
                     END IF
            CASE: R is equal to or better than N and worse than or equal to
              B:
                     R will become the new vertex of the next simplex
            CASE: R is worse than N:
                     IF R is worse than W:
                     THEN
                              Calculate CW using
                                        CW = G - 0.5(G - W)
                              Get response at CW
                              CW will become the new vertex of the next
                                simplex
                     ELSE
                              Calculate CR using
                                        CR = G + 0.5(G - W)
                              Get response at CR
                              CR will become the new vertex of the next
                                simplex
                     END IF
            END CASE
            Step 5: In order to construct the next simplex replace the
              rejected vertex (W) of the previous simplex by the vertex accepted
              in step 4
          ELSE:
            The optimum is the best of the last simplex
          END IF
END LOOP
END
```

Table 2.33 Evolution of the modified simplex for the worked example.

Simplex number	*Vertex number*	X_1	X_2	*Y*	*Order*	*Times retained*	*Rejected*
1	1	6.0	5.4	50	W		r
	2	15.0	5.4	59	N		
	3	6.0	6.0	63	B	1	
G = (B + N)/2		10.5	5.7				
R = G + (G – W)	4	15.0	6.0	71	R > B		
E = G + 2(G – W)	5	19.5	6.3	78			
Simplex number	**Vertex number**	$\mathbf{X_1}$	$\mathbf{X_2}$	**Y**	**Order**	**Times retained**	**Rejected**
2	5	19.5	6.3	78	B	1	
	2	15.0	5.4	59	W		r
	3	6.0	6.0	63	N		
G = (B + N)/2		12.8	6.2				
R = G + (G – W)	6	10.5	6.9	71	N < R < B		
Simplex number	**Vertex number**	$\mathbf{X_1}$	$\mathbf{X_2}$	**Y**	**Order**	**Times retained**	**Rejected**
3	5	19.5	6.3	78	B	2	
	6	10.5	6.9	71	N		
	3	6.0	6.0	63	W		r
G = (B + N)/2		15.0	6.6				
R = G + (G – W)	7	24.0	7.2	89	R > B		
E = G + 2(G – W)	8	33.0	7.8	92	E > B		
Simplex number	**Vertex number**	$\mathbf{X_1}$	$\mathbf{X_2}$	**Y**	**Order**	**Times retained**	**Rejected**
4	5	19.5	6.3	78	N		
	6	10.5	6.9	71	W		r
	8	33.0	7.8	92	B	1	
G = (B + N)/2		26.3	7.1				
R = G + (G – W)	9	42.0	7.2	93	R > B		
E = G + 2(G – W)	10	57.8	7.4	89	E < B		

Two consecutive expansions allows the simplex to approach fast to a maximum.

Simplex number	Vertex number	X_1	X_2	Y	Order	Times retained	Rejected
5	5	19.5	6.3	78	W		r
	9	42.0	7.2	93	B	1	
	8	33.0	7.8	92	N		
G = (B + N)/2		37.5	7.5				
R = G + (G – W)	11	55.5	8.7	81	W < R < N		
CR = G + 0.5(G – W)	12	46.5	8.1	90			

Simplex number	Vertex number	X_1	X_2	Y	Order	Times retained	Rejected
6	12	46.5	8.1	90	W		[r]
	9	42.0	7.2	93	B	2	
	8	33.0	7.8	92	N		r
G = (B + W)/2		44.3	7.7				
R = G + (G – N)	13	55.5	7.5	90	R=N*		

Vertex 12 is the point accepted from simplex 5, but now is the worst point, so we must reject the 'next to the worst' N.
Vertex 13 is as bad as vertex 12; the 'next to the worst' (rejected) vertex N*, so we do not try more points.

Simplex number	Vertex number	X_1	X_2	Y	Order	Times retained	Rejected
7	12	46.5	8.1	90	W		r
	9	42.0	7.2	93	B	3	
	13	55.5	7.5	90	W		
G = (B + N)/2		48.8	7.4				
R = G + (G – W)	14	51.0	6.6	88	R < W		
CW = G – 0.5(G – W)	15	47.6	7.7	92			

We have two 'worst' points; reject the oldest one (vertex number 12).

Simplex number	Vertex number	X_1	X_2	Y	Order	Times retained	Rejected
8	15	47.6	7.7	92	N		
	9	42.0	7.2	93	B	4	
	13	55.5	7.5	90	W		r
G = (B + N)/2		44.8	7.5				
R = G + (G – W)	16	34.1	7.4	92	R=N		

Table 2.33 *(continued)*.

Simplex number	*Vertex number*	X_1	X_2	*Y*	*Order*	*Times retained*	*Rejected*
Simplex number	**Vertex number**	$\mathbf{X_1}$	$\mathbf{X_2}$	**Y**	**Order**	**Times retained**	**Rejected**
9	15	47.6	7.7	92	W		r
	9	42.0	7.2	93	B	5	
	16	34.1	7.4	92	W		
G = (B + N)/2		38.1	7.3				
R = G + (G – W)	17	28.5	6.9	89	R < W		
CW = G – 0.5(G – W)	18	42.8	7.5	93			
Simplex number	**Vertex number**	$\mathbf{X_1}$	$\mathbf{X_2}$	**Y**	**Order**	**Times retained**	**Rejected**
10	18	42.8	7.5	93	B	1	r
	9	42.0	7.2	93	B	6	
	16	34.1	7.4	92	W		
G = (B + N)/2		38.1	7.3				
R = G + (G – W)	19	33.3	7.1	91	R < W		
CW = G – 0.5(G – W)	20	40.4	7.4	93			
Two consecutive contractions reduce the size of the simplex design.							
Simplex number	**Vertex number**	$\mathbf{X_1}$	$\mathbf{X_2}$	**Y**	**Order**	**Times retained**	**Rejected**
11	18	42.8	7.5	93	B	2	
	9	42.0	7.2	93	B	7	
	20	40.5	7.4	93	B	1	

The simplex has a small size (no practical differences between levels in each factor) and the three responses are equal, so we stop the process here.

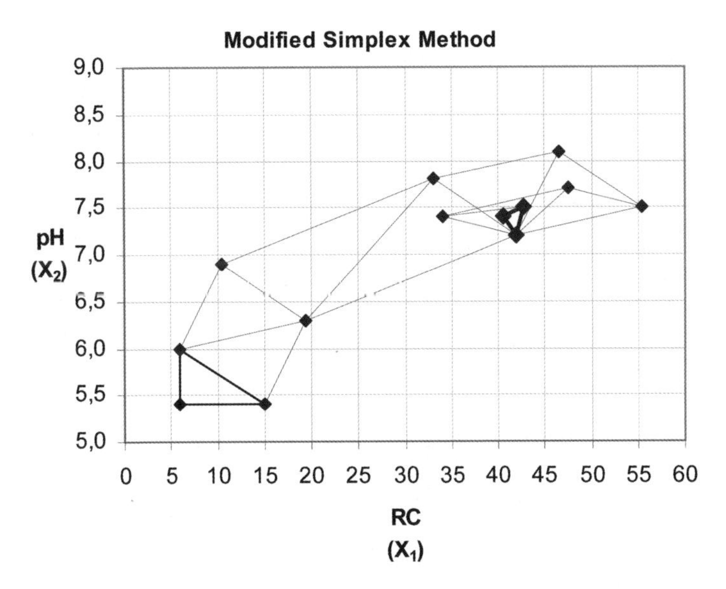

Figure 2.16 Evolution of the modified simplex in the example.

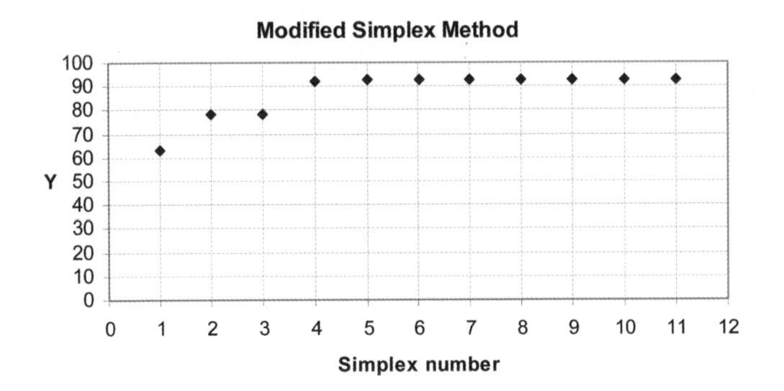

Figure 2.17 Evolution of the best response in the example (modified simplex).

There are three main criteria to stop the process:

1. Simplex too small, no practical differences between level factors.
2. Responses are very similar.
3. You achieved a good enough response and your budget was spent.

If conditions 1 and 2 are fulfilled simultaneously, it is probable that you have achieved the optimum (perhaps a local optimum).

Table 2.34 Examples of practical applications of experimental designs in atomic spectrometry.

Experimental design	*Ref.*	*Analyte(s)*	*Matrix/sample*	*Technique*	*Comments*
Full factorial	[19]	Cr	Aqueous solutions, waste water, dam water, carrot, parsley and lettuce	FAAS	The optimisation of the preconcentration variables was carried out using a full factorial design, from which it was deduced that the most important parameter was the concentration of the eluent
	[20]	Al	Juice and soft drinks	SS-ETAAS	Five variables were studied to 'optimise' the ultrasound slurry pseudo-digestion (preintensification and intensification times, acid concentrations in the mixtures and temperature)
	[21]	Cd, Cr, Cu, Pb	Seafood	FAAS and ETAAS	Several systematic errors were investigated, mainly related to the sample pretreatment steps (microwave acid digestion, ultrasound acid leaching and slurry sampling), calibration mode and number of repeats. Experimental designs and principal components were used
	[22]	Ca, Fe, Mg, Mn, Zn	Beans	ICP-OES	The sample treatment stage, focused microwave aqua regia digestion of beans, was studied to determine the metals simultaneously
	[23]	Cr, Mo, Ti, V	Diesel and fuel oil	ICP-OES	Samples were converted to slurries and a 2^3 factorial design was deployed to study the effects of nitric acid, sample amount and presence of oxygen in the Ar flow
	[24]	Ni	Table salt and baking soda	FAAS	Homocysteine linked to agarose was presented as a novel technique to adsorb Ni. The effect of five variables on the recovery was studied simultaneously by factorial designs
	[25]	Ba, Ca, Cu, K, Mg, Na, P, Zn	Milk	ICP-OES	A two-level full factorial design was made to evaluate the optimum experimental conditions to reduce the residual carbon and the final acidity of the digests
	[26]	Organic species of Pb and Sn	Mussels, gasoline	FAAS	A full experimental design was employed to set the best conditions of a heated electrospray interface intended to couple quartz furnace AAS with HPLC
	[27]	Sb	Soils	ICP-OES	Relevant variables influencing vapour generation of Sb(III) with bromide were screened using an experimental design. Then they were optimised using a simplex algorithm

[28]	Se	Cosmetic, dietetic products	ICP-OES	Relevant variables influencing vapour generation of Se with bromide were screened using an experimental design. Then they were optimised using a simplex algorithm
[29]	As	Homeopathic drugs	ICP-OES	Relevant variables influencing vapour generation of As with bromide ions in acid media were screened using an experimental design. Subsequently, they were optimised via simplex algorithms
[30]	Cd, Cu, Cr, Ni, Pb, V	Waste lubricant oils	ICP-OES	Five instrumental ICP variables were optimised via a full experimental design to measure heavy metals using a kerosene dilution method
[31]	Butyl- and phenyltin	Sediments, water	ICP-OES	The critical parameters determining the instrumental hyphenation between solid-phase microextraction, gas chromatography and ICP were optimised
[32]	Ba, Ca, Co, Cr, Cs, Cu, Fe, Mg, Mn, Ni, Pb, Rb, Sr, Ti, Zn	Tea leaves	ICP-MS, ICP-OES, FAAS	Experimental designs and principal components analysis were combined to study systematic errors in each stage of the methodologies
[33]	Al, Ar, Ba, Cd, Co, Cr, Cu, Fe, Mg, Mn, Na, Ni, Ti, Zn	Aqueous standards	ICP-OES	The long-term stabilities of the ICP radiofrequency power and nebuliser gas flow rate were assessed by experimental designs. 3-way multivariate studies were also carried out. 3-way PARAFAC studies were also employed
[34]	Hg	Blood serum	ETAAS	A trapping technique for the determination of Hg as hydride was optimised using a 4-variable full design
[35]	B, Ba, Cd, Cr, Cu, Fe, Mn, Ni, Si, Sr	Water	ICP-OES	The polynomial effects of Ca and Mg on the metals were modelled by a 2-level full experimental design. Then, a mathematical model to correct for their interferences was developed
[36]	Fe, Pb	Seawater, urine, mineral water, soil, physiological solutions	FAAS	A preconcentration procedure based on the precipitation of the metals was performed using full experimental designs
[37]	Pb	Water	FAAS	The complexation and sorption process on an ester copolymer were studied by factorial designs (pH, shaking time and reagent concentration)

Table 2.34 *(continued)*.

Experimental design	*Ref.*	*Analyte(s)*	*Matrix/sample*	*Technique*	*Comments*
	[38]	Cd	Drinking water	FAAS	Cd precipitation in a knotted reactor was studied with a 3-variable full design
	[39]	S	Sulfides	ICP-MS	An experimental design was applied to study which variables determined the measurement of total reduced sulfur species in natural waters and sediments. Then, a simplex optimisation of relevant variables was made
	[40]	Organic and inorganic species of Sb	Seawater	Atomic fluorescence	Sb species were separated in an anion-exchange column, then hydrides were generated. The chromatographic separation and measurement parameters were studied simultaneously using face-centred full designs
Response surface	[41]	Al, As, Ba, Cd, Ce, Cr, Cu, Mg, Mn, Ni, Pb, Sc, Se, Sr, Zn	Aqueous solutions	ICP-OES	The signal-to-noise ratio on an axially viewed ICP was studied for different lines of the elements and the best conditions were sought
Fractional factorial	[42]	Al, Ba, Cu, Cd, Cr, Fe, Mg, Mn, Pb, Zn	Nuts	ICP-OES	Seven factors (three acid volumes, digestion time, predigestion, temperatures and sample weight) ruling two wet digestion procedures were developed and optimised
	[43]	Species of Hg	Water	Atomic fluorescence	Hydride generation coupled with purge and cryogenic trapping was implemented after an experimental design where purge flow, time and amount of borohydride were studied simultaneously
	[44]	Cu	Tap and river water	FAAS	An agarose-based anion exchanger loaded with chromotropic acid was employed for Cu preconcentration. Five variables were studied simultaneously by the experimental design
	[45]	Hg	Human red blood cells	FAAS	Cold vapour was generated in acid media to relate the amount of Hg in blood cells with the use of dental restorative material made with amalgams. Robustness of the methodology was studied using fractional factorial designs

	[46]	Cd, Co, Cr, Cu, Mn, Ni, Pb, Zn	Sewage sludge	ICP-OES	A traditional ultrasound-assisted digestion was studied by a design, whereas a novel 'all-glass sonotrode' was optimised using three experimental designs
	[47]	Organotin species	Sediments, freshwater, wastewater	ICP-OES	A coupling between gas chromatography and ICP was made and the crucial part (analyte passing through the transfer line) was studied by a experimental design
	[48]	Ge	Coal and lignite	ICP-OES	The chemical vaporisation of Fe with chloride ions was studied by means of fractional designs. Relevant variables were optimised by simplex algorithms
Central composite	[49]	Mg, Pb	Aqueous solutions	ICP-OES	The plasma characteristics of an axially viewed ICP where amounts of vapours or aerosols were injected were investigated. Experimental designs and empirical modelling (multivariate linear regression) were combined to select robust conditions
	[50]	As	Gasoline, diesel and naphtha	ETAAS	Experimental designs were employed to define the optimum microemulsion composition (formed by using propan-1-ol and nitric acid) as well as the temperature programme
	[51]	As	Water	ETAAS	Five permanent modifiers were studied (central composite design at five levels)
	[52]	Al, Ca, Co, Cr, Cu, Fe, K, Mg, Mn, Na, Ni, Pb, Zn	Environmental matrices	ICP-OES	The best conditions to operate an axially viewed ICP were set employing experimental designs. The multivariate effect of carrier gas flow and rf power on several analytical figures of merit was studied. Multivariate regression and principal component analysis were also used to model the system
	[53]	Organic and inorganic species of Se	Yeast	Atomic fluorescence	A pressurised liquid extraction device and hydride generation were coupled. The design was used to investigate the dependence of extraction time, temperature and solvent on the recovery
	[54]	Methylmercury and Hg(II)	Cockle and reference materials	ICP-OES	A desirability function was defined to optimise the simultaneous derivatisation solid-phase microextraction of Hg and methylmercury in order to couple it with microwave extraction, gas chromatography and ICP.

Table 2.34 *(continued)*.

Experimental design	*Ref.*	*Analyte(s)*	*Matrix/sample*	*Technique*	*Comments*
	[55]	Se	Not-fat milk powder, oyster tissue, human urine, sediment	ICP-OES	Eight factors were studied to evaluate the performance of a new hydride generation nebuliser system
	[56]	Butyltin	Mussel tissue, marine sediment	ETAAS	HPLC and hydride generation–*in situ* trapping ETAAS were combined, the experimental conditions for the hydride generation being set after experimental designs
Plackett–Burman	[57]	Ni	Food samples	FAAS	The critical variables defining a combination of flow injection and FAAS was studied
	[58]	Al, Cr, Pb	Macroalgae	ETAAS	25 parameters were screened via the experimental design and a final methodology was proposed
	[59]	Cr, Co	Seafood	FAAS	A dynamic ultrasound-assisted sample treatment and online minicolumn preconcentration were studied by experimental design
	[60]	As, Ca, Cd, Co, K, Li, Mg, Na, Ni, Sr	Human hair	ICP-OES	A novel pressurised-liquid extraction method was studied (concentration of the solvent, temperature, static time, pressure, extraction steps, particle size, diatomaceous/sample ratio and flush volume) to carry out sample pretreatment
	[61]	Sb	Soils, fly ash and sediments	ETAAS	Several experimental designs were deployed at different concentration levels of Sb to develop multivariate regression models capable of handling typical interferences which can not be solved easily by traditional methods
	[62]	Cd	Solid and semisolid milk products	FAAS	The experimental design was used to set the conditions to implement a dynamic continuous, ultrasound-assisted acid extraction
	[63]	Sb	Water, acid digests	ETAAS	Several experimental designs were made at different concentrations of Sb and major concomitants (which typically affect its atomic signal) in order to develop multivariate regression models

[64]	Cd	Water	FAAS	A laboratory-made electrolytic cell was designed as an electrolytic generator of the molecular hydride of Cd. The influence of several parameters on the recorded signal was evaluated by the experimental design and subsequently optimised univariately
[65]	Cd	Meat	FAAS	Experimental designs were used to set the operational conditions of a flow injection method which combines continuous acid extraction, preconcentration and flame measurement
[66]	Cd	Legumes and dried fruit	FAAS	The continuous leaching procedure (a minicolumn packed with a chelating resin) and concentration step were established using the experimental design
[67]	Zn	Meat	FAAS	A continuous ultrasound-assisted extraction system connected to a flow injection manifold was implemented and relevant parameters were studied by the experimental design
[68]	Cd	Vegetables	FAAS	A minicolumn containing a chelating resin was implemented in an on-line flow injection device. The experimental design was used to set the operational conditions
[69]	Cu, Fe	Seafood	FAAS	A continuous ultrasound-assisted device was studied by an experimental design in order to determine the leaching parameters
[70]	Cd, Pb	Mussel	FAAS	The leaching conditions of a ultrasound-assisted device were studied to obtain the best experimental settings
[71]	As, Bi, Se, Te	Coal, fly ash, slag	Atomic fluorescence	Microwave acid extraction was implemented to recover metals from solid materials. The operational conditions needed to form the hydrides were set after analyses by experimental designs
[72]	Co	Coal fly ash, soils, sediments	ETAAS	The most important variables determining the slurry extraction of Co were searched for. Two liquid media and three agitation modes were studied
[73]	Ca	Seafood	FAAS	The experimental conditions influencing an ultrasound-assisted continuous leaching process were studied using an experimental design

Table 2.34 *(continued)*.

Experimental design	*Ref.*	*Analyte(s)*	*Matrix/sample*	*Technique*	*Comments*
	[74]	Fe	Milk powder and infant formula	FAAS	The experimental design evaluated the factors influencing a flow-injection manifold including an ultrasound-assisted step
	[75]	Cr	Mussels	FAAS	A continuous ultrasound-assisted device was studied and the operational experimental conditions were established
	[76]	Cu, Ti, Zn	Cu–Ti–Zn alloys	Glow discharge optical emission	The parameters of different sampling methods were set after studies based on experimental designs
	[77]	Mn	Mussels, tuna, sardine, clams	FAAS	A flow-injection on-line manifold where a leaching procedure occurs was implemented with the aid of experimental designs
	[78]	Fe	Mussels	FAAS	An ultrasound-assisted acid extraction was studied with the designs, after which the setup was established
	[79]	Cu	Mussels	FAAS	An ultrasound-assisted acid extraction was studied with the designs, after which the setup was established
	[80]	Cu	Seawater	FAAS	An ultrasound-assisted acid extraction was studied with the designs, after which the setup was established
	[81]	Ca, Cd, Cr, Cu, Fe, Mg, Mn, Pb, Zn	Seafood	FAAS, ETAAS	Seven factors were studied to assess their importance on the acid microwave-assisted leaching of metallic elements
	[82]	Cd	Seawater	FAAS	Experimental designs were used to set up a field flow preconcentration system (six important variables were considered) coupled to a minicolumn concentration system filled with Chelite P
Taguchi	[83]	Cu	Water	FAAS	Four variables at three levels each were considered to establish an electrodialysis setup to separate Cu ions from the solutions
	[84]	Cd, Cr	Marine sediments	ETAAS	Four variables at three levels each were studied to reduce noise during metal measurement. The effects of the variables determining the acid digestion and their interactions were studied

Doehlert	[85]	Pb	Tap and mineral water, serum and synthetic seawater	FAAS	Multiwall C nanotubes were proposed to absorb lead and act as a preconcentration step. The variables influencing the absorption were studied. The absorption step was coupled on-line to an FAAS device
	[86]	Ni	Saline oil-refinery effluents	FAAS	A four-variable Doehlert design was used to optimise a preconcentration procedure involving centrifugation, a surfactant, methanol and pH buffer
	[87]	As	Naphtha	ETAAS	The best settings for the furnace programme were first studied and then optimised with Doehlert designs
	[88]	Na, K, Ca, Al, Fe	Water	ICP-OES	A 5-level Doehlert design provided multivariate insight into the complex relations between interfering effects and matrix composition. Models were developed through multiple linear regression
	[89]	V	Petroleum condensates	ETAAS	Doehlert designs were used to optimise the temperature program of the ETAAS measurement of emulsions of samples
D-optimal	[90]	Hg, Sn	Sediments	FAAS, ICP-MS	A flow injection–hydride generation–quartz furnace system was studied by experimental designs considering concentration of HCl, % of $NaBH_4$ and furnace temperature. A design was developed specifically to measure Sn in sediments by microwave-assisted digestion and ICP
	[91]	Fe	Seawater	ETAAS	A desirability function was developed to optimise two sets of variables each with practical constraints
	[92]	Al	Seawater	ETAAS	Seven variables were studied simultaneously by experimental designs with the aim of decreasing the detection limit
Three-level orthogonal	[93]	As, Hg	Medicinal herbs	Atomic fluorescence	The experimental conditions for the acid digestion (closed-vessel microwave device) of samples were established. Hydride generation was carried out on the digests before atomising them
Full and fractional factorial	[94]	Hg	Atmospheric particulate	ICP-OES	Experimental parameters concerning plasma conditions, amalgamation, desorption and mercury vapour generation were studied by the experimental design
	[95]	Cu, Cd, K, Fe, Zn, Pb	Human hair	FAAS	Mutual interferences between the analytes and between them and some other elements were modelled with the aid of experimental designs

Table 2.34 *(continued)*.

Experimental design	*Ref.*	*Analyte(s)*	*Matrix/sample*	*Technique*	*Comments*
	[96]	Cd, Zn	Bentonite	FAAS	The adsorption and ultrasonic desorption of heavy metals on bentonites was studied by factorial designs
	[97]	Organic species of P	Polymeric systems	ETAAS	Ashing temperature, digestion time and amount of reagents were studied by experimental designs to prevent volatilisation of organic P compounds present in minor concentrations in polymer materials
	[98]	As	Gasoline	FAAS	Five factors for hydride generation were studied to develop a method to determine As in gasolines. A central composite design was used to develop the response surface and, thus, optimise the extraction procedure
Full factorial and response surface	[99]	Al	Soft drinks	ETAAS	Conventional and multivariate methods were used to establish the best pyrolysis and atomisation temperatures and the chemical modifier (centred full factorial designs). A comparison is presented
Full factorial and central composite	[100]	Sb	Poly(ethylene terephthalate)	FAAS	A screening design detected significant instrumental and chemical variables to volatilise and measure Sb. They were optimised using response surfaces derived from central composite designs. Findings were confirmed using artificial neural networks
	[101]	Al, Fe, K, Li, Na, Mn, Zn	Human hair	ICP-OES	A desirability function was defined to study all elements simultaneously according to three critical variables (sheath gas flow, pump speed and auxiliary gas flow)
	[102]	Se	Electrolytic manganese	ICP-OES	A flow injection–hydride generation procedure was defined using experimental designs. Several chemical and ICP operational variables were simultaneously considered, the relevant ones optimised with a central composite design
	[103]	Sn	Sediments	ETAAS	The experimental conditions to extract Sn from solid sediments were optimised with two sequential experimental designs

Full and Plackett–Burman	[104]	Ca, Cu, Fe, Mg, Zn	Animal feed	FAAS	Experimental designs were used to select the extraction conditions of a dynamic ultrasound-assisted system. Different setups were defined for macro- and micronutrients.
Full factorial and Doehlert	[105]	Mn	Seawater	FAAS	A 2-level full design and a 2-Doehlert matrix design were used to optimise a preconcentration procedure where interaction between the four relevant variables could not be disregarded
	[106]	Cu, V	Seawater	ICP-OES	Four variables were optimised to preconcentrate V and Cu by combining full factorial and Doehlert matrix designs
	[107]	Cd, Pb	Drinking water	FAAS	Experimental designs were used to study the main variables affecting a preconcentration procedure using cloud point extraction. A multiple response function was set to obtain the experimental conditions
	[108]	Cd	Drinking water	FAAS	The optimisation of a preconcentration system involving a column of polyurethane foam and a complexant was made using two-level full factorial and Doehlert designs
	[109]	Zn	Seawater	ICP-OES	Five variables were considered in a full factorial and a Doehlert matrix design to study an on-line preconcentration procedure
Fractional factorial and Doehlert	[110]	Al	Chocolate and candies	ETAAS	Five variables were studied to optimise ultrasound sample treatment before the measurement stage and their interactions were studied
Fractional and central composite	[111]	Hg and organic species	Seawater	ICP-OES	A desirability function was established to optimise the working conditions of a headspace solid-phase microextraction step, combined with capillary gas chromatography and ICP
	[112]	Ag, Au, Cd, Cu, Ni, Zn	Fish	ICP-OES	A fractional factorial design plus two added points in the centre of the domain selected the relevant variables, which were set using an orthogonal central composite design. A desirability function was defined
	[113]	Al, Ba, Ca, Cu, Fe, K, Mg, Mn, Pb, S, Si, V, Zn	Soils, sediments	ICP-OES	A centrifugation step to pass solutions with different pH through the solid sample was studied via experimental designs. Relevant variables were optimised with a central composite design

Table 2.34 *(continued)*.

Experimental design	*Ref.*	*Analyte(s)*	*Matrix/sample*	*Technique*	*Comments*
Fractional factorial and surface response	[114]	Cd	Mineral, tap and river water	TS-FF-AAS	On-line coupling of a C-nanotubes-based preconcentration system with a flow injection manifold. Variables considered most relevant to control the process were evaluated and optimised using experimental designs and response surface methods
	[115]	Co	Aqueous	ETAAS	A solid -phase preconcentration system was implemented. Several chemical and flow-related parameters potentially influencing the enrichment factor were studied, of which only three were revealed to be significant
Fractional factorial and Box–Behnken	[116]	Cd	Drinking water	FAAS	The wall of a knotted reactor was covered with a chelating agent which retains Cd. A factorial design was employed to ascertain the main factors influencing the system and then the final experimental conditions were determined using a Box–Behnken design
Fractional factorial, with star and centre points	[117]	Mn	Seawater	ETAAS	The multivariate interferences of Na, K, Mg and Ca on Mn determination were studied by experimental designs and multivariate regression
Plackett–Burman and central composite	[118]	As, Cd. Pb	Fish	ETAAS	Ultrasound pseudo-digestion sample treatment was studied to extract metals from fish muscle tissue. Relevant variables were finally optimised with a central composite design
	[119]	Al, Ba, Cd, Cr, Cu, Fe, Mn, Pb, Sn, V, Zn	Mussel	ICP-OES	Sample treatment was studied by the saturated fractional design considering volumes and concentrations of acids, temperatures, ramp time and hold time for the microwave heating. An optimised programme was set after the central composite study
	[120]	Al, As, Cd, Co, Cu, Fe, Hg, Li, Mn, Pb, Se, Sr, V, Zn	Marine biological materials	ICP-OES	Pressurised liquid extraction was applied to treat biological materials. Seven variables were studied simultaneously to find the best digestion conditions. Different variables were critical for different sets of metals

[121]	As, Sb	Seawater	Axial ICP-OES	A sample treatment device composed of continuous hydride generation, pre-redissolution with thiourea and preconcentration by coprecipitation with La in a knotted reactor was studied
[122]	Cr, Cu, Ni	Honey	ETAAS	The conditions to determine the three metals were assessed by experimental design and, finally, optimised by response surfaces developed after applying central composite designs
[123]	As, Al, Cd, Cr, Cu, Fe, Mn, Ni, Pb, Zn	Mussels	ICP-OES	The experimental setup for seven variables determining ultrasound bath-assisted enzymatic hydrolysis was optimised using a central composite design after a previous Plackett–Burman design
[124]	Cd, Cr, Cu, Pb, Zn	Mussels	ETAAS	The operating conditions of a new blending device which mixed the mussel tissue with the extractant were set after an experimental design and optimised with a central composite design
[125]	As, Bi, Sb, Se, Te	Atmospheric particulate matter	Atomic fluorescence	Hydride generation conditions were assessed by Plackett–Burman designs. Relevant variables were then optimised using a star central composite design
[126]	Al, Ba, Cd, Cr, Cu, Fe, Mg, Mn, Zn	Vegetables	ICP-OES	Two open-vessel wet digestion methods using mixtures of nitric and sulfuric acids and H_2O_2 were developed and optimised to determine metals in plants
[127]	Al, Ba, Ca, Cd, Cu, Fe, K, Mg, Mn, Na, P, Pb, S, Si, Sr, Ti, V, Zn	Soils, sediments	ICP-OES	An extraction method using centrifugation to pass solutions with different pH through the sample was studied via experimental designs
[128]	Bi	Seawater, hot-spring water, coal, coal fly ash, slag	ETAAS	Nine variables influencing the hydride generation and the flow injection device were studied simultaneously by a Plackett–Burman design. The relevant variables were optimised by a star central composite design
[129]	Cu, Fe	River water	FAAS	Screening studies were made by Plackett–Burman designs to ascertain the main variables affecting the solvent extraction and then optimised using a central composite design

Table 2.34 (*continued*).

Experimental design	*Ref.*	*Analyte(s)*	*Matrix/sample*	*Technique*	*Comments*
	[130]	As, Bi, Ge, Hg, Se	Coal fly ash	ETAAS	Experimental designs were used to evaluate the effects of several parameters influencing the vapour generation efficiency. Relevant variables were optimised with a central composite design
	[131]	Hg	Marine sediments, soils, coal	ETAAS	Experimental designs were used to evaluate the effects of several parameters influencing cold vapour generation. Relevant variables were optimised with a central composite design
	[132]	Organic Sn	Sediments	ETAAS	Evaluation of the main effects of a two-phase extraction method was made by Plackett–Burman experimental designs and then optimised with a fractional central composite design
	[133]	As, Ca, Cd, Co, Cr, Cu, Fe, hg, Mg, Mn, Pb, Se, Zn	Seafood	ETAAS, FAAS	Experimental designs were used to evaluate the effects of several parameters influencing acid leaching. Relevant variables were optimised with a central composite design
	[134]	As, Cd, Sb, Se	Fresh water, hot-spring water	ETAAS	The effect of several variables on the efficiency of the hydride generation procedure employed to extract the analytes was studied using experimental designs. Optimum conditions were found after a central composite matrix was deployed
	[135]	Cd, Cr, Hg, Pb, Se	Human hair	ETAAS	Experimental designs were used to evaluate a leaching procedure implying an ultrasound treatment. The experimental conditions for relevant variables were optimised with a central composite design
	[136]	Ca, Cu, Fe, Mg, Mn, Zn	Human hair	FAAS, ETAAS	The variables implied on an ultrasound-assisted acid leaching procedure were evaluated by experimental designs. The relevant variables were subsequently optimised by a central composite design and response surface

Plackett–Burman and Box–Behnken	[137]	As, Se	Coal fly ash	ETAAS	A coal fly ash leaching procedure was studied. Plackett–Burman designs were used to screen several operational variables and decide which were more relevant. These were optimised by a Box–Behnken approach
Uniform shell Doehlert	[138]	Al, Fe	Water	ETAAS	The experimental design was devoted to study the sample treatment in order to develop a model to remove metals selectively from water using resins (3 factors were considered)
Doehlert and response surface	[139]	Mn	Citrus leaves, bovine liver, dogfish liver	ICP-OES	Doehlert designs and response surface models were developed to evaluate the effects and interactions of five variables on the extraction of Mn in a flow-injection device.
Full factorial and Doehlert	[140]	Cu	Water	Thermospray flame furnace AAS	An on-line coupling of a flow-injection system with the AAS technique was proposed. The full factorial design was used to ascertain the relevant variables which were then optimised using a Doehlert design (pH and concentration of the complexing agent)
Central composite, response surface	[141]	As, Bi, Ge, Hg, Se, Sn, Sb, Te	Aqueous standards	ICP-OES	Vapour generation was studied for each metal and similarities determined by principal components analysis. Response surfaces were plotted after multiple linear regression was made on the scores of a principal component analysis
The sample matrix is aqueous standards	[142]	As and Se species	Aqueous standards	ETAAS	A combination of reversed-phase chromatography and ETAAS was proposed. A central composite design studied the chromatographic conditions. Response curves were deployed for each analyte in order to select the best experimental conditions and evaluate robustness

Table 2.35 Examples of practical applications of simplex optimisation in atomic spectrometry.

Analyte	*Reference*	*Comments*	*Matrix*	*Technique*	*Simplex mode*
Al	[145]	Optimisation of the slurry nebulisation conditions was performed for the analysis of zirconia powder	Zirconia powder	ICP-OES	Simplex
	[146]	Four procedures to determine trace impurities in zirconia powder were developed. Operational characteristics of one of them were optimised by simplex	Zirconia powder	ICP-OES	Simplex
	[147]	Analysis of cement after optimisation of the operational variables (forward power, sample uptake rate, flow rates of intermediate plasma gas, outer plasma gas, aerosol carrier and spray chamber, nebuliser and type of torch)	Cement	ICP-OES	Modified simplex
As	[148]	Optimisation of hydride generation parameters from slurry samples after ultrasonication and ozonation for the direct determination of trace amounts of As(III) and total inorganic As by their *in situ* trapping	Environmental samples (marine sediment, soil, rock salt, waste water) and biological samples (human hair, urine, lobster, liver, muscle, beer and wort, tablets)	HG-ETAAS	Simplex
	[149]	Determination of As in homeopathic drugs by bromide volatilisation–ICP was carried out after the main and interactive effects of the variables affecting the method were evaluated by factorial designs. The optimal conditions (bromide, sulfuric acid) were optimised by simplex	Homeopathic drugs	ICP-OES	Modified simplex

	[150]	Presents graphic representations of multifactor simplex optimisation		ICP-OES	Modified simplex
	[151]	Simplex optimisation of the instrumental conditions (ashing and atomisation temperature, modifier concentration and atomisation ramping time) to determine As	Environmental samples of marine origin	ETAAS	Simplex
	[152]	Reaction conditions were optimised by simplex to determine As, Sb and Sn using a new continuous hydride generator	Aqueous	HG-FAAS	Simplex
	[153]	Optimisation of batch HGAAS to find robust, compromise conditions to level off the sensitivity of As species (related to Balkan endemic nephropathy)	Urine	HG-FAAS	Simplex
	[154]	Optimisation of operational characteristics (viewing height, rf plasma power and nebuliser argon flow) to carry out a multielemental analysis of slurry samples. Comparison of three optimisation procedures	Sediments, rocks, sewage sludge	ICP-OES	Modified simplex
	[155]	Simultaneous speciation of redox species of As and Se using an anion-exchange microbore column coupled with a microconcentric nebuliser. Experimental designs coupled to simplex to optimise ICP conditions	Natural water, soil	ICP-MS	Simplex
Au	[156]	Optimum experimental conditions for both conventional and fast (standardless) furnace programmes were selected to determine Au	Geological samples	ETAAS	Simplex

Table 2.35 (*continued*).

Analyte	*Reference*	*Comments*	*Matrix*	*Technique*	*Simplex mode*
	[157]	An atomised on-line flow-injection (FI) preconcentration system for determination of trace gold in ore samples was studied by simplex	Ore samples	FI-ETAAS	Simplex
	[158]	Trace gold quantities were measured after on-line preconcentration with flow-injection atomic absorption spectrometry. Experimental conditions were set using simplex	Ore samples	FAAS	Scaled simplex method
B	[145]	Optimisation of the slurry nebulisation conditions was performed for the analysis of zirconia powder	Zirconia powder	ICP-OES	Simplex
	[159]	B was determined at sub-$\mu g\,g^{-1}$ levels in biological tissues after simplex optimisation of gas flow, power and observation height	Biological tissues	ICP-OES	Modified simplex
Ba	[160]	Evaluation of three nebulisers for use in microwave-induced plasma optical emission spectrometry	Feminatal tablets, iodide tablets, lichen, lobster hepatopancreas, human hair, soya bean, flour	ICP-OES	Simplex
Bi	[161]	On-line preconcentration and simultaneous determination of heavy metals after optimising experimental conditions (preconcentration flow rate, eluent flow rate, weight of solid phase and eluent loop volume)	Natural water	ICP-OES (CCD)	Super-modified simplex
C	[162]	Determination of the residual carbon content in biological and environmental samples by microwave–ICP	Digested biological samples, environmental samples	MIP-OES	Simplex

Ca	[145]	Optimisation of the slurry nebulisation conditions was performed for the analysis of zirconia powder	Zirconia powder	ICP-OES	Simplex
	[146]	Four procedures to determine trace impurities in zirconia powder were developed. Operational characteristics of one of them were optimised by simplex	Zirconia powder	ICP-OES	Simplex
	[147]	Analysis of cement after optimisation of the operational variables (forward power, sample uptake rate, flow rates of intermediate plasma gas, outer plasma gas, aerosol carrier and spray chamber, nebuliser and type of torch)	Cement	ICP-OES	Modified simplex
	[160]	Evaluation of three nebulisers for use in microwave-induced plasma optical emission spectrometry	Feminatal tablets, iodide tablets, lichen, lobster hepatopancreas, human hair, soya bean, flour	ICP-OES	Simplex
	[163]	Characterisation and optimisation of four parameters defining the performance of a 22 mm torch were made by simplex. Several criteria for optimisation were studied		ICP-OES	Simplex
	[164]	A comparison between linear-flow and a tangential-flow torches was made after simplex optimisation of two flow rates, rf power and height above the load coil		ICP-OES	Simplex
	[165]	Simplex strategy to optimise flame conditions (acetylene and air flow rates)		FAAS	Modified simplex

Table 2.35 (*continued*).

Analyte	*Reference*	*Comments*	*Matrix*	*Technique*	*Simplex mode*
	[166]	Optimising signal-to-noise ratio in flame AAS using sequential simplex optimisation		FAAS	Simplex
	[167]	Optimisation of six experimental variables to develop a new high–temperature plasma sources for spectrochemical analysis		ICP-OES	Modified simplex
	[168]	Optimisation of flow rates of the cooling gas, top and bottom electrode gas, sample liquid flow rate, nebuliser gas and dc current in a variable-diameter multi-electrode dc plasma		DC-ICP-OES	Modified simplex
	[169]	Operating ICP conditions: rf power, intermediate flow, nebuliser flow, peristaltic pump and sample uptake were optimised using an automated simplex programme		ICP-OES	Modified simplex
Cd	[160]	Evaluation of three nebulisers for use in microwave induced plasma optical emission spectrometry	Feminatal tablets, iodide tablets, lichen, lobster hepatopancreas, human hair, soya bean, flour	ICP-OES	Simplex
	[161]	On-line preconcentration and simultaneous determination of heavy metals after optimising experimental conditions (preconcentration flow rate, eluent flow rate, weight of solid phase and eluent loop volume)	Natural water	ICP-OES (CCD)	Super-modified simplex

	[170]	Multielemental separation and determination of Cu, Co, Cd and Ni in tap water and high-salinity media by CGA (colloidal gas aphron) coflotation and simplex optimisation	Tap water, synthetic seawater	FAAS	Simplex
	[171]	Multielemental separation of Cu, Co, Cd and Ni in natural waters by means of colloidal gas aphron coflotation	Natural waters	FAAS	Simplex
	[172]	Separation and preconcentration of Cd in natural water using a liquid membrane system with 2-acetylpyridine benzoylhydrazone as carrier. Transport processes across the membrane were optimised	Natural water	FAAS	Modified simplex
	[173]	Determination of Cd and Pb in environmental samples, optimisation of the experimental conditions by simplex	Environmental samples	ETAAS	Simplex
	[174]	Ultra-trace determination of Cd by vapour generation atomic fluorescence spectrometry	Potable waters, sewage sludge (domestic), sewage sludge (industrial)	VG-AFS	Simplex
	[175]	Atomic absorption determination of traces of Cd in urine after electrodeposition on to a tungsten wire	Urine, river water	AAS	Modified and weighted centroid simplex
	[176]	Simplex-optimised automatic on-line column preconcentration system for determination of Cd	Seawater	ETAAS	Simplex
Cl	[177]	Plasma operating conditions were set using a simplex for low-level determination of non-metals (Cl, Br, I, S, P) in waste oils using prominent spectral lines in the 130–190 nm range	Waste oils	ICP-OES	Simplex

Table 2.35 (*continued*).

Analyte	*Reference*	*Comments*	*Matrix*	*Technique*	*Simplex mode*
Co	[161]	On-line preconcentration and simultaneous determination of heavy metals after optimising experimental conditions (preconcentration flow rate, eluent flow rate, weight of solid phase and eluent loop volume)	Natural water	ICP-OES (CCD)	Super-modified simplex
	[170]	Multielemental separation and determination of Cu, Co, Cd and Ni in tap water and high salinity media by CGA (colloidal gas aphron) coflotation and simplex optimisation	Tap water, synthetic seawater	FAAS	Simplex
	[171]	Multielemental separation of Cu, Co, Cd and Ni in natural waters by means of colloidal gas aphron coflotation	Natural waters	AAS	Simplex
	[178]	Simultaneous determination of Co and Mn in urine. Method development using a simplex optimisation approach	Urine	GF-AAS	Simplex
Cr	[165]	Simplex strategy to optimise atomisation flames		FAAS	Modified simplex
	[179]	Preconcentration of Cr(VI) by solvent extraction was optimised using a regular simplex	Seawater, waste water	FI-ETAAS	Simplex
	[180]	Experimental conditions to determine Cr without flames were optimised by simplex. The effects of matrix concomitants were also studied		ETAAS	Simplex

Cu	[145]	Optimisation of the slurry nebulisation conditions was performed to quantify zirconia	Zirconia powder	ICP-OES	Simplex
	[146]	Four procedures to determine trace impurities in zirconia powder were developed. Operational characteristics of one of them were optimised by simplex	Zirconia powder	ICP-OES	Simplex
	[154]	Optimisation of operational characteristics (viewing height, rf plasma power and nebuliser argon flow) to carry out a multielemental analysis of slurry samples. Comparison of three optimisation procedures	Sediments, rocks, sewage sludge	ICP-OES	Modified simplex
	[160]	Evaluation of three nebulisers for use in microwave-induced plasma optical emission spectrometry	Feminatal tablets, iodide tablets, lichen, lobster hepatopancreas, human hair, soya bean, flour	ICP-OES	Simplex
	[161]	On-line preconcentration and simultaneous determination of heavy metals after optimising experimental conditions (preconcentration flow rate, eluent flow rate, weight of solid phase and eluent loop volume)	Natural water	ICP-OES (CCD)	Super-modified simplex
	[165]	Simplex strategy to optimise atomisation flames		AAS	Modified simplex
	[168]	Optimisation of flow rates of the cooling gas, top and bottom electrode gas, sample liquid flow rate, nebuliser gas and dc current in a variable-diameter multi-electrode dc plasma		DC-ICP-OES	Modified simplex

Table 2.35 (*continued*).

Analyte	*Reference*	*Comments*	*Matrix*	*Technique*	*Simplex mode*
	[170]	Multielemental separation and determination of Cu, Co, Cd and Ni in tap water and high-salinity media by CGA (colloidal gas aphron) coflotation and simplex optimisation	Tap water, synthetic seawater	FAAS	Simplex
	[171]	Multielemental separation of copper, cobalt, cadmium and nickel in natural waters by means of colloidal gas aphron coflotation	Natural waters	FAAS	Simplex
	[181]	Determination of Fe, Cu and Zn in tinned mussels. Univariate setups were found more satisfactory than those from simplex optimisation	Tinned mussels	ICP-OES	Simplex
	[182]	The overall response of a simultaneous multielement determination was optimised. A univariate search procedure was also evaluated		FAAS	Modified simplex
	[183]	Determination of Cu in seawater based on a liquid membrane preconcentration system. Optimisation of transport processes throughout the membrane	Seawater	FAAS	Modified simplex
	[184]	Optimisation of lamp control parameters in glow discharge optical emission spectroscopy for the analysis of Cu–Ti–Zn alloy using simplex	Pure zinc and zinc alloys	GD-OES	Simplex
Fe	[145]	Optimisation of the slurry nebulisation conditions was performed for the analysis of zirconia powder	Zirconia powder	ICP-OES	Simplex

[146]	Four procedures to determine trace impurities in zirconia powder were developed. Operational characteristics of one of them were optimised by simplex	Zirconia powder	ICP-OES	Simplex
[147]	Analysis of cement after optimisation of the operational variables (forward power, sample uptake rate, flow rates of intermediate plasma gas, outer plasma gas, aerosol carrier and spray chamber, nebuliser and type of torch)	Cement	ICP-OES	Modified simplex
[160]	Evaluation of three nebulisers for use in microwave-induced plasma optical emission spectrometry	Feminatal tablets, iodide tablets, lichen, lobster hepatopancreas, human hair, soya bean, flour	ICP-OES	Simplex
[161]	On-line preconcentration and simultaneous determination of heavy metals after optimising experimental conditions (preconcentration flow rate, eluent flow rate, weight of solid phase and eluent loop volume)	Natural water	ICP-OES (CCD)	Super-modified simplex
[163]	Characterisation and optimisation of four parameters defining the performance of a 22 mm torch were made by simplex. Several criteria for optimisation were studied		ICP-OES	Simplex
[165]	Simplex strategy to optimise atomisation flames		AAS	Modified simplex
[168]	Optimisation of flow rates of the cooling gas, top and bottom electrode gas, sample liquid flow rate, nebuliser gas and dc current in a variable-diameter multi-electrode dc plasma		DC-ICP-OES	Modified simplex

Table 2.35 *(continued)*.

Analyte	*Reference*	*Comments*	*Matrix*	*Technique*	*Simplex mode*
	[182]	Simplex optimisation of the overall response of a simultaneous multi-element flame atomic absorption spectrometer (air to fuel ratio, slit width, height above the burner head and four hollow-cathode lamp currents). Cu, Fe, Mn and Zn were measured		AAS	Modified simplex
	[181]	The overall response of a simultaneous multielement determination was optimised. A univariate search procedure was also evaluated	Tinned mussels	ICP-OES	Modified simplex
	[185]	Experimental conditions to determine Fe in water–methanol were simplex optimised (ashing temperature, ashing time), ramp and atomisation temperature)	Atmospheric water	ETAAS	Modified simplex
Ge	[186]	The three most influential factors (sulfuric acid volume, sodium chloride concentration and sample volume) on the chemical vaporisation of Ge were optimised	Coal, lignite and ash	ICP-OES	Modified simplex
	[187]	Optimisation and empirical modelling of HG–ICP-AES analytical technique; simplex and artificial neural networks were combined		HG-ICP-OES	Simplex
	[188]	Application of factorial designs and simplex optimisation in the development of flow injection–hydride generation–graphite furnace atomic absorption spectrometry (FI–HG–GFAAS) procedures		FI-HG-ETAAS	Simplex

Hg	[189]	Experimental conditions to determine methylmercury and inorganic Hg by slurry sampling cold vapour atomic absorption spectrometry were determined by simplex (APDC concentration, agitation time, silica amount and agitation time), after an experimental design underlined their importance	Water samples (drinking and pond water samples)	SS-FI-CV-ETAAS	Simplex
	[190]	Simplex methodology to measure methylmercury and inorganic Hg by high-performance liquid chromatography–cold vapour atomic fluorescence spectrometry	Sediments	HPLC-CV-AFS	Modified simplex
	[191]	Determination of Hg species in gas condensates by on-line coupled high-performance liquid chromatography and cold-vapour atomic absorption spectrometry	Gas condensates	CVAAS	Modified simplex
	[192]	Determination of Hg(II), monomethylmercury cation, dimethylmercury and diethylmercury by hydride generation, cryogenic trapping and atomic absorption spectrometric detection	Estuarine samples of the marsh grass *Spartina alterniflora* and eelgrass	AAS	Simplex
	[193]	Optimisation of cold vapour atomic absorption spectrometric determination of Hg with and without amalgamation by subsequent use of complete and fractional factorial designs with simplex	Tap water, synthetic seawater	CV-FAAS	Modified simplex

Table 2.35 (*continued*).

Analyte	*Reference*	*Comments*	*Matrix*	*Technique*	*Simplex mode*
I	[160]	Evaluation of three nebulisers for use in microwave-induced plasma optical emission spectrometry	Feminatal tablets, iodide tablets, lichen, lobster hepatopancreas, human hair, soya bean, flour	ICP-OES	Simplex
	[177]	Plasma operating conditions were set using a simplex for low-level determination of non-metals (Cl, Br, I, S, P) in waste oils using prominent spectral lines in the 130–190 nm range	Waste oils	ICP-OES	Simplex
K	[147]	Analysis of cement after optimisation of the operational variables (forward power, sample uptake rate, flow rates of intermediate plasma gas, outer plasma gas, aerosol carrier and spray chamber, nebuliser and type of torch)	Cement	ICP-OES	Modified simplex
Lanthanides	[194]	A separation procedure was developed after a simplex optimisation of a nitric acid digestion to determine lanthanides and yttrium in rare earth ores and concentrates	Rare earth ores	ICP-OES	Simplex
Mg	[145]	Optimisation of the slurry nebulisation conditions was performed for the analysis of zirconia powder	Zirconia powder	ICP-OES	Simplex
	[146]	Four procedures to determine trace impurities in zirconia powder were developed. Operational characteristics of one of them were optimised by simplex	Zirconia powder	ICP-OES	Simplex

	[147]	Analysis of cement after optimisation of the operational variables (forward power, sample uptake rate, flow rates of intermediate plasma gas, outer plasma gas, aerosol carrier and spray chamber, nebuliser and type of torch)	Cement	ICP-OES	Modified simplex
	[160]	Evaluation of three nebulisers for use in microwave induced plasma optical emission spectrometry	Feminatal tablets, iodide tablets, lichen, lobster hepatopancreas, human hair, soya bean, flour	ICP-OES	Simplex
	[163]	Characterisation and optimisation of four parameters defining the performance of a 22 mm torch were made by simplex. Several criteria for optimisation were studied		ICP-OES	Simplex
	[164]	Automated simplex optimisation for monochromatic imaging ICP		ICP-OES	Simplex
	[165]	Simplex strategy to optimise atomisation flames		FAAS	Modified simplex
	[168]	Optimisation of flow rates of the cooling gas, top and bottom electrode gas, sample liquid flow rate, nebuliser gas and dc current in a variable-diameter multi-electrode dc plasma		DC-ICP-OES	Modified simplex
	[169]	Operating ICP conditions: rf power, intermediate flow, nebuliser flow, peristaltic pump and sample uptake were optimised using an automated simplex programme		ICP-OES	Modified simplex
Mn	[145]	Optimisation of the slurry nebulisation conditions was performed for the analysis of zirconia powder	Zirconia powder	ICP-OES	Simplex

Table 2.35 (*continued*).

Analyte	*Reference*	*Comments*	*Matrix*	*Technique*	*Simplex mode*
	[146]	Four procedures to determine trace impurities in zirconia powder were developed. Operational characteristics of one of them were optimised by simplex	Zirconia powder	ICP-OES	Simplex
	[147]	Analysis of cement after optimisation of the operational variables (forward power, sample uptake rate, flow rates of intermediate plasma gas, outer plasma gas, aerosol carrier and spray chamber, nebuliser and type of torch)	Cement	ICP-OES	Modified simplex
	[154]	Optimisation of operational characteristics (viewing height, rf plasma power and nebuliser argon flow) to measure environmental slurry samples. Comparison of three optimisation procedures	Sediments, rocks, sewage sludge	ICP-OES	Modified Simplex
	[160]	Evaluation of three nebulisers for use in microwave-induced plasma optical emission spectrometry	Feminatal tablets, iodide tablets, lichen, lobster hepatopancreas, human hair, soya bean, flour	ICP-OES	Simplex
	[165]	Simplex strategy to optimise atomisation flames		FAAS	Modified simplex
	[168]	Optimisation of flow rates of the cooling gas, top and bottom electrode gas, sample liquid flow rate, nebuliser gas and dc current in a variable-diameter multi-electrode dc plasma		DC-ICP-OES	Modified simplex

	[178]	Simultaneous determination of cobalt and manganese in urine by electrothermal atomic absorption spectrometry. Extraction procedure optimised using simplex	Urine	ETAAS	Simplex
	[182]	Simplex optimisation of the overall response of a simultaneous multi-element flame atomic absorption spectrometer (air to fuel ratio, slit width, height above the burner head and four hollow cathode lamp currents). Cu, Fe, Mn and Zn were measured		FAAS	Modified simplex
Mo	[195]	Experimental designs revealed the critical variables (HNO_3 and HCl concentrations) and they were optimised by two simplex to measure Mo by slurry extraction–ETAAS	Solid environmental samples: coal fly ash, sediment, soil and urban dust	SS-ETAAS	Simplex and modified simplex
Na	[145]	Optimisation of the slurry nebulisation conditions was performed for the analysis of zirconia powder	Zirconia powder	ICP-OES	Simplex
	[146]	Four procedures to determine trace impurities in zirconia powder were developed. Operational characteristics of one of them were optimised by simplex	Zirconia powder	ICP-OES	Simplex
	[147]	Analysis of cement after optimisation of the operational variables (forward power, sample uptake rate, flow rates of intermediate plasma gas, outer plasma gas, aerosol carrier and spray chamber, nebuliser and type of torch)	Cement	ICP-OES	Modified simplex

Table 2.35 *(continued)*.

Analyte	*Reference*	*Comments*	*Matrix*	*Technique*	*Simplex mode*
Ni	[161]	On-line preconcentration and simultaneous determination of heavy metals after optimising experimental conditions (preconcentration flow rate, eluent flow rate, weight of solid phase and eluent loop volume)	Natural water	ICP-OES (CCD)	Super-modified simplex
	[170]	Multielemental separation and determination of Cu, Co, Cd and Ni in tap water and high-salinity media by CGA (colloidal gas aphron) coflotation and simplex optimisation	Tap water, synthetic seawater	FAAS	Simplex
	[171]	Multielemental separation of Cu, Co, Cd and Ni in natural waters by means of colloidal gas aphron coflotation	Natural waters	FAAS	Simplex
	[196]	The main variables affecting the separation and enrichment of Ni from natural waters using a liquid membrane were optimised	River water and seawater	FAAS, ICP-MS, AdCSV	Modified simplex
	[197]	A permeation liquid membrane system for determination of nickel in seawater. Transport processes were optimised	Seawater	GFAAS, FAAS, AdCSV	Modified simplex
	[198]	Computer-assisted simplex optimisation of a graphite furnace was accomplished to measure Ni using an on-line preconcentration system	Seawater	ETAAS	Modified simplex
P	[147]	Analysis of cement after optimisation of the operational variables (forward power, sample uptake rate, flow rates	Cement	ICP-OES	Modified simplex

		of intermediate plasma gas, outer plasma gas, aerosol carrier and spray chamber, nebuliser and type of torch)			
	[177]	Plasma operating conditions were set using a simplex for low-level determination of non-metals (Cl, Br, I, S, P) in waste oils using prominent spectral lines in the 130–190 nm range	Waste oils	ICP-OES	Simplex
Pb	[154]	Optimisation of operational characteristics (viewing height, rf plasma power and nebuliser argon flow) to measure environmental slurry samples. Comparison of three optimisation procedures	Sediments, rocks, sewage sludge	ICP-OES	Modified Simplex
	[160]	Evaluation of three nebulisers for use in microwave induced plasma optical emission spectrometry	Feminatal tablets, iodide tablets, lichen, lobster hepatopancreas, human hair, soya bean, flour	ICP-OES	Simplex
	[161]	On-line preconcentration and simultaneous determination of heavy metals after optimising experimental conditions (preconcentration flow rate, eluent flow rate, weight of solid phase and eluent loop volume)	Natural water	ICP-OES (CCD)	Super-modified simplex
	[165]	Simplex strategy to optimise atomisation flames		AAS	Modified simplex
	[173]	Determination of Pb and Cd by atomic absorption methods optimised with simplex	Environmental samples	ETAAS	Simplex
	[199]	An approach to determine Pb by vapour generation atomic	Standard reference water	VG-FAAS	Simplex

Table 2.35 *(continued)*.

Analyte	*Reference*	*Comments*	*Matrix*	*Technique*	*Simplex mode*
		absorption spectrometry was presented and its conditions were optimised by simplex			
	[200]	Simplex optimisation of the digestion conditions for the acid attack of river sediments	River sediments		Simplex
	[201]	Three noise reduction techniques were applied to determine Pb at low concentration levels		ETAAS	Modified simplex
S	[147]	Analysis of cement after optimisation of the operational variables (forward power, sample uptake rate, flow rates of intermediate plasma gas, outer plasma gas, aerosol carrier and spray chamber, nebuliser and type of torch)	Cement	ICP-OES	Modified simplex
	[177]	Plasma operating conditions were set using a simplex for low-level determination of non-metals (Cl, Br, I, S, P) in waste oils using prominent spectral lines in the 130–190 nm range	Waste oils	ICP-OES	Simplex
	[202]	Determination of total reduced sulfur species in natural waters and volatile sulfides in sediments was performed after experimental design and simplex optimisation of relevant variables (octopole bias and helium flow rate)	Natural water and sediments	ICP-MS	Simplex

Sb	[152]	Reaction conditions were optimised by simplex to determine As, Sb and Sn using a new continuous hydride generator	Aqueous	HG-FAAS	Simplex
	[203]	Vaporisation of Sb(III) with bromide was optimised using experimental design and simplex (volumes of sulfuric acid, KBr and solution)	Soils	ICP-OES	Modified simplex
	[204]	Simplex optimisation of experimental conditions to determine Sb in environmental samples	Pine needle, pond weed sample	ETAAS	Modified simplex
	[205]	Simplex optimisation of experimental conditions to determine Sb in environmental samples	Pine needle, pond weed sample	ETAAS	Simplex
	[206]	Simplex was used to optimise the best conditions to generate, transport and atomise Sb volatile species to determine total and inorganic Sb species at ultratrace levels by hydride generation and *in situ* trapping flame atomic absorption spectrometry	Soil, sediment, coal fly ash, sewage, river water	HG-IAT-FAAS	Simplex
	[207]	Direct generation of stibine from slurries and its determination by ETAAS was studied using simplex multivariate optimisation (concentrations of HCl and $NaBH_4$ and Ar flow rate). Previous experimental designs were made to select those variables	Soils, sediments, coal fly ash and coals	HG-ETAAS	Modified simplex
Se	[155]	Simultaneous speciation of redox species of As and Se using an anion-exchange microbore column coupled with a microconcentric nebuliser.	Natural water	ICP-MS	Simplex

Table 2.35 (*continued*).

Analyte	*Reference*	*Comments*	*Matrix*	*Technique*	*Simplex mode*
		Experimental designs coupled to simplex to optimise ICP conditions			
	[208]	Experimental designs revealed three main factors affecting the volatilisation efficiency of two volatile species of Se, which were optimised with simplex	Shampoos and dietetic capsules	ICP-OES	Modified simplex
Si	[147]	Analysis of cement after optimisation of the operational variables (forward power, sample uptake rate, flow rates of intermediate plasma gas, outer plasma gas, aerosol carrier and spray chamber, nebuliser and type of torch)	Cement	ICP-OES	Modified simplex
	[168]	Optimisation of flow rates of the cooling gas, top and bottom electrode gas, sample liquid flow rate, nebuliser gas and dc current in a variable-diameter multi-electrode dc plasma		DC-ICP-OES	Modified simplex
Sn	[152]	A new continuous hydride generator for the determination of As, Sb and Sn by hydride generation. Reaction conditions were optimised by simplex		HG-FAAS	Simplex
	[209]	The experimental conditions for a time-based device used for the determination of tin by hydride generation flow injection atomic	Biological materials	HG-FI-AAS	Simplex

		absorption techniques were optimised by simplex (sample and reagent concentrations and volumes, carrier gas flow rate and atomiser conditions)			
Sr	[147]	Analysis of cement after optimisation of the operational variables (forward power, sample uptake rate, flow rates of intermediate plasma gas, outer plasma gas, aerosol carrier and spray chamber, nebuliser and type of torch)	Cement	ICP-OES	Modified simplex
	[160]	Evaluation of three nebulisers for use in microwave-induced plasma optical emission spectrometry	Feminatal tablets, iodide tablets, lichen, lobster hepatopancreas, human hair, soya bean, flour	ICP-OES	Simplex
Ti	[145]	Optimisation of the slurry nebulisation conditions was performed for the analysis of zirconia powder	Zirconia powder	ICP-OES	Simplex
	[146]	Four procedures to determine trace impurities in zirconia powder were developed. Operational characteristics of one of them were optimised by simplex	Zirconia powder	ICP-OES	Simplex
	[147]	Analysis of cement after optimisation of the operational variables (forward power, sample uptake rate, flow rates of intermediate plasma gas, outer plasma gas, aerosol carrier and spray chamber, nebuliser and type of torch)	Cement	ICP-OES	Modified simplex

Table 2.35 (*continued*).

Analyte	*Reference*	*Comments*	*Matrix*	*Technique*	*Simplex mode*
V	[145]	Optimisation of the slurry nebulisation conditions was performed for the analysis of zirconia powder	Zirconia powder	ICP-OES	Simplex
	[146]	Four procedures to determine trace impurities in zirconia powder were developed. Operational characteristics of one of them were optimised by simplex	Zirconia powder	ICP-OES	Simplex
	[168]	Optimisation of flow rates of the cooling gas, top and bottom electrode gas, sample liquid flow rate, nebuliser gas and dc current in a variable-diameter multi-electrode dc plasma		DC-ICP-OES	Modified simplex
W	[168]	Optimisation of flow rates of the cooling gas, top and bottom electrode gas, sample liquid flow rate, nebuliser gas and dc current in a variable-diameter multi-electrode dc plasma		DC-ICP-OES	Modified simplex
Y	[145]	Optimisation of the slurry nebulisation conditions was performed for the analysis of zirconia powder	Zirconia powder	ICP-OES	Simplex
	[146]	Four procedures to determine trace impurities in zirconia powder were developed. Operational characteristics of one of them were optimised by simplex	Zirconia powder	ICP-OES	Simplex

	[194]	A separation procedure was developed after a simplex optimisation of a nitric digestion to determine lanthanides and yttrium in rare earth ores and concentrates	Rare earth ores	ICP-OES	Simplex
Zn	[154]	Optimisation of operational characteristics (viewing height, rf plasma power and nebuliser argon flow) to measure environmental slurry samples. Comparison of three optimisation procedures	Sediments, rocks, sewage sludge	ICP-OES	Modified simplex
	[160]	Evaluation of three nebulisers for use in microwave-induced plasma optical emission spectrometry	Feminatal tablets, iodide tablets, lichen, lobster hepatopancreas, human hair, soya bean, flour	ICP-OES	Simplex
	[161]	On-line preconcentration and simultaneous determination of heavy metals after optimising experimental conditions (preconcentration flow rate, eluent flow rate, weight of solid phase and eluent loop volume)	Natural water	ICP-OES (CCD)	Super-modified simplex
	[165]	Simplex strategy to optimise atomisation flames		FAAS	Modified simplex
	[182]	Simplex optimisation of the overall response of a simultaneous multi-element flame atomic absorption spectrometer (air to fuel ratio, slit width, height above the burner head and four hollow-cathode lamp currents). Cu, Fe, Mn and Zn were measured		FAAS	Modified simplex

Table 2.35 (*continued*).

Analyte	*Reference*	*Comments*	*Matrix*	*Technique*	*Simplex mode*
	[181]	The overall response of a simultaneous multielement determination was optimised. A univariate search procedure was also evaluated	Tinned mussels	ICP-OES	Modified simplex
Zr	[168]	Optimisation of flow rates of the cooling gas, top and bottom electrode gas, sample liquid flow rate, nebuliser gas and dc current in a variable-diameter multi-electrode dc plasma		DC-ICP-OES	Modified simplex

Table 2.33 summarises the evolution of the modified simplex until the optimum is reached in the worked example.

Figure 2.16 displays the evolution of the modified simplex in the worked example. Note that only 11 simplexes were carried out.

Figure 2.17 summarises the evolution of the vertex B for each of the simplexes. Observe that the optimum (Y = 93%) was already obtained in simplex number 5 but we needed to continue until simplex number 11 to confirm it. In this example, 20 experiments were carried out, as in the previous original simplex method; nevertheless, in general the modified simplex approaches the zone of the optimum faster (*i.e.* using fewer experiments).

2.5 Examples of Practical Applications

Several thousand papers using experimental designs have been published. We summarise in Table 2.34 the most important in our field from 1999 to 2008. Quite surprisingly, the number of papers using simplex methods to optimize analytical procedures is not so large and, so, our review started at 1990 (see Table 2.35). For older references on simplex applications see Grotti [143] and Wienke *et al.* [144].

References

1. G. E. P. Box, J. S. Hunter and W. G. Hunter, *Statistics for Experimenters. Design, Innovation and Discovery*, 2nd edn, Wiley-Interscience, New York, 2005.
2. E. Morgan, *Chemometrics: Experimental Design*, Wiley, New York, 1991.
3. S. N. Deming and S. L. Morgan, *Experimental Designs: a Chemometric Approach*, Elsevier, Amsterdam, 1987.
4. T. Mohammadi, A. Moheb, M. Sadrzadeh and A. Razni, Separation of copper ions by electrodialysis using Taguchi experimental design, *Desalinination*, 169(1), 2004, 131–139.
5. J. G. Vlachogiannis, G. V. Vlachonis. Taguchi's method in a marine sediments heavy metal determinations, *J. Environ. Anal. Chem.*, 85(8), 2005, 553–565.
6. G. Taguchi, S. Chowdhury, Y. Wu, *Orthogonal Arrays and Linear Graphs*, American Supplier Institute Press, Dearborn, MI, 1987.
7. R. N. Kacker, K. Tsui. Interaction graphs: graphical aids for planning experiments, *J. Qual. Technol.*, 22, 1990, 1–14.
8. G. Taguchi. *System of Experimental Design*, UNIPUB/Kraus International Publications, White Plains, NY, and American Supplier Press, 1987.
9. L. Huwang, C. F. J. Wu and C. H. Yen, The idle column method: design constructions, properties and comparisons, *Technometrics*, 44(4), 2002, 347–355.

10. R. H. Myers, A. I. Khuri and G. G. Vining, Response surface alternatives to the Taguchi robust parameter design approach, *Am. Stat.*, 46(2), 1992, 131–139.
11. G. Taguchi and S. Chowdhury, *Taguchi's Quality Engineering Handbook*, Wiley-Interscience, New York, 2004.
12. S. H. Park, *Robust Design and Analysis for Quality Engineering*, Chapman and Hall, London, 1996.
13. M. J. Nash, J. E. Maskall and S. J. Hill, Methodologies for determination of antimony in terrestrial environmental samples, *J. Environ. Monit.*, 2, 2000, 97–109.
14. V. Czitrom, One-factor-at-a-time versus designed experiments, *Am. Stat.*, 53(2), 1999, 126–131.
15. G. E. P. Box and K. B. Wilson, On the experimental attainment of optimum conditions, *J. R. Stat. Soc., Ser. B*, 13(1), 1951, 1–45.
16. W. Spendley, G. R. Hext and F. R. Himsworth, Sequential application of simplex designs in optimisation and evolutionary operation, *Technometrics*, 4(4), 1962, 441–461.
17. J. A. Nelder and R. Mead, A simplex method for function minimisation, *Comput. J.*, 8, 1965, 42–52.
18. M. Soylak, I. Narin, M. Almeida Becerra and S. L. Costa Ferreira, Factorial design in the optimisation of preconcentration procedure for lead determination by FAAS, *Talanta*, 65(4), 2005, 895–899.
19. E. Kenduezler, O. Yalcinkaya, S. Baytak and A. R. Tuerker, Application of full factorial design for the preconcentration of chromium by solid phase extraction with Amberlyst 36 resin, *Microchim. Acta*, 160(4), 2008, 389–395.
20. N. Jalbani, T. G. Kazi, B. M. Arain, M. K. Jamali, H. I. Afridi and R. A. Sarfraz, Application of factorial design in optimisation of ultrasonic-assisted extraction of aluminum in juices and soft drinks, *Talanta*, 70(2), 2006, 307–314.
21. A. Moreda-Piñeiro, P. Bermejo-Barrera and A. Bermejo-Barrera, Chemometric investigation of systematic error in the analysis of biological materials by flame and electrothermal atomic absorption spectrometry, *Anal. Chim. Acta*, 560(1–2), 2006, 143–152.
22. L. M. Costa, M. G. A. Korn, J. T. Castro, W. P. C. Santos, E. V. Carvalho and A. R. A. Nogueira, Factorial design used for microwave-assisted digestion of bean samples, *Quim. Nova*, 29(1), 2006, 149–152.
23. R. M. De Souza, B. M. Mathias, I. S. Scarminio, C. L. P. Da Silveira and R. Q. Aucelio, Comparison between two sample emulsification procedures for the determination of Mo, Cr, V and Ti in diesel and fuel oil by ICP-OES along with factorial design, *Microchim. Acta*, 153(3–4), 2006, 219–225.
24. P. Hashemi and Z. Rahmani, A novel homocystine–agarose adsorbent for separation and preconcentration of nickel in table salt and baking soda using factorial design optimisation of the experimental conditions, *Talanta*, 68(5), 2006, 1677–1682.

25. D. M. Santos, M. M Pedroso, L. M. Costa, A. R. A. Nogueira and J. A. Nobrega, A new procedure for bovine milk digestion in a focused microwave oven: gradual sample addition to pre-heated acid, *Talanta*, 65(2), 2005, 505–510.
26. P. Rychlovsky, P. Cernoch and M. Sklenickova, Application of a heated electrospray interface for on-line connection of the AAS detector with HPLC for detection of organotin and organolead compounds, *Anal. Bioanal. Chem.*, 374(5), 2002, 955–962.
27. A. López-Molinero, O. Mendoza, A. Callizo, P. Chamorro and J. R. Castillo, Chemical vapor generation for sample introduction into inductively coupled plasma atomic emission spectroscopy: vaporisation of antimony(III) with bromide, *Analyst*, 127(10), 2002, 1386–1391.
28. A. López-Molinero, R. Giménez, P. Otal, A. Callizo, P. Chamorro and J. R. Castillo, New sensitive determination of selenium by bromide volatilisation inductively coupled plasma atomic emission spectrometry, *J. Anal. At. Spectrom.*, 17(4), 2002, 352–357.
29. A. López-Molinero, A. Villareal, C. Velilla, D. Andia and J. R. Castillo, Determination of arsenic in homeopathic drugs by bromide volatilisation–inductively coupled plasma atomic emission spectrometry, *J. AOAC Int.*, 85(1), 2002, 31–35.
30. T. Kuokkanen, P. Peramaki, I. Valimaki and H. Ronkkomaki, Determination of heavy metals in waste lubricating oils by inductively coupled plasma-optical emission spectrometry, *Int. J. Environ. Anal. Chem.*, 81(2), 2001, 89–100.
31. S. Aguerre, C. Pecheyran, G. Lespes, E. Krupp, O. F. X. Donard and M. Potin-Gautier, Optimisation of the hyphenation between solid-phase microextraction, capillary gas chromatography and inductively coupled plasma atomic emission spectrometry for the routine speciation of organotin compounds in the environment, *J. Anal. At. Spectrom.*, 16(12), 2001, 1429–1433.
32. A. Moreda-Piñeiro, A. Marcos, A. Fisher and S. J. Hill, Chemometrics approaches for the study of systematic error in inductively coupled plasma atomic emission spectrometry and mass spectrometry, *J. Anal. At. Spectrom.*, 16(4), 2001, 350–359.
33. A. Marcos, M. Foulkes and S. J. Hill, Application of a multi-way method to study long-term stability in ICP-AES, *J. Anal. At. Spectrom.*, 16(2), 2001, 105–114.
34. B. Izgi, C. Demir and S. Gucer, Application of factorial design for mercury determination by trapping and graphite furnace atomic absorption spectrometry, *Spectrochim. Acta, Part B*, 55(7), 2000, 971–977.
35. M. Villanueva, M. Pomares, M. Catases and J. Díaz, Application of factorial designs for the description and correction of combined matrix effects in ICP-AES, *Quim. Anal.*, 19(1), 2000, 39–42.
36. S. Saracoglu, M. Soylak, D. S. Kacar-Peker, L. Elci, W. N. L. dos-Santos, V. A. Lemos and S. L. C. Ferreira, A pre-concentration procedure using coprecipitation for determination of lead and iron in several samples

using flame atomic absorption spectrometry, *Anal. Chim. Acta*, 575(1), 2006, 133–137.

37. M. Soylak, I. Narin, M. a. Bezerra and S. L. C. Ferreira, Factorial design in the optimisation procedure for lead determination by FAAS, *Talanta*, 65(4), 2005, 895–899.
38. S. Cerutti, S. L. C. Ferreira, J. A. Gásquez, R. A. Olsina and L. D. Martínez, Optimisation of the preconcentration system of cadmium with 1-(2-thiazolylazo)-*p*-cresol using a knotted reactor and flame atomic absorption spectrometric detection, *J. Hazard. Mater.*, 112(3), 2004, 279–283.
39. M. Colón, M. Iglesias and M. Hidalgo, Development of a new method for sulfide determination by vapour generator-inductively coupled plasma-mass spectrometry, *Spectrochim. Acta, Part B*, 62, 2007, 470–475.
40. I. De Gregori, W. Quiroz, H. Pinochet, F. Pannier and M. Potin-Gautier, Simultaneous speciation analysis of Sb(III), Sb(V) and $(CH_3)_3SbC_{12}$ by high performance liquid chromatography-hydride generation-atomic fluorescence spectrometry detection (HPLC-HG-AFS): application to antimony speciation in sea water, *J. Chromatogr. A*, 1091(1–2), 2005, 94–101.
41. M. Chausseau, E. Poussel and J. M. Mermet, Signal and signal-to-background ratio response surface using axially viewed inductively coupled plasma multichannel detection-based emission spectrometry, *J. Anal. At. Spectrom.*, 15, 2000, 1293–1301.
42. A. A. Momen, G. A. Zachariadis, A. N. Anthemidis and J. A. Stratis, Use of fractional factorial design for optimisation of digestion procedures followed by multielement determination of essential and non-essential elements in nuts using ICP-OES technique, *Talanta*, 71(1), 2007, 443–451.
43. T. Stoichev, R. C. Rodríguez-Martin-Doimeadios, D. Amouroux, N. Molenat and O. F. X. Donard, Application of cryofocusing hydride generation and atomic fluorescence detection for dissolved mercury species determination in natural water samples, *J. Environ. Monit.*, 4, 2002, 517–521.
44. P. Hashemi, S. Bagheri and M. R. Fat'hi, Factorial design for optimisation of experimental variables in preconcentration of copper by a chromotropic acid loaded Q-Sepharose adsorbent, *Talanta*, 68(1), 2005, 72–78.
45. O. Ertas and H. Tezel, A validated cold vapour-AAS method for determining mercury in human red blood cells, *J. Pharm. Biomed. Anal.*, 36(4), 2004, 893–897.
46. D. Hristozov, C. E. Domini, V. Kmetov, V. Stefanova, D. Georgieva and A. Canals, Direct ultrasound-assisted extraction of heavy metals from sewage sludge samples for ICP-OES analysis, *Anal. Chim. Acta*, 516(1–2), 2004, 187–196.
47. S. Aguerre, C. Pecheyran and G. Lespes, Validation, using a chemometric approach, of gas chromatography-inductively coupled plasma-atomic emission spectrometry (GC-ICP-AES) for organotin determination, *Anal. Bioanal. Chem.*, 376(2), 2003, 226–235.

48. A. López-Molinero, A. Villareal, D. Andia, C. Velilla and J. R. Castillo, Volatile germanium tetrachloride for sample introduction and germanium determination by inductively coupled plasma atomic emission spectroscopy, *J. Anal. At. Spectrom.*, 16(7), 2001, 744–749.
49. M. Grotti, C. Lagomarsino and J. M. Mermet, Effect of operating conditions on excitation temperature and electron number density in axially-viewed ICP-OES with introduction of vapors or aerosols, *J. Anal. At. Spectrom.*, 21(9), 2006, 963–969.
50. G. P. Brandao, R. C. Campos, A. S. Luna, E. V. R. Castro and H. C. Jesús, Determination of arsenic in diesel, gasoline and naphtha by graphite furnace atomic absorption spectrometry using microemulsion medium for sample stabilisation, *Anal. Bioanal. Chem.*, 385(8), 2006, 1562–1569.
51. A. K. Avila, T. O. Araújo, P. R. G. Couto and R. M. H. Borges, Experimental design applied to the optimisation of pyrolysis and atomisation temperatures for As measurement in water samples by GFAAS, *Metrologia*, 42(5), 2005, 368–375.
52. M. Grotti, C. Lagomarsino, F. Soggia and R. Frache, Multivariate optimisation of an axially viewed inductively coupled plasma multichannel-based emission spectrometer for the analysis of environmental samples, *Ann. Chim. (Rome)*, 95(1–2), 2005, 37–51.
53. J. Gómez-Ariza, M.-A. Caro-de-la-Torre, I. Giráldez and E. Morales, Speciation analysis of selenium compounds in yeasts using pressurized liquid extraction and liquid chromatography–microwave-assisted digestion-hydride generation-atomic fluorescence spectrometry, *Anal. Chim. Acta*, 524(1–2), 2004, 305–314.
54. R. Rodil, A. M. Carro, R. A. Lorenzo, M. Abuín and R. Cela, Methylmercury determination in biological samples by derivatisation, solid-phase microextraction and gas chromatography with microwave-induced plasma atomic emission spectrometry, *J. Chromatogr. A*, 963(1–2), 2002, 313–323.
55. N. Carrión, M. Murillo, E. Montiel and D. Díaz, Development of a direct hydride generation nebuliser for the determination of selenium by inductively coupled plasma optical emission spectrometry. *Spectrochim. Acta, Part B*, 58(8), 2003, 1375–1389.
56. M. Grotti, P. Rivaro and R. Frache, Determination of butyltin compounds by high-performance liquid chromatography–hydride generation-electrothermal atomisation atomic absorption spectrometry, *J. Anal. At. Spectrom.*, 16(3), 2001, 270–274.
57. M. C. Yebra, S. Cancela and R. M. Cespón, Automatic determination of nickel in foods by flame atomic absorption spectrometry, *Food Chem.*, 108(2), 2008, 774–778.
58. M. El Ati-Hellal, F. Hellal, M. Dachraoui and A. Hedhili, Plackett–Burman designs in the pretreatment of macroalgae for Pb, Cr and Al determination by GF-AAS, *C. R. Chim.*, 10(9), 2007, 839–849.
59. Yebra-Biurrun and S. Cancela-Pérez, Continuous approach for ultrasound-assisted acid extraction-minicolumn preconcentration of

chromium and cobalt from seafood samples prior to flame atomic absorption spectrometry, *Anal. Sci.*, 23(8), 2007, 993–996.
60. J. Moreda-Piñeiro, E. Alonso-Rodríguez, P. López-Mahía, S. Muniategui-Lorenzo, D. Prada-Rodríguez, A. Moreda-Piñeiro and P. Bermejo-Barrera, Determination of major and trace elements in human scalp hair by pressurized-liquid extraction with acetic acid and inductively coupled plasma-optical-emission spectrometry, *Anal. Bioanal. Chem.*, 388(2), 2007, 441–449.
61. M. Felipe-Sotelo, M. J. Cal-Prieto, M. P. Gómez-Carracedo, J. M. Andrade, A. Carlosena and D. Prada, Handling complex effects in slurry-sampling–electrothermal atomic absorption spectrometry by multivariate calibration, *Anal. Chim. Acta*, 571(2), 2006, 315–323.
62. S. Cancela and M. C. Yebra, Flow-injection flame atomic absorption spectrometric determination of trace amounts of cadmium in solid and semisolid milk products coupling a continuous ultrasound-assisted extraction system with the online preconcentration on a chelating aminomethylphosphoric acid resin, *J. AOAC Int.*, 89(1), 2006, 185–191.
63. M. Felipe-Sotelo, M. J. Cal-Prieto, J. Ferré, R. Boqué, J. M. Andrade and A. Carlosena, Linear PLS regression to cope with interferences of major concomitants in the determination of antimony by ETAAS, *J. Anal. At. Spectrom.*, 21(1), 2006, 61–68.
64. M. H. Arbab-Zavar, M. Chamsaz, A. Youssefi and M. Aliakbari, Electrochemical hydride generation atomic absorption spectrometry for determination of cadmium, *Anal. Chim. Acta*, 546(1), 2005, 126–132.
65. S. Cancela-Pérez and M. C. Yebra-Biurrun, Flow injection determination of Cd in meat samples using a continuous lixiviation/preconcentration system coupled to a flame AAS, *At. Spectrosc.*, 26(3), 2005, 110–116.
66. M. C. Yebra and S. Cancela, Continuous ultrasound-assisted extraction of cadmium from legumes and dried fruit samples coupled with on-line preconcentration-flame atomic absorption spectrometry, *Anal. Bioanal. Chem.*, 382(4), 2005, 1093–1098.
67. M. C. Yebra-Biurrun, A. Moreno-Cid and S. Cancela-Pérez, Fast on-line ultrasound-assisted extraction coupled to a flow injection-atomic absorption spectrometric system for zinc determination in meat samples, *Talanta*, 66(3), 2005, 691–695.
68. M. C. Yebra, S. Cancela and A. Moreno-Cid, Continuous ultrasound-assisted extraction of cadmium from vegetable samples with on-line preconcentration coupled to a flow injection-flame atomic spectrometric system, *Int. J. Environ. Anal. Chem.*, 85(4–5), 2005, 305–313.
69. M. C. Yebra, A. Moreno-Cid and S. Cancela, Flow injection determination of copper and iron in seafoods by a continuous ultrasound-assisted extraction system coupled to FAAS, *Int. J. Environ. Anal. Chem.*, 85(4–5), 2005, 315–323.
70. M. C. Yebra-Biurrun, S. Cancela-Pérez and A. Moreno-Cid-Barinaga, Coupling continuous ultrasound-assisted extraction, preconcentration and

flame atomic absorption spectrometric detection for the determination of cadmium and lead in mussel samples, *Anal. Chim. Acta*, 533(1), 2005, 51–56.
71. C. Moscoso-Pérez, J. Moreda-Piñeiro, P. López-Mahía, S. Muniategui-Lorenzo, E. Fernández-Fernández and D. Prada-Rodríguez, As, Bi, Se(IV) and Te(IV) determination in acid extracts of raw materials and by-products from coal-fired power plants by hydride generation-atomic fluorescence spectrometry, *At. Spectrosc.*, 25(5), 2004, 211–216.
72. M. Felipe-Sotelo, A. Carlosena, J. M Andrade, E. Fernández, P. López-Mahía, S. Muniategui and D. Prada, Development of a slurry-extraction procedure for direct determination of cobalt by electrothermal atomic absorption spectrometry in complex environmental samples, *Anal. Chim. Acta*, 522(2), 2004, 259–266.
73. A. Moreno-Cid and M. C. Yebra, Continuous ultrasound-assisted extraction coupled to a flow injection-flame atomic absorption spectrometric system for calcium determination in seafood samples, *Anal. Bioanal. Chem.*, 379(1), 2004, 77–82.
74. M. C. Yebra, A. Moreno-Cid, R. Cespón and S. Cancela, Preparation of a soluble solid sample by a continuous ultrasound assisted dissolution system for the flow-injection atomic absorption spectrometric determination of iron in milk powder and infant formula, *Talanta*, 62(2), 2004, 403–406.
75. M. C. Yebra, A. Moreno-Cid, R. M. Cespón and S. Cancela, Flame AAS determination of total chromium in mussel samples using a continuous ultrasound-assisted extraction system connected to an on-line flow injection manifold, *At. Spectrosc.*, 24(1), 2003, 31–36.
76. B. Koklic and M. Veber, Influence of sampling to the homogeneity of Cu–Ti–Zn alloy samples for the analyses with glow discharge optical emission spectroscopy, *Accredit. Qual. Assur.*, 8(3–4), 2003, 146–149.
77. M. C. Yebra and A. Moreno-Cid, On-line determination of manganese in solid seafood samples by flame atomic absorption spectrometry, *Anal. Chim. Acta*, 477(1), 2003, 149–155.
78. M. C. Yebra and A. Moreno-Cid, Continuous ultrasound-assisted extraction of iron from mussel samples coupled to a flow injection-atomic spectrometric system, *J. Anal. At. Spectrom.*, 17(10), 2002, 1425–1428.
79. A. Moreno-Cid and M. C. Yebra, Flow injection determination of copper in mussels by flame atomic absorption spectrometry after on-line continuous ultrasound-assisted extraction, *Spectrochim. Acta, Part B*, 57(5), 2002, 967–974.
80. M. C. Yebra and A. Moreno-Cid, Optimisation of a field flow preconcentration system by experimental design for the determination of copper in sea water by flow-injection-atomic absorption spectrometry, *Spectrochim. Acta, Part B*, 57(1), 2002, 85–93.
81. P. Bermejo-Barrera, A. Moreda-Piñeiro, O. Muñiz-Navarro, A. M. J. Gómez-Fernández and A. Bermejo-Barrera, Optimisation of a microwave-pseudo-digestion procedure by experimental designs for the determination of trace elements in seafood products by atomic absorption spectrometry, *Spectrochim. Acta, Part B*, 55(8), 2000, 1351–1371.

82. M. C. Yebra-Biurrun, A. Moreno-Cid and L. Puig, Minicolumn field preconcentration and flow-injection flame atomic absorption spectrometric determination of cadmium in seawater, *Anal. Chim. Acta*, 524(1–2), 2004, 73–77.
83. T. Mohammadi, A. Moheb, M. Sadrzadeh and A. Razni, Separation of copper ions by electrodialysis using Taguchi experimental design, *Desalination*, 169(1), 2004, 21–31.
84. J. G. Vlachogiannis and G. V. Vlachonis, Taguchi's method in a marine sediment's heavy metal determination, *Int. J. Environ. Anal. Chem.*, 85(8), 2005, 553–565.
85. A. F. Barbosa, M. G. Segatelli, A. C. Pereira, A. De Santana Santos, L. T. Kubota, P. O. Luccas and C. R. T. Tarley, Solid-phase extraction system for Pb(II) ions enrichment based on multiwall carbon nanotubes coupled on-line to flame atomic absorption spectrometry, *Talanta*, 71(4), 2007, 1512–1519.
86. M. A. Bezerra, A. L. B. Conceicao and S. L. C. Ferreira, Doehlert matrix for optimisation of procedure for determination of nickel in saline oil-refinery effluents by use of flame atomic absorption spectrometry after preconcentration by cloud-point extraction, *Anal. Bioanal. Chem.*, 378(3), 2004, 798–803.
87. M. V. Reboucas, S. L. C. Ferreira and B. De-Barros-Neto, Arsenic determination in naphtha by electrothermal atomic absorption spectrometry after preconcentration using multiple injections, *J. Anal. At. Spectrom.*, 18(10), 2003, 1267–1273.
88. M. Grotti, E. Magi and R. Frache, Multivariate investigation of matrix effects in inductively coupled plasma atomic emission spectrometry using pneumatic or ultrasonic nebulisation, *J. Anal. At. Spectrom.*, 15(1), 2000, 89–95.
89. R. E. Santelli, M. A. Becerra, A. S. Freire, E. P. Oliveira and A. F. Batista-de-Carvalho, Non-volatile vanadium determination in petroleum condensate, diesel and gasoline prepared as detergent emulsions using GFAAS, *Fuel*, 87(8–9), 2008, 1617–1622.
90. P. Navarro, J. Raposo, G. Arana and N. Etxebarría, Optimisation of microwave assisted digestion of sediments and determination of Sn and Hg, *Anal. Chim. Acta*, 566(1), 2006, 37–44.
91. S. Salomon, P. Giamarchi and A. Le-Bihan, Desirability approach for optimisation of electrothermal atomic absorption spectrometry factors in iron determinations, *Analusis*, 28(7), 2000, 575–586.
92. S. Salomon, P. Giamarchi, A. Le-Bihan, H. Becker-Ross and U. Heitmann, Improvements in the determination of nanomolar concentrations of aluminum in seawater by electrothermal atomic absorption spectrometry, *Spectrochim. Acta, Part B*, 55(8), 2000, 1337–1350.
93. Z. Long, J. Xin and X. Hou, Determination of arsenic and mercury in Chinese medicinal herbs by atomic fluorescence spectrometry with closed-vessel microwave digestion, *Spectrosc. Lett.*, 37(3), 2004, 263–274.

94. M. Murillo, N. Carrión, J. Chirinos, A. Gammiero and E. Fassano, Optimisation of experimental parameters for the determination of mercury by MIP/AES, *Talanta*, 54(2), 2001, 389–395.
95. V. V. Polonnikova, O. M. Karpukova, A. N. Smagunova, E. I. Ivanova and O. A. Proidakova, Mutual interferences of elements in the atomic-absorption analysis of human hairs, *J. Anal. Chem.*, 55(1), 2000, 29–33.
96. O. Lacin, B. Bayrak, O. Korkut and E. Sayan, Modeling of adsorption and ultrasonic desorption of cadmium(II) and zinc(II) on local bentonite, *J. Colloid Interface Sci.*, 292(2), 2005, 330–335.
97. A. P. Krushevska, K. Klimash, J. F. Smith, E. A. Williams, P. J. McCloskey and V. Ravikumar, Determination of phosphorus in polymeric systems using an ashing procedure and inductively coupled plasma atomic emission spectrometry, *J. Anal. At. Spectrom.*, 19(9), 2004, 1186–1191.
98. J. M. Trindade, A. L. Marques, G. S. Lopes, E. P. Marques and J. Zhang, Arsenic determination in gasoline by hydride generation atomic absorption spectroscopy combined with a factorial experimental design approach, *Fuel*, 85(14–15), 2006, 2155–2161.
99. F. R. Amorim, C. Bof, M. B. Franco, J. B. B. Silva and C. C. Nascentes, Comparative study of conventional and multivariate methods for aluminium determination in soft drinks by graphite furnace atomic absorption spectrometry, *Microchem. J.*, 82(2), 2006, 168–173.
100. A. López-Molinero, P. Calatayud, D. Sipiera, R. Falcón, D. Liñán and J. R. Castillo, Determination of antimony in poly(ethylene terephthalate) by volatile bromide generation flame atomic absorption spectrometry, *Microchim. Acta*, 158(3–4), 2007, 247–253.
101. F. Bianchi, M. Maffini, E. Marengo and C. Mucchino, Experimental design optimisation for the ICP-AES determination of Li, Na, K, Al, Fe, Mn and Zn in human serum, *J. Pharm. Biomed. Anal.*, 43(2), 2007, 659–665.
102. N. Etxebarría, R. Antolín, G. Borge, T. Posada and J. C. Raposo, Optimisation of flow-injection-hydride generation inductively coupled plasma spectrometric determination of selenium in electrolytic manganese, *Talanta*, 65(5), 2005, 1209–1214.
103. I. Arambarri, R. García and E. Millán, Optimisation of tin determination in aqua regia–HF extracts from sediments by electrothermal atomic absorption spectrometry using experimental design, *Analyst*, 125(11), 2000, 2084–2088.
104. F. Priego-Capote and M. D. Luque-de-Castro, Dynamic ultrasound-assisted leaching of essential macro and micronutrient metal elements from animal feeds prior to flame atomic absorption spectrometry, *Anal. Bioanal. Chem.*, 378(5), 2004, 1376–1381.
105. A. C. Ferreira, M. G. A. Korn and S. L. C. Ferreira, Multivariate optimisation in preconcentration procedure for manganese determination in seawater samples by FAAS, *Microchim. Acta*, 146(3–4), 2004, 271–278.
106. S. L. C. Ferreira, A. S. Queiroz, M. S. Fernandes and D. C. Dos-Santos, Application of factorial designs and Doehlert matrix in optimisation of

experimental variables associated with the preconcentration and determination of vanadium and copper in seawater by inductively coupled plasma optical emission spectrometry, *Spectrochim. Acta, Part B*, 57(12), 2002, 1939–1950.

107. A. P. Lindomar, H. S. Ferreira, W. N. L. Dos-Santos and S. L. C. Fereira, Simultaneuous pre-concentration procedure for the determination of cadmium and lead in drinking water employing sequential multi-element flame atomic absorption spectrometry, *Microchem. J.*, 87(1), 2007, 77–80.
108. W. N. L. Dos-Santos, J. L. O. Costa, R. G. O. Araújo, D. S. De-Jesús and A. C. S. Costa, An on-line pre-concentration system for determination of cadmium in drinking water using FAAS, *J. Hazard. Mater.*, 137(3), 2006, 1357–1361.
109. M. Zougagh, P. C. Rudner, A. García-de-Torres and J. M. Cano-Pavón, Application of Doehlert matrix and factorial designs in the optimisation of experimental variables associated with the on-line preconcentration and determination of zinc by flow injection inductively coupled plasma atomic emission spectrometry, *J. Anal. At. Spectrom.*, 15(12), 2000, 1589–1594.
110. N. Jalbani, T. G. Kazi, M. K. Jamali, M. B. Arain, H. I. Afridi, S. T. Sheerazi and R. Ansari, Application of fractional factorial design and Doehlert matrix in the optimisation of experimental variables associated with the ultrasonic-assisted acid digestion of chocolate samples for aluminium determination by atomic absorption spectrometry, *J. AOAC Int.*, 90(6), 2007, 1682–1688.
111. A. M. Carro, I. Neira, R. Rodil and R. A. Lorenzo, Speciation of mercury compounds by gas chromatography with atomic emission detection. Simultaneous optimisation of a headspace solid-phase microextraction and derivatisation procedure by use of chemometric techniques, *Chromatographia*, 56(11/12), 2002, 733–738.
112. E. Pena-Vázquez, J. Villanueva-Alonso and P. Bermejo-Barrera, Optimisation of a vapor generation method for metal determination using ICP-OES, *J. Anal. At. Spectrom.*, 22(6), 2007, 642–649.
113. R. Santamaría-Fernández, A. Moreda-Piñeiro and S. J. Hill, Optimisation of a multielement sequential extraction method employing an experimental design approach for metal partitioning in soils and sediments, *J. Environ. Monit.*, 4(2), 2002, 330–336.
114. C. R. Teixeira-Tarley, A. F. Barbosa, M. Gava-Segatelli, E. Costa-Figueiredo and P. Orival-Luccas, Highly improved sensitivity of TS-FF-AAS for Cd(II) determination at ng L^{-1} levels using a simple flow injection minicolumn preconcentration system with multiwall carbon nanotubes, *J. Anal. At. Spectrom.*, 21(11), 2006, 1305–1313.
115. G. D. Matos, C. R. T. Tarley, S. L. C. Ferreira and M. A. Z. Arruda, Use of experimental design in the optimisation of a solid phase preconcentration system for cobalt determination by GFAAS, *Eclet. Quim.*, 30(1), 2005, 65–74.

116. A. S. Souza, W. N dos Santos and S. L. C. Ferreira, Application of Box–Behnken design in the optimisation of an on-line pre-concentration system using knotted reactor for cadmium determination by flame atomic absorption spectrometry, *Spectrochim. Acta, Part B*, 60(5), 2005, 737–742.
117. M. Grotti, R. Leardi, C. Gnecco and R. Frache, Determination of manganese by graphite furnace atomic absorption spectrometry: matrix effects control by multiple linear regression model, *Spectrochim. Acta, Part B.*, 54(5), 1999, 845–851.
118. M. B. Arain, T. G. Kazi, M. K. Jamali, N. Jalbani, H. I. Afridi, R. A. Sarfraz and A. Q. Shah, Determination of toxic elements in muscle tissues of five fish species using ultrasound-assisted pseudodigestion by electrothermal atomic absorption spectrophotometry: optimisation study, *Spectrosc. Lett.*, 40(6), 2007, 861–878.
119. E. M. Seco-Gesto, A. Moreda-Piñeiro, A. Bermejo-Barrera and P. Bermejo-Barrera, Multielement determination in raft mussels by fast microwave-assisted acid leaching and inductively coupled plasma-optical emission spectrometry, *Talanta*, 72(3), 2007, 1178–1185.
120. J. Moreda-Piñeiro, E. Alonso-Rodríguez, P. López-Mahía, S. Muniategui-Lorenzo, D. Prada-Rodríguez, A. Moreda-Piñeiro, A. Bermejo-Barrera and P. Bermejo-Barrera, Pressurized liquid extraction as a novel sample pretreatment for trace element leaching from biological material, *Anal. Chim. Acta*, 572(2), 2006, 172–179.
121. E. Pena-Vázquez, A. Bermejo-Barrera and P. Bermejo-Barrera, Use of lanthanum hydroxide as a trapping agent to determine hydrides by HG-ICP-OES, *J. Anal. At. Spectrom.*, 20(12), 2005, 1344–1349.
122. J. C. Rodríguez-García, J. Barciela-García, C. Herrero-Latorre, S. García-Martín and R. M. Peña-Crecente, Direct and combined methods for the determination of chromium, copper and nickel in honey by electrothermal atomic absorption spectroscopy, *J. Agric. Food Chem.*, 53(17), 2005, 6616–6623.
123. C. Pena-Farfal, A. Moreda-Piñeiro, A. Bermejo-Barrera, P. Bermejo-Barrera, H. Pinochet-Cancino and I. De-Gregori-Henríquez, Ultrasound bath-assisted enzymatic hydrolysis procedures as sample pretreatment for multielement determination in mussels by inductively coupled plasma atomic emission spectrometry, *Anal. Chem.*, 76(13), 2004, 3541–3547.
124. R. Santamaría-Fernández, S. Santiago-Rivas, A. Moreda-Piñeiro, A. Bermejo-Barrera, P. Bermejo-Barrera and S. J. Hill, Blending procedure for the cytosolic preparation of mussel samples for AAS determination of Cd, Cr, Cu, Pb and Zn bound to low molecular weight compounds, *At. Spectrosc.*, 25(1), 2004, 37–43.
125. C. Moscoso-Pérez, J. Moreda-Piñeiro, P. López-Mahía, S. Muniategui-Lorenzo, E. Fernández-Fernández and D. Prada-Rodríguez, Hydride generation atomic fluorescence spectrometric determination of As, Bi, Sb, Se(IV) and Te(IV) in aqua regia extracts from atmospheric particulate matter using multivariate optimisation, *Anal. Chim. Acta*, 526(2), 2004, 185–192.

126. A. A. Momen, G. A. Zachariadis, A. N. Anthemidis and J. A. Stratis, Optimisation and comparison of two digestion methods for multielement analysis of certified reference plant materials by ICP-AES. Application of Plackett–Burman and central composite designs, *Microchim. Acta*, 160(4), 2008, 397–403.
127. R. Santamaría-Fernández, M. R. Cave and S. J. Hill, The effect of humic acids on the sequential extraction of metals in soils and sediments using ICP-AES and chemometric analysis, *J. Environ. Monit.*, 5(6), 2003, 929–934.
128. C. Moscoso-Pérez, J. Moreda-Piñeiro, P. López-Mahía, S. Muniategui, E. Fernández-Fernández and D. Prada-Rodríguez, Bismuth determination in environmental samples by hydride generation-electrothermal atomic absorption spectrometry, *Talanta*, 1(5), 2003, 633–642.
129. P. Bermejo-Barrera, A. Moreda-Piñeiro, R. González-Iglesias and A. Bermejo-Barrera, Multivariate optimisation of solvent extraction with 1,1,1-trifluoroacetylacetonates for the determination of total and labile Cu and Fe in river surface water by flame atomic absorption spectrometry, *Spectrochim. Acta, Part B*, 57(12), 2002, 1951–1966.
130. J. Moreda-Piñeiro, P. López-Mahía, S. Muniategui-Lorenzo, E. Fernández-Fernández and D. Prada-Rodríguez, Direct As, Bi, Ge, Hg and Se(IV) cold vapor/hydride generation from coal fly ash slurry samples and determination by electrothermal atomic absorption spectrometry, *Spectrochim. Acta, Part B*, 57(5), 2002, 883–895.
131. J. Moreda-Piñeiro, P. López-Mahía, S. Muniategui-Lorenzo, E. Fernández-Fernández and D. Prada-Rodríguez, Direct mercury determination in aqueous slurries of environmental and biological samples by cold vapor generation-electrothermal atomic absorption spectrometry, *Anal. Chim. Acta*, 460(1), 2002, 111–122.
132. I. Arambarri, R. García and E. Millán, Application of experimental design in a method for screening sediments for global determination of organic tin by electrothermal atomic absorption spectrometry (ETAAS), *Fresenius' J. Anal. Chem.*, 371(7), 2001, 955–960.
133. P. Bermejo-Barrera, O. Muñiz-Naveiro, A. Moreda-Piñeiro and A. Bermejo-Barrera, The multivariate optimisation of ultrasonic bath-induced acid leaching for the determination of trace elements in seafood products by atomic absorption spectrometry, *Anal. Chim. Acta*, 439(2), 2001, 211–227.
134. J. Moreda-Piñeiro, C. Moscoso-Pérez, P. López-Mahía, S. Muniategui-Lorenzo, E. Fernández-Fernández and D. Prada-Rodríguez, Multivariate optimisation of hydride generation procedures for single element determinations of As, Cd, Sb and Se in natural waters by electrothermal atomic absorption spectrometry, *Talanta*, 53(4), 2001, 871–883.
135. P. Bermejo-Barrera, A. Moreda. Piñeiro and A. Bermejo-Barrera, Factorial designs for Cd, Cr, Hg, Pb and Se ultrasound-assisted acid leaching from human hair followed by atomic absorption spectrometric determination, *J. Anal. At. Spectrom.*, 15(2), 2000, 121–130.

136. P. Bermejo-Barrera, O. Muñiz-Naveiro, A. Moreda-Piñeiro and A. Bermejo-Barrera, Experimental designs in the optimisation of ultrasonic bath-acid-leaching procedures for the determination of trace elements in human hair samples by atomic absorption spectrometry, *Forensic Sci. Int.*, 107(1–3), 2000, 105–120.
137. R. Otero-Rey, M. Mato-Fernández, J. Moreda-Piñeiro, E. Alonso-Rodríguez, S. Muniategui-Lorenzo, P. López-Mahía and D. Prada-Rodríguez, Influence of several experimental parameters on As and Se leaching from coal fly ash samples, *Anal. Chim. Acta*, 531(2), 2005, 299–305.
138. P. Vanloot, J. L. Boudenne, C. Brach-Papa, M. Sergent and B. Coulomb, An experimental design to optimise the flow extraction parameters for the selective removal of Fe(III) and Al(III) in aqueous samples using salicylic acid grafted on Amberlite XAD- and final determination by GF-AAS, *J. Hazard. Mater.*, 147(1–2), 2007, 463–470.
139. M. Zougagh, A. García-de-Torres and J. M. Cano-Pavón, On-line separation and determination of trace amounts of manganese in biological samples by flow injection inductively coupled plasma atomic emission spectrometry, *Anal. Lett.*, 36(6), 2003, 1115–1130.
140. C. R. T. Tarley, E. C. Figueiredo and G. D. Matos, Thermospray flame furnace-AAS determination of copper after on-line sorbent preconcentration using a system optimised by experimental designs, *Anal. Sci.*, 21(11), 2005, 1337–1342.
141. M. Grotti, C. Lagomarsino and R. Frache, Multivariate study in chemical vapor generation for simultaneous determination of arsenic, antimony, bismuth, germanium, tin, selenium, tellurium and mercury by inductively coupled plasma optical emission spectrometry, *J. Anal. At. Spectrom.*, 20(12), 2005, 1365–1373.
142. B. Do, S. Robinet, D. Pradeau and F. Guyon, Speciation of arsenic and selenium compounds by ion-pair reversed-phase chromatography with electrothermal atomic absorption spectrometry. Application of experimental design for chromatographic optimisation, *J. Chromatogr. A*, 918(1), 2001, 87–98.
143. A. M. Grotti, Improving the analytical performances of inductively coupled plasma optical emission spectrometry by multivariate analysis techniques, *Ann. Chim. (Rome)*, 94, 2004, 1–15.
144. D. Wienke, C. Lucasius and G. Kateman, Multicriteria target vector optimisation of analytical procedures using a genetic algorithm: Part I. Theory, numerical simulations and application to atomic emission spectroscopy, *Anal. Chim. Acta*, 265, 1992, 211–225.
145. R. Lobinski, W. Van Borm, J. A. C. Broekaert, P. Tschoepel and G. Toelg, Optimisation of slurry nebulisation inductively-coupled plasma emission spectrometry for the analysis of zirconia powder, *Fresenius' J. Anal. Chem.*, 342(7), 1992, 563–568.
146. R. Lobinski, J. A. C. Broekaert, P. Tschoepel and G. Toelg, Inductively-coupled plasma atomic emission spectroscopic determination of trace

impurities in zirconia powder, *Fresenius' J. Anal. Chem.*, 342(7), 1992, 569–580.

147. L. Marjanovic, R. I. McCrindle, B. M. Botha and J. H. Potgieter, Analysis of cement by inductively coupled plasma optical emission spectrometry using slurry nebulisation, *J. Anal. At. Spectrom.*, 15, 2000, 983–985.
148. H. Matusiewicz and M. Mroczkowska, Hydride generation from slurry samples after ultrasonication and ozonation for the direct determination of trace amounts of As (III) and total inorganic arsenic by their *in situ* trapping followed by graphite furnace atomic absorption spectrometry, *J. Anal. At. Spectrom.*, 18, 2003, 751–761.
149. A. López-Molinero, A. Villareal, C. Velilla, D. Andia and J. R. Castillo, Determination of arsenic in homeopathic drugs by bromide volatilisation-inductively cooupled plasma-atomic emission spectrometry, *J. AOAC Int.*, 85(1), 2002, 31–35.
150. A. López-Molinero, Possibilities for graphic representation of multifactor simplex optimisation, *Anal. Chim. Acta*, 297(3), 1994, 417–425.
151. S. A. Pergantis, W. R. Cullen and A. P. Wade, Simplex optimisation of conditions for the determination of arsenic in environmental samples by using electrothermal atomic absorption spectrometry, *Talanta*, 41(2), 1994, 205–209.
152. X. Ch. Le, W. R. Cullen, K. J. Reimer and I. D. Brindie, A new continous hybride generator for the determination of arsenic, antimony and tin by hydride generation atomic absorption spectrometry, *Anal. Chim. Acta*, 258(2), 1992, 307–315.
153. R. B. Georgieva, P. K. Petrov, P. S. Dimitrov and D. L. Tsalev, Observations on toxicologically relevant arsenic in urine in adult offspring of families with Balkan endemic nephropathy and controls by batch hydride generation atomic absorption spectrometry, *Int. J. Environ. Anal. Chem.*, 87(9), 2007, 673–685.
154. L. Ebdon, M. Foulkes and K. O'Hanlon, Optimised simultaneous multielement analysis of environmental slurry samples by inductively coupled plasma atomic emission spectrometry using a segmented array charge-coupled device detector, *Anal. Chim. Acta*, 311, 1995, 123–134.
155. A. Woller, H. Garraud, J. Boisson, A. M. Dorthe, P. Fodor and O. F. X. Donard, Simultaneuous speciation of redox species of arsenic and selenium using an anion-exchange microbore column coupled with a micro-concentric nebuliser and an inductively coupled plasma mass spectrometer as detector, *J. Anal. At. Spectrom.*, 13, 1998, 141–149.
156. W.-M. Yang and Z.-M. Ni, The possibility of standardless analysis in graphite furnace atomic absorption spectrometry: determination of gold in geological samples, *Spectrochim. Acta, Part B*, 51(1), 1996, 65–73.
157. P. Di, D. E. Davey. An atomized online preconcentration system for determination of trace gold in ore samples, *Talanta*, 42(8), 1995, 1081–1088.
158. P. Di and D. E. Davey, Trace gold determination by online preconcentration with flow injection atomic absorption spectrometry, *Talanta*, 41(4), 1994, 565–571.

159. D. Pollmann, J. A. C. Broekaert, F. Leis, P. Tschoepel and G. Tölg, Determination of boron in biological tissues by inductively coupled plasma optical emission spectrometry (ICP-OES), *Fresenius' J. Anal. Chem.*, 346(4), 1993, 441–445.
160. H. Matusiewicz, M. Ślachciñski, M. Hidalgo and A. Canals, Evaluation of various nebulisers for use in microwave induced plasma optical emission spectrometry, *J. Anal. At. Spectrom.*, 22, 2007, 1174–1178.
161. H. Karami, M. F. Mousavi, Y. Yamini and M. Shamsipur, On-line preconcentration and simultaneous determination of heavy metal ions by inductively coupled plasma-atomic emission spectrometry, *Anal. Chim. Acta*, 509(1), 2004, 89–94.
162. H. Matusiewicz, B. Golik and A. Suszka, Determination of the residual carbon content in biological and environmental samples by microwave-induced-plasma atomic emission spectrometry, *Chem. Anal. (Warsaw)*, 44(3B), 1999, 559–566.
163. J. A. Horner and G. M. Hieftje, Characterisation of a 22 mm torch for ICP-AES, *Appl. Spectrosc.*, 53(6), 1999, 713–718.
164. N. N. Sesi, P. J. Galley and G. M. Hieftje, Evaluation of a linear-flow torch for inductively coupled plasma atomic emission spectrometry, *J. Anal. At. Spectrom.*, 8, 1993, 65–70.
165. J. Echeverría, M. T. Arcos, M. J. Fernández and J. Garrido Segovia, Simplex method strategy for optimisation in flames atomic absorption spectrometry, *Quím. Anal.*, 11(1), 1992, 25–33.
166. R. J. Stolzberg, Optimizing signal-to-noise ratio in flame atomic absorption spectrophotometry using sequential simplex optimisation, *J. Chem. Educ.*, 76(6), 1999, 834–838.
167. G. H. Lee, Development of new high temperature plasma sources for spectrochemical analysis: multivariate optimisation by the modified sequential simplex method, *Bull. Korean Chem. Soc.*, 14(2), 1993, 275–281.
168. J. A. McGuire and E. H. Piepmeier, The characterisation and simplex optimisation of a variable-diameter, multi-electrode, direct current plasma for atomic emission spectroscopy, *Can. J. Appl. Spectrosc.*, 36(6), 1991, 127–139.
169. P. J. Galley, J. A. Horner and G. M. Hieftje, Automated simplex optimisation for monochromatic imaging inductively coupled plasma atomic emission spectroscopy, *Spectrochim. Acta, Part B*, 50(1), 1995, 87–107.
170. L. M. Cabezón, M. Caballero, J. M. Díaz, R. Cela and J. A. Pérez-Bustamante, Multielemental separation and determination of some heavy metals (copper, cobalt, cadmium and nickel) in tap water and high salinity media by CGA (colloidal gas aphron)-coflotation, *Analusis*, 19(4), 1991, 123–127.
171. J. M. Díaz, M. Caballero, J. A. Perez-Bustamante and R. Cela, Multielemental separation of copper, cobalt, cadmium and nickel in natural waters by means of colloidal gas aphron coflotation, *Analyst*, 115(9), 1990, 1201–1205.

172. M. D. Granado-Castro, M. D. Galindo-Riaño and M. García-Vargas, Separation and preconcentration of cadmium ions in natural water using a liquid membrane system with 2-acetylpyridine benzoylhydrazone as carrier by flame atomic absorption spectrometry, *Spectrochim. Acta, Part B*, 59(4), 2004, 577–583.
173. C. D. Stalikas, G. A. Pilidis and M. I. Karayannis, Determination of lead and cadmium in environmental samples by simplex optimised atomic absorption methods, *J. Anal. At. Spectrom.*, 11(8), 1996, 595–599.
174. L. Ebdon, Ph. Goodall, S. J. Hill, P. B. Stockwell and K. C. Thompson, Ultra-trace determination of cadmium by vapor generation atomic fluorescence spectrometry, *J. Anal. At. Spectrom.*, 8(5), 1993, 723–729.
175. G. Zhang, J. Jinghua, H. Dexue and P. Xiang, Atomic absorption determination of traces of cadmium in urine after electrodeposition onto a tungsten wire, *Talanta*, 40(3), 1993, 409–413.
176. E. I. Vereda Alonso, L. P. Gil, M. T. Siles Cordero, A. García de Torres and J. M. Cano Pavón, Automatic on-line column preconcentration system for determination of cadmium by electrothermal atomic absorption spectrometry, *J. Anal. At. Spectrom.*, 16, 2001, 293–295.
177. K. Krengel-Rothensee, U. Richter and P. Heitland, Low-level determination of non-metals (Cl, Br, I, S, P) in waste oils by inductively coupled plasma optical emission spectrometry using prominent spectral lines in the 130–190 nm range, *J. Anal. At. Spectrom.*, 14, 1999, 699–702.
178. B. S. Iversen, A. Panayi, J. P. Camblor and E. Sabbioni, Simultaneous determination of cobalt and manganese in urine by electrothermal atomic absorption specttrometry. Method development using a simplex optimisation approach, *J. Anal. At. Spectrom.*, 11(8), 1996, 591–594.
179. S. Ch. Nielsen, S. Sturup, H. Spliid and E. H. Hansen, Selective flow injection analysis of ultra-trace amounts of Cr(VI), preconcentration of it by solvent extraction and determination by electrothermal atomic absorption spectrometry (ETAAS), *Talanta*, 49(5), 1999, 1027–1044.
180. A. Bassimane, C. Porte and A. Delacroix, Optimisation of chromium analysis by atomic absorption without flame and interference study, *Analusis*, 25(5), 1997, 168–171.
181. F. J. Copa-Rodríguez and M. I. Basadre-Pampín, Determination of iron, copper and zinc in tinned mussels by inductively coupled plasma atomic emission spectrometry (ICP-AES), *Fresenius' J. Anal. Chem.*, 348(5–6), 1994, 390–395.
182. K. S. Farah and J. Sneddon, Optimisation of a simultaneous multielement atomic absorption spectrometer, *Talanta*, 40(6), 1993, 879–882.
183. C. Mendiguchia, C. Moreno and M. García-Vargas, Determination of copper in seawater based on a liquid membrane preconcentration system, *Anal. Chim. Acta*, 460(1), 2002, 35–40.
184. B. Koklic, M. Veber and J. Zupan, Optimisation of lamp control parameters in glow discharge optical emission spectroscopy for the analysis of

copper–titanium–zinc alloy using the simplex method, *J. Anal. At. Spectrom.*, 18(2), 2003, 157–160.
185. A. M. Sofikitis, J. L. Colin, K. V. Desboeufs and R. Losno, Iron analysis in atmospheric water samples by atomic absorption spectroscopy (AAS) in water–methanol, *Anal. Bioanal. Chem.*, 378(2), 2004, 460–464.
186. A. López-Molinero, A. Villareal, D. Andia, C. Velilla and J. R. Castillo, Volatile germanium tetrachloride for sample introduction and germanium determination by inductively coupled plasma atomic emission spectroscopy, *J. Anal. At. Spectrom.*, 16(7), 2001, 744–749.
187. J. G. Magallanes, P. Smichowski and J. Marrero, Optimisation and empirical modeling of HG-ICP-AES analytical technique through artificial neural networks, *J. Chem. Inf. Comput. Sci.*, 41(3), 2001, 824–829.
188. B. Hilligsoe and E. H. Hansen, Application of factorial designs and simplex optimisation in the development of flow injection-hydride generation-graphite furnace atomic absorption spectrometry (FI-HG-GFAAS) procedures as demonstrated for the determination of trace levels of germanium, *Fresenius' J. Anal. Chem.*, 358(7–8), 1997, 775–780.
189. S. Río Segade and J. F. Tyson, Determination of methylmercury and inorganic mercury in water samples by slurry sampling cold vapor atomic absorption spectrometry in a flow injection system after preconcentration on silica C18 modified, *Talanta*, 71(4), 2007, 1696–1702.
190. E. Ramalhosa, S. Río Segade, E. Pereira, C. Vale and A. Duarte, Simple methodology for methylmercury and inorganic mercury determinations by high-performance liquid chromatography-cold vapour atomic fluorescence spectrometry, *Anal. Chim. Acta*, 448(1–2), 2001, 135–143.
191. C. Schickling and J. A. C. Broekaert, Determination of mercury species in gas condensates by online coupled high-performance liquid chromatography and cold-vapor atomic absorption spectrometry, *Appl. Organomet. Chem.*, 9(1), 1995, 29–36.
192. R. Puk and J. H. Weber, Determination of mercury (II), monomethylmercury cation, dimethylmercury and diethylmercury by hydride generation, cryogenic trappling and atomic absorption spectrometric detection, *Anal. Chim. Acta*, 292(1–2), 1994, 175–183.
193. G. A. Zachariadis and J. A. Stratis, Optimisation of cold vapour atomic absorption spectrometric determination of mercury with and without amalgamation by subsequent use of complete and fractional factorial designs with univariate and modified simplex methods, *J. Anal. At. Spectrom.*, 6(3), 1991, 239–245.
194. M. I. Rucandio, Determination of lanthanides and yttrium in rare earth ores and concentrates by inductively coupled plasma atomic emission spectrometry, *Anal. Chim. Acta*, 264(2), 1992, 333–344.
195. M. Felipe-Sotelo, M. J. Cal-Prieto, A. Carlosena, J. M. Andrade, E. Fernández and D. Prada, Multivariate optimisation for molybdenum

determination in environmental solid samples by slurry extraction–ETAAS, *Anal. Chim. Acta*, 553, 2005, 208–213.
196. F. C. Domínguez-Lledó, M. D. Galindo-Riaño, I. C. Díaz-López, M. García-Vargas and M. D. Granado-Castro, Applicability of a liquid membrane in enrichment and determination of nickel traces from natural waters, *Anal. Bioanal. Chem.*, 389, 2007, 653–659.
197. A. Aouarram, M. D. Galindo-Riaño, M. García-Vargas, M. Stitou and F. El Yousfi, A permeation liquid membrane system for determination of nickel in seawater, *Talanta*, 71, 2007, 165–170.
198. M. T. Siles Cordero, E. I. Vereda Alonso, P. Canada Rudner, A. García de Torres and J. M. Cano Pavón, Computer-assisted simplex optimisation of an on-line preconcentration system for determination of nickel in seawater by electrothermal atomic absorption spectrometry, *J. Anal. At. Spectrom.*, 14(7), 1999, 1033–1037.
199. L. Ebdon, P. Goodall, S. J. Hill, P. Stockwell and K. C. Thompson, Approach to the determination of lead by vapor generation atomic absorption spectrometry, *J. Anal. At. Spectrom.*, 9(12), 1994, 1417–1421.
200. J. M. Casas, R. Rubio and G. Rauret, Simplex optimisation of acid attack on river sediments for determination of lead, *Quím. Anal.*, 9(2), 1990, 163–170.
201. A. Economou, P. R. Fielden, A. J. Packham, E. J. Woollaston and R. D. Snook, Comparison of data smoothing techniques as applied to electrothermal atomic absorption spectrometry, *Analyst*, 122(11), 1997, 1331–1334.
202. M. Colón, M. Iglesias and M. Hidalgo, Development of a new method for sulfide determination by vapor generator-inductively coupled plasma-mass spectrometry, *Spectrochim. Acta, Part B*, 62, 2007, 470–475.
203. A. López-Molinero, O. Mendoza, A. Callizo, P. Chamorro and J. R. Castillo, Chemical vapor generation for sample introduction into inductively coupled plasma atomic emission spectroscopy: vaporisation on antimony (III) with bromide, *Analyst*, 127(10), 2002, 1386–1391.
204. I. Koch, Ch. F. Harrington, K. J. Reimer and W. R. Cullen, Simplex optimisation of conditions for the determination of antimony in environmental samples by using electrothermal atomic absorption spectrometry, *Talanta*, 44(7), 1997, 1241–1251.
205. I. Koch, C. F. Harrington, K. J. Reimer and W. R. Cullen, Simplex optimisation of conditions for the determination of antimony in environmental samples by using electrothermal atomic absorption spectrometry, *Talanta*, 44(5), 1997, 771–780.
206. H. Matusiewicz and M. Krawczyk, Determination of total antimony and inorganic antimony species by hydride generation *in situ* trapping flame atomic absorption spectrometry: a new way to (ultra)trace speciation analysis, *J. Anal. At. Spectrom.*, 23, 2008, 43–53.
207. M. J. Cal-Prieto, M. Felipe-Sotelo, A. Carlosena and J. M. Andrade, Direct generation of stibine from slurries and its determination

by ETAAS using multivariate optimisation, *At. Spectrosc.*, 26(3), 2005, 94–101.
208. A. López-Molinero, R. Giménez, P. Otal, A. Callizo, P. Chamorro and J. R. Castillo, New sensitive determination of selenium by bromide volatilisation inductively coupled plasma atomic emission spectrometry, *J. Anal. At. Spectrom.*, 17, 2002, 352–357.
209. M. Burguera, J. L. Burguera, C. Rivas, P. Carrero, R. Brunetto and M. Gallignani, Time-based device used for the determination of tin by hydride generation flow injection atomic absorption techniques, *Anal. Chim. Acta*, 308(1–3), 1995, 339–348.

CHAPTER 3

Ordinary Multiple Linear Regression and Principal Components Regression

JOAN FERRÉ-BALDRICH AND RICARD BOQUÉ-MARTÍ

Department of Analytical and Organic Chemistry, University Rovira i Virgili, Tarragona, Spain

3.1 Introduction

3.1.1 Multivariate Calibration in Quantitative Analysis

Most analytical problems require some of the constituents of a sample to be identified (qualitative analysis) or their concentrations to be determined (quantitative analysis). Quantitative analysis assumes that the measurands, usually concentrations of the constituents of interest in a sample, are related to the quantities (signals) measured using the technique with which the sample was analysed. In atomic spectroscopy, typical measured signals are absorbance and intensity of emission. These are used to predict the quantities of interest in new unknown samples using a validated mathematical model. The term '*unknown sample*' is used here to designate a sample to be analysed, not considered at the calibration stage.

Since the quality of the analytical result depends on the quality of every step of the analytical procedure, it is obvious that the establishment of a valid (useful) mathematical model is as important as performing adequately other steps of the analytical procedure such as sampling and sample pretreatment.

RSC Analytical Spectroscopy Monographs No. 10
Basic Chemometric Techniques in Atomic Spectroscopy
Edited by Jose Manuel Andrade-Garda

Published by the Royal Society of Chemistry, www.rsc.org

For this reason, it is of interest to learn the diverse types of calibration, together with their mathematical/statistical assumptions, the methods for validating these models and the possibilities of outlier detection. The objective is to select the calibration method that will be most suited for the type of analysis one is carrying out.

A variety of mathematical models can be used to establish appropriate relationships between instrumental responses and chemical measurands. Quantitative analysis in single-element atomic absorption spectroscopy is typically based on a single measured signal that is converted to concentration of the analyte of interest via the calibration line:

$$r_i = s_{0,k} + s_{1,k}c_{i,k} \tag{3.1}$$

where r_i is the single instrumental response obtained for a sample (*e.g.* absorbance), $c_{i,k}$ is the concentration of the analyte in the sample and $s_{0,k}$ and $s_{1,k}$ are the estimates of the coefficients of the model. The subscript *i* indicates that this equation holds for the sample *i*. A different sample may have a different concentration and, hence, produce a different instrumental response. The subscript *k* indicates that this equation is valid for analyte *k*. The coefficients will be different for another analyte. The coefficients in the model are estimated by solving a system of equations. These equations are obtained from a set of calibration samples for which the instrumental responses and the quantities of the chemical variables (here analyte concentrations) are known. Once the coefficients have been estimated, the concentration in a sample *i* is predicted from its response, *r*, as

$$c_{i,k} = b_{0,k} + r_i \times b_{1,k} \tag{3.2}$$

where $b_0 = -s_{0,k}/s_{1,k}$ and $b_1 = 1/s_{1,k}$. Equation (3.1) is called the classical (direct) calibration model and, for prediction, the equation must be reversed into eqn (3.2). It is also possible to use the calibration data to fit eqn (3.2) from the very beginning (see Figure 3.1). This process is called *inverse* calibration. Statistically it is different to fit the model in eqn (3.1) than to fit the model in eqn (3.2). The reason is that the ordinary least-squares (OLS) method (the most widely used method for estimating the model coefficients) requires that the

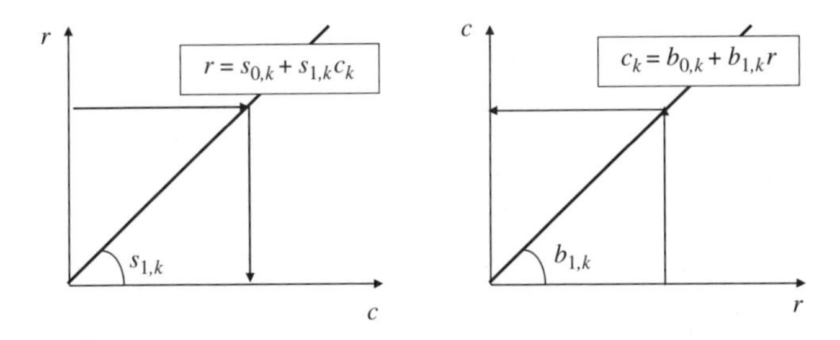

Figure 3.1 Univariate calibration. Direct model (left) and inverse model (right). The arrow indicates the direction in which the prediction is made.

random errors in the y be much larger than the errors in the x. This is usually achieved when we plot the concentrations of the standards c_k on the abscissa and the instrumental responses r on the ordinate. Hence eqn (3.1) is preferred. In multivariate calibration, however, we will see that it is also possible to fit both the classical model and the inverse model, but that the inverse model is preferred.

The application of a univariate calibration model requires the instrumental response to depend only on the concentration of the analyte of interest. In fact, the response is also allowed to contain a constant background contribution, which must be the same for all standards and unknown samples. This constant background is modelled in the constant term of eqn (3.1). Apart from that constant background, all selectivity problems must be removed before measurement. If this is to be achieved, the analysis of mixtures requires either a method to separate the analyte from the interferents (with the drawback that manipulating samples in the laboratory is a source of error and increases costs) or a highly selective instrument. Although univariate calibration is the most common situation, it is also true that unexpected interferences may contribute to the measured signal (such in the analysis of complex natural samples, environmental samples, clinical samples, *etc*.), thus introducing bias in the result unless additional work is done to achieve the necessary high selectivity. An alternative to the use of single point measurements and univariate calibration is the use of multiple responses simultaneously (multivariate signals), which may overcome most of the limitations of univariate calibration.

The most common, straightforward multivariate calibration model is the natural extent of the univariate calibration, the linear equation for which is

$$c_{i,k} = b_{0,k} + r_{i,1}b_{1,k} + \ldots + r_{i,J}b_{J,k} \tag{3.3}$$

where $r_{i,1}, \ldots, r_{i,j}, \ldots, r_{i,J}$ are J instrumental measurements/responses (*e.g*. absorbances at selected wavelengths or absorbances at different times of the atomic peak) made on a sample i and $b_{0,k}, b_{1,k}, \ldots, b_{J,k}$ are the model coefficients. Note that eqn (3.3) is an extension of eqn (3.2) to several responses.

An important advantage of multivariate calibration models over the univariate ones is that, since many measurements are obtained from the same sample, the signal from the analytes and from the interferences can be separated mathematically, so concentrations can be determined without the need for highly selective measurements for the analyte. This advantage has been termed the first-order advantage and eqn (3.3) is also called the first-order calibration model. The term 'first-order' means that the response from a sample is a vector (a first-order tensor) [1]. This nomenclature and the advantages of first-order calibration have been well described [2]. To use this advantage, however, there is one major requirement: the multivariate measurements of the calibration samples must contain contributions from the constituents that will be found in the future unknown samples to be predicted. In this way, the model is trained to distinguish the signal of the analyte from the signal of the interferences. Multivariate calibration has other additional advantages. One is that more than one property (*i.e.* the concentrations of several analytes) can be determined from the same instrumental response (*e.g.* the same spectrum) by using different coefficients b_k in eqn (3.3). Moreover, it is possible to detect whether the measured signal contains non-modelled contributions that could lead to severe bias, which makes it possible to detect outliers in prediction. This constitutes an important advantage of multivariate calibration over univariate calibration, in which outlier detection in prediction is not possible unless the prediction has abnormally high or low values which lack physical/chemical meaning.

For all the mentioned reasons, there is an ongoing tendency in spectroscopic studies to manipulate samples less and perform fewer experiments but to obtain more data in each of them and use more sophisticated mathematical techniques than simple univariate calibration. Hence multivariate calibration methods are being increasingly used in laboratories where instruments providing multivariate responses are of general use. Sometimes, these models may give less precise or less accurate results than those given by the traditional method of (univariate) analysis, but they are much quicker and cheaper than classical approaches.

As in univariate calibration, the coefficients in eqn (3.3) are estimated by solving a set of equations which relate the responses of the calibration samples to their known analyte concentrations. The equation used for calculating the coefficients distinguishes the different regression methods [3–6]. In this book, we will comment on four multivariate calibration techniques: classical least squares (CLS), inverse least squares (ILS), principal component regression (PCR) and partial least squares (PLS). The underlying theory, advantages and limitations of CLS, ILS and PCR are discussed in the following sections. The discussion is used to present the setup and the advantages of multivariate calibration. This serves to introduce the most advanced method, PLS, which, owing to its importance, is discussed extensively in Chapter 4. Note that both PCR and PLS are based on the same concepts and yield the same type of results. In order to avoid unnecessary repetitions, the typical graphical outputs from PCR will not be discussed in this chapter, since these will be discussed for PLS in Chapter 4.

3.1.2 Notation

Multivariate calibration can be used with any suitable multivariate data to model any property. However, the terminology used here associates instrumental responses with sample spectra and the property of interest with the analyte concentration since the bulk of this book deals with this type of data.

Matrices are shown in bold capital letters (*e.g.* **R**), column vectors in bold lowercase letters (*e.g.* **c**) (row vectors are transposed column vectors) and scalars in italic characters (*e.g.* c_k). True values are indicated by Greek characters or the subscript 'true'. Calculated or measured values are indicated by Roman characters. The *hat* (^), used in the literature to indicate calculated, has been dropped to simplify the notation; whether the magnitude is measured or calculated can be deduced from the context. The running indexes in multivariate calibration are as follows: $k=1$ to K analytes are present in $i=1$ to I

calibration samples whose instrumental responses are measured using $j=1$ to J sensors (in this book we will also refer to '*signals*' or '*variables*'; all these terms should be considered synonymous).

3.2 Basics of Multivariate Regression

3.2.1 The Multiple Linear Regression Model

This section introduces the regression theory that is needed for the establishment of the calibration models in the forthcoming sections and chapters. The multivariate linear models considered in this chapter relate several independent variables (x) to one dependent variable (y) in the form of a first-order polynomial:

$$y_i = \beta_0 + \beta_1 x_{i,1} + \ldots + \beta_J x_{i,J} + \varepsilon_i \tag{3.4}$$

where $\beta_0, \ldots, \beta_J$ are the *true* model coefficients (they are unknown and must be estimated from a set of calibration standards), ε_i indicates random error and the subscript i indicates that this equation holds for a given sample i. By defining the vector of x-variables $\mathbf{x}_i=[1, x_{i,1}, \ldots, x_{i,J}]^{\mathrm{T}}$ and the vector of regression coefficients $\boldsymbol{\beta}=[\beta_0, \beta_1, \ldots, \beta_J]^{\mathrm{T}}$, eqn (3.4) can be written as

$$y_i = \mathbf{x}_i^{\mathrm{T}}\boldsymbol{\beta} + \varepsilon_i \tag{3.5}$$

In order to estimate the coefficients, I sets of observations on $x_1, \ldots, x_J$ and y are collected. These observations are obtained from a set of I calibration standards, for which the concentrations and spectra are known. The application of eqn (3.5) to these I standards yields a system of equations:

$$\mathbf{y} = \mathbf{X}\boldsymbol{\beta} + \boldsymbol{\varepsilon} \tag{3.6}$$

where $\mathbf{y}$ ($I \times 1$) is the vector containing the observed y_i ($i=1, \ldots, I$), $\mathbf{X}$ is the $I \times (J+1)$ matrix that contains, in the ith row, the measured variables for sample i ($\mathbf{x}_i=[1, x_{i,1}, \ldots, x_{i,J}]^{\mathrm{T}}$) and $\boldsymbol{\varepsilon}$ ($I \times 1$) is the vector of non-observable random errors. Note that $\mathbf{X}$ specifies the form of the model and that as the model in eqn (3.4) contains a constant term (or 'intercept') β_0, matrix $\mathbf{X}$ should include a column of ones (see Figure 3.2).

$$\begin{bmatrix} y_1 \\ \vdots \\ y_i \\ \vdots \\ y_I \end{bmatrix} = \begin{bmatrix} 1 & x_{11} & \cdots & x_{1J} \\ \vdots & \vdots & \cdots & \vdots \\ 1 & x_{i1} & \cdots & x_{iJ} \\ \vdots & \vdots & \cdots & \vdots \\ 1 & x_{I1} & \cdots & x_{IJ} \end{bmatrix} \begin{bmatrix} \beta_0 \\ \beta_1 \\ \vdots \\ \beta_J \end{bmatrix} + \begin{bmatrix} \varepsilon_1 \\ \vdots \\ \varepsilon_i \\ \vdots \\ \varepsilon_I \end{bmatrix}$$

Figure 3.2 Matrix representation of eqn (3.6) for the model in eqn (3.4).

3.2.2 Estimation of the Model Coefficients

An estimate of $\boldsymbol{\beta}$ in eqn (3.6) can be obtained as

$$\mathbf{b} = \mathbf{X}^{+}\mathbf{y} \tag{3.7}$$

where $\mathbf{b} = [b_0, \ldots, b_j, \ldots, b_J]^{\mathrm{T}}$ is the vector of estimated regression coefficients and $\mathbf{X}^{+}$ is the so-called *pseudoinverse* of $\mathbf{X}$. Equation (3.7) provides an optimal solution if certain assumptions about the model and the model error ε_i are fulfilled [7]. The pseudoinverse is calculated in a different way for each regression method, thus producing different estimated coefficients **b**. In multiple linear regression (MLR), the pseudoinverse is calculated with the OLS solution and in PCR is calculated as [4,8]

$$\mathbf{X}^{+} = \mathbf{P}\mathbf{D}^{-1}\mathbf{U}^{\mathrm{T}} \tag{3.8}$$

where **U**, **D** and **P** are generated by the singular-value decomposition (SVD) of **X** (see Appendix). In MLR, **U**, **D** and **P** are used as such, whereas in PCR, **U**, **P** and **D** are truncated to have size $\mathbf{U}(I \times A)$, $\mathbf{P}(J \times A)$ and $\mathbf{D}(A \times A)$, where A is the number of latent variables that are retained (see Section 3.4.2). This truncation provides a low-dimension approximation of the data which retains the relevant information and has less noise. For systems in which **X** has full column rank (*i.e.* independent columns and at least as many rows as columns), eqn (3.8) is the same as $\mathbf{X}^{+} = (\mathbf{X}^{\mathrm{T}}\mathbf{X})^{-1}\mathbf{X}^{\mathrm{T}}$ and eqn (3.7) becomes the popular OLS solution equation:

$$\mathbf{b} = (\mathbf{X}^{\mathrm{T}}\mathbf{X})^{-1}\mathbf{X}^{\mathrm{T}}\mathbf{y} \tag{3.9}$$

Note that, in order to apply eqn (3.9), the symmetric square matrix $\mathbf{X}^{\mathrm{T}}\mathbf{X}$ must be invertible (non-singular, *i.e.* $\det(\mathbf{X}^{\mathrm{T}}\mathbf{X}) \neq 0$). If **X** is not of full rank, then the parameter estimates **b** cannot be found with eqn (3.9) but they can still be found with equations (3.7) and (3.8). See, for example, Ben-Israel and Greville [9] for an extensive discussion on how to solve the system of equations in a variety of situations.

It is also important to note that the variance–covariance matrix of the OLS estimated coefficients is

$$\mathrm{var}(\mathbf{b}) = (\mathbf{X}^{\mathrm{T}}\mathbf{X})^{-1}\sigma^2 = \begin{pmatrix} var(b_0) & cov(b_0, b_1) & \ldots & cov(b_0, b_J) \\ cov(b_0, b_1) & var(b_1) & \ldots & cov(b_1, b_J) \\ \vdots & \vdots & \ddots & \vdots \\ cov(b_J, b_0) & cov(b_J, b_1) & \ldots & var(b_J) \end{pmatrix} \tag{3.10}$$

where σ^2 is the variance of the random error. Equation 3.10 indicates that the variance of the coefficients will be larger whenever the elements of $(\mathbf{X}^{\mathrm{T}}\mathbf{X})^{-1}$ are larger. This will happen when the columns of **X** are collinear (see Section 3.2.4, where collinearity and its implications are discussed in more detail). As in univariate regression, the variance of the coefficients is used to decide on the statistical significance of the coefficients and to calculate their confidence intervals. The variance of the coefficients is also important because it propagates to the variance of the predictions obtained from the model.

3.2.3 Prediction

Once the model coefficients have been estimated, the prediction from a vector of input variables $\mathbf{x} = [1, x_1, \ldots, x_j, \ldots, x_J]^T$ is

$$\hat{y} = b_0 + b_1 x_1 + \ldots + b_j x_j + \ldots + b_J x_J = \mathbf{x}^T \mathbf{b} \quad (3.11)$$

3.2.4 The Collinearity Problem in Multivariate Regression

Collinearity (also called multicollinearity or ill-conditioning) is defined as the approximate linear dependence of, at least, one of the columns in **X** with another or other column(s) of **X**. Singularity occurs when the variables are perfectly correlated. In the case of singularity, the OLS solution [eqn (3.9)] cannot be calculated since $\det(\mathbf{X}^T\mathbf{X}) = 0$. However, in the case of collinearity, the OLS can be calculated, but the coefficients are estimated poorly: small relative changes in the observed **y** due to random errors may lead to very different estimated coefficients **b**. This instability is translated into large (inflated) variances and covariances of the coefficients of the variables involved in the linear dependencies [*i.e.* at least one diagonal element of the $(\mathbf{X}^T\mathbf{X})^{-1}$ matrix is large in eqn (3.10)]. These large variances and covariances sometimes cause unacceptable signs of the coefficients, numerical values which are too large and also statistical and interpretation problems. Collinearity makes it more difficult to interpret the impact of each x-variable on the y because correlated estimates cannot be interpreted separately. Also, a t-test can indicate statistical insignificance of the coefficients owing to large variances of the regression coefficient estimates [10]. Interestingly, collinearity can exist in models with a good fit (a high multiple correlation coefficient) and it is not a problem if we are only interested in predictions at combinations of xs which are similar to those in the calibration data. However, collinearity is a serious and critical problem when predictions are made at different combinations of the xs or extrapolation beyond the range of the data. Different measures for evaluating the extent of the collinearity problem have been proposed for the MLR models: the correlation matrix of the x-variables, the eigenvalues of the matrix of x-variables (and related measures such as the condition indices and the condition number), the tolerance and variance inflation factors of the estimated coefficients and the variance–decomposition proportions. These diagnostic measures are discussed elsewhere [11] and not too many details will be given here. The *variance inflation factors* can be calculated easily and have been recommended as a general diagnostic measure of collinearity. The variance inflation factor (VIF_j) of the regression coefficient b_j is calculated as [12]

$$\mathrm{VIF}_j = (1 - \mathrm{R}_j^2)^{-1} \quad (3.12)$$

where R_j^2 is the multiple coefficient of determination of the variable x_j regressed on all the other terms in the model. The VIF_j ranges from 1 (non-correlated coefficients) to ∞ (perfect correlation). Values larger than 1 indicate that the variable is affected by collinearity. Values larger than 10 indicate that the correlation among the variables is so high that the coefficient is likely to be poorly estimated [13].

3.3 Multivariate Direct Models

3.3.1 Classical Least Squares

3.3.1.1 Model Formulation

This quantification method is based on the linear additive model applied to many instrumental responses. The model assumes that, for a sample *i*, the signal (measured response) at sensor *j* results from the sum of the responses of all the analytes (*K*) that contribute to the signal in this sensor (the term sensor will be used to refer globally to wavelength, time of the atomic peak, *m*/*z* ratio, *etc.*):

$$r_{i,j} = r_{j,0} + r_{i,j,1} + \ldots + r_{i,j,k} + \ldots + r_{i,j,K} + \varepsilon_{i,j} \tag{3.13}$$

where $r_{i,j}$ is the response measured at the *j*th sensor and $r_{i,j,k}$ is the response of analyte *k* in sample *i* in that sensor *j*. For example, the absorbance measured at 350 nm is the sum of the absorbances at 350 nm of the *K* analytes in the sample. The measured response also contains a constant term $r_{j,0}$ (the background) and a random term $\varepsilon_{i,j}$. This error is assumed to derive from the measurement of the responses and to be independent and normally distributed [*i.e.* $\varepsilon_{i,j} \sim N(0,\sigma^2)$]. In addition, the model assumes that the response of each analyte is proportional to its concentration. Hence, for analyte *k*:

$$r_{i,j,k} = s_{j,k} c_{i,k} \tag{3.14}$$

where $s_{j,k}$ is the sensitivity of the sensor *j* for analyte *k* (the response at unit concentration, which is the traditional 'absorptivity' term in spectral measurements) and $c_{i,k}$ is the concentration of analyte *k* in the sample. Hence the model equation for the response at sensor *j* of a sample *i* of *K* analytes is

$$r_{i,j} = r_{0,j} + \sum_{k=1}^{K} s_{j,k}\, c_{i,k} + \varepsilon_{i,j} \tag{3.15}$$

For a series of *J* measured responses, eqn (3.15) becomes

$$\mathbf{r}_i = \mathbf{r}_{i,0} + \mathbf{S}\mathbf{c}_{i,\text{true}} + \boldsymbol{\varepsilon}_i \tag{3.16}$$

where $\mathbf{r}_i = [r_{i,1}, \ldots, r_{i,j}, \ldots, r_{i,J}]^T$ is the $J \times 1$ response vector of sample *i* measured at the *J* sensors, $\mathbf{c}_{i,\text{true}}$ is the $K \times 1$ vector of all the analyte concentrations in the sample that give a response, $\mathbf{S} = [\mathbf{s}_1, \ldots, \mathbf{s}_k, \ldots, \mathbf{s}_K]$ is a

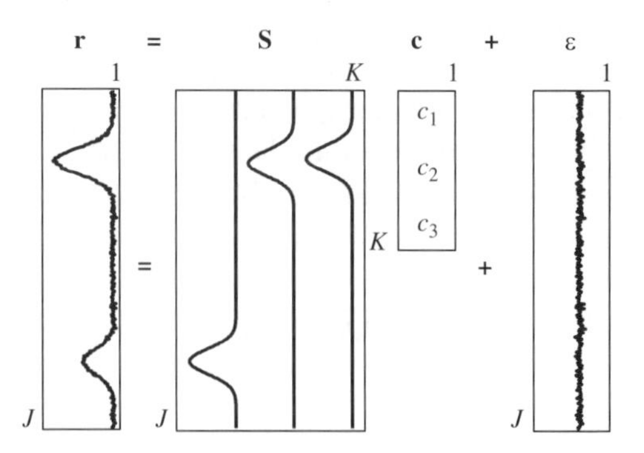

Figure 3.3 Graphical and matrix representations of the CLS model.

$J \times K$ matrix of sensitivities and $\boldsymbol{\varepsilon}_i$ is a $J \times 1$ vector of errors. For the case of absorbance spectroscopy, eqn (3.16) is the application of the Lambert–Beer law to many wavelengths: $\mathbf{r}_i$ is the absorbance spectrum at J selected wavelengths, which is written as a linear additive function of the concentrations of the chemical constituents. Then the columns of **S** are the pure-component spectra at unit concentration and unit pathlength (absorptivity–pathlength products). Note that the model assumes also that there are no interaction terms in the spectrum between the various components of the sample. Usually the measured spectra are background corrected (for example, by subtracting the response of a blank containing only the sample matrix from each measured signal, without the analytes of interest), so that eqn (3.16) is rewritten as (see Figure 3.3)

$$\mathbf{r}_i = \mathbf{S}\mathbf{c}_{i,\text{true}} + \boldsymbol{\varepsilon}_i \tag{3.17}$$

Note that $\mathbf{c}_{i,\text{true}}$ 'counts' how many times each column of **S** can be found in **r**. Hence, if the columns in **S** are the spectra at unit concentration and unit pathlength, $\mathbf{c}_{i,\text{true}}$ is the concentration of the analytes in the sample, expressed in the same concentration units as those used in the columns of **S**.

3.3.1.2 Calibration

Equation (3.17) is the fundamental equation of the CLS model that allows calibration and prediction. The calibration step consists of calculating **S**, which is the matrix of coefficients that will allow the quantification of future samples. **S** is found by entering the spectra and the known concentration of a set of calibration samples in eqn (3.17). These calibration samples, which can be either pure standards or mixtures of the analytes, must contain in total *all* the analytes that will be found in future samples to be predicted. Then, eqn (3.17) for I calibration samples becomes

$$\mathbf{R}^{\text{T}} = \mathbf{S}\mathbf{C}^{\text{T}} + \mathbf{E} \tag{3.18}$$

where $\mathbf{R}$ ($I \times J$) contains the signals (atomic absorbances) measured at J sensors (recall that sensors means wavelengths, times of the atomic peak, *etc.*) for I calibration samples of individual components or mixtures, $\mathbf{C}$ ($I \times K$) are the known concentrations of the K analytes in the calibration samples and $\mathbf{E}$ ($I \times J$) is the matrix of spectral errors. Equation (3.18) is of the form of eqn (3.6), so $\mathbf{S}$ is estimated by OLS solution as

$$\mathbf{S} = (\mathbf{C}^{+}\mathbf{R})^{\mathrm{T}} \tag{3.19}$$

where $\mathbf{C}^{+}$ is the pseudoinverse of $\mathbf{C}$. As a particular case, if the number of calibration samples and constituents is the same, $\mathbf{C}$ is square ($K \times K$) and $\mathbf{S}$ can be estimated as

$$\mathbf{S} = (\mathbf{C}^{-1}\mathbf{R})^{\mathrm{T}} \tag{3.20}$$

In addition, if each calibration sample contains only one analyte, then $\mathbf{R}$ contains already the spectra of the pure components and $\mathbf{C}$ ($K \times K$) is a diagonal matrix (*i.e.* non-zero values only in the main diagonal). Hence eqn (3.20) simply calculates $\mathbf{S}$ by dividing the spectrum of each pure calibration sample by its analyte concentration. To obtain a better estimate of $\mathbf{S}$, the number of calibration samples is usually larger than the number of components, so eqn (3.19) is used. We still have two further requirements in order to use the previous equations. First, the relative amounts of constituents in at least K calibration samples must change from one sample to another. This means that, unlike the common practice in laboratories when the calibration standards are prepared, the dilutions of one concentrated calibration sample cannot be used alone. Second, in order to obtain the necessary number of linearly independent equations, the number of wavelengths must be equal to or larger than the number of constituents in the mixtures ($J \geq K$). Usually the entire spectrum is used.

3.3.1.3 Prediction

Note that eqn (3.17) has the same form as eqn (3.6). Hence, by introducing in $\mathbf{r}$ the vector of instrumental responses (*e.g.* a spectrum) of an unknown sample and in $\mathbf{S}$ the matrix of sensitivities that we obtained in the calibration step, the OLS solution [see the similarity with eqn (3.7)] is

$$\mathbf{c} = \mathbf{S}^{+}\mathbf{r} \tag{3.21}$$

and it gives the concentrations of the K analytes in the unknown sample (*i.e.* of the K analytes whose responses were included in $\mathbf{S}$ at the calibration stage) in a single step. Note also that the concentration of analyte k (the kth element of $\mathbf{c}$) is obtained by multiplying the kth row of $\mathbf{S}^{+}$ by $\mathbf{r}$:

$$c_k = \mathbf{S}^{+}_{k\text{-row}}\mathbf{r} = \mathbf{r}^{\mathrm{T}}\mathbf{S}^{\mathrm{T}+}{}_{k\text{-colum}} \tag{3.22}$$

since $(\mathbf{S}^+)^T = \mathbf{S}^{T+}$. By renaming the kth column of $\mathbf{S}^{T+}$ as the vector of coefficients:

$$\mathbf{b}_{k,\mathrm{CLS}} = \mathbf{S}^{T+}_{k\text{-col}} \tag{3.23}$$

it turns out that the prediction of the concentration of analyte k in a sample is obtained as

$$c_k = r_1 b_{1,k,\mathrm{CLS}} + \ldots + r_J b_{J,k,\mathrm{CLS}} \tag{3.24}$$

which is the general equation of the linear multivariate calibration shown in eqn (3.3) after accounting for the intercept $b_{0,k}$. This interesting result shows that the prediction of the concentration of a single analyte in a mixture can be done if we are able to find the adequate regression coefficients. In this case, these coefficients were found for all the analytes simultaneously (matrix **S**). We will see, in Section 3.4, that it is possible to find the coefficients for only one analyte of interest in the mixture without the need to know the pure spectra of all the constituents in the mixture.

3.3.1.4 Considerations About CLS

The main advantage of multivariate calibration based on CLS with respect to univariate calibration is that CLS does not require selective measurements. Selectivity is obtained mathematically by solving a system of equations, without the requirement for chemical or instrumental separations that are so often needed in univariate calibration. In addition, the model can use a large number of sensors to obtain a signal-averaging effect [4], which is beneficial for the precision of the predicted concentration, making it less susceptible to the noise in the data. Finally, for the case of spectroscopic data, the Lambert–Bouguer–Beer's law provides a sound foundation for the predictive model.

The main limitation of this model [6,14] is that it assumes that the measured response at a given sensor is due entirely to the constituents considered in the calibration step, whose spectra are included in the matrix of sensitivities, **S**. Hence, in the prediction step, the response of the unknown sample is decomposed *only* in the contributions that are found in **S**. If the response of the unknown contains some contributions from constituents that have not been included in **S** (in addition to background problems and baseline effects), biased predicted concentrations may be obtained, since the system will try to assign this signal to the components in **S**. For this reason, this model can only be used for systems of known qualitative composition (*e.g.* gas-phase spectroscopy, some process monitoring or pharmaceutical samples), in which the signal of all the pure constituents giving rise to a response can be known. For the same reason, CLS is not useful for mixtures where interaction between constituents or deviations from the Lambert-Beer law (nonlinear calibration curves) occur.

An additional limitation is related to spectral overlap and collinearity. Since the sought for concentrations are the estimated coefficients of the model in eqn (3.17), the OLS solution used to estimate these coefficients suffers from the collinearity problem indicated in Section 3.2.4. We have to be aware of this problem because it happens whenever the analytes included in the model have similar spectra (*i.e.* the columns of **S** are similar), which is often termed high overlap of the pure component spectra. Recall that collinearity gives an unstable system of equations, which means that small variations in the data produce large variations in the solutions of the systems of equations. In this case, it means that two repeated measured spectra **r** may produce very different calculated concentrations in eqn (3.21) only because of small changes in the random error in **r**. This translates into large uncertainties associated with the predicted concentrations and even misleading results (*e.g.* negative concentrations for analytes that indeed are present in the sample). The collinearity problem also sets a practical limit to the number of analytes that can be resolved at the same time. Although eqn (3.17) can be solved in principle for any number of analytes (provided that a sufficient amount of data have been measured), the fact is that the larger the number of analytes that are present in the mixture, the larger is the number of columns in the matrix **S** and the larger the collinearity in the system. In technical terms, it is said that a reduction in the amount of signal of each analyte that can be used for prediction occurred. This corresponds to the concept of *net analyte signal* [15], which will be commented on in Chapter 4. The effects of collinearity can be reduced by correctly choosing the sensors to be used in the regression, although collinearity cannot be eliminated unless completely selective sensors are available and by limiting the number of analytes in the mixtures. The criteria for variable selection are usually based on some measure of orthogonality in the **S** matrix, such as the selectivity defined by Lorber [15]. Fortunately, these limitations are partially avoided with the use of inverse multivariate calibration models.

3.4 Multivariate Inverse Models

The multivariate quantitative spectroscopic analysis of samples with complex matrices can be performed using *inverse* calibration methods, such as ILS, PCR and PLS. The term '*inverse*' means that the concentration of the analyte of interest is modelled as a function of the instrumental measurements, using an empirical relationship with no theoretical foundation (as the Lambert–Bouguer–Beer's law was for the methods explained in the paragraphs above). Therefore, we can formulate our calibration like eqn (3.3) and, in contrast to the CLS model, it can be calculated without knowing the concentrations of all the constituents in the calibration set. The calibration step requires only the instrumental response and the reference value of the property of interest (*e.g.* concentration) in the calibration samples. An important advantage of this approach is that unknown interferents may be present in the calibration samples. For this reason, inverse models are more suited than CLS for complex samples.

3.4.1 Inverse Least Squares (ILS)

3.4.1.1 Model Formulation

ILS is a least-squares method that assumes the inverse calibration model given in eqn (3.4). For this reason it is often also termed *multiple linear regression* (MLR). In this model, the concentration of the analyte of interest, k, in sample i is regressed as a linear combination of the instrumental measurements at J selected sensors [5,16–19]:

$$c_{i,k} = \beta_{0,k} + \beta_{1,k} r_{i,1} + \ldots + \beta_{J,k} r_{i,J} + \varepsilon_{i,k} = \mathbf{r}_i^{\mathrm{T}} \boldsymbol{\beta}_k + \varepsilon_{i,k} \tag{3.25}$$

where the error $\varepsilon_{i,k}$ is assumed to derive from uncertainties in the determination of the concentration in the calibration samples whereas no error is assumed in the absorbance values. Note that this is eqn (3.4) by using the concentration as dependent variable (y) and the instrumental responses as independent variables (x). Hence all the mathematical equations given in Section 3.2 are applied here, simply by updating them with instrumental responses and concentrations.

3.4.1.2 Calibration

Calibration consists of estimating the $J+1$ coefficients of the model by solving a system of I ($I \geq J+1$) equations. For I calibration samples, the model becomes

$$\mathbf{c}_k = \mathbf{R}\boldsymbol{\beta}_k + \boldsymbol{\varepsilon}_k \tag{3.26}$$

where $\mathbf{c}_k$ contains the concentration of analyte k in those samples and the rows of $\mathbf{R}$ are the instrumental responses (*e.g.* spectra) of the calibration samples (plus a column of ones to account for the constant term). The coefficients are estimated as

$$\mathbf{b}_{k,\mathrm{ILS}} = \mathbf{R}^{+}\mathbf{c}_k \tag{3.27}$$

where the pseudoinverse is $\mathbf{R}^{+}$ calculated as in eqn (3.8).

3.4.1.3 Prediction

The predicted concentration of analyte k in a sample from its vector of instrumental responses $\mathbf{r}^{\mathrm{T}} = [1\ \mathrm{r}_1,\ \ldots\ \mathrm{r}_J]$ is given by

$$c_k = \mathbf{r}^{\mathrm{T}}\mathbf{b}_{k,\mathrm{ILS}} \tag{3.28}$$

3.4.1.4 Considerations About ILS

The number of spectral variables and the collinearity problem are the two main drawbacks of the ILS model when applied to spectroscopic data.

The number of variables (sensors, wavelengths, times of the atomisation peak, *m*/*z* ratios, *etc.*) should be restricted to a subset of the usually large number of variables we can measure. The reason is, on the one hand, mathematical: there must be at least as many equations as unknowns in order to obtain the OLS solution to the system of equations. Since each coefficient for a particular variable response is an unknown and we need at least as many equations as unknowns, and since each equation corresponds to a calibration sample, then the number of calibration samples must not be lower than the number of sensors. Hence, using too many sensors would require obtaining the reference values for a large number of calibration samples, which may be costly and tedious. Second, the constituent of interest must absorb at the selected variables in order to obtain accurate predictions. Prediction improves if sensors related to the constituent of interest are added to the model, but it degrades if the sensors include excessive noise, which is unique to the training set. Such noise, which is unlikely to vary in exactly the same manner in the unknown samples and in the calibration samples, will degrade the prediction accuracy for unknown samples (this is called *the overfitting problem* and it will be addressed also in Chapter 4). On the other hand, collinear variables (*i.e.* the correlation between the columns of **R**) must be avoided since they cause large variances for the coefficients $\mathbf{b}_k$ and for the predicted concentration in unknown samples c_k. Collinearity is particularly relevant in molecular spectroscopy, where the number of responses can easily be high and the absorbances at adjacent wavelengths tend to increase and decrease together in the calibration samples. The large variance (low precision) of the coefficients produces large variances in the predictions. Ideally, there is a crossover point between selecting enough variables for an accurate OLS solution and selecting few enough that the calibration is not affected by the noisiest measurements and the collinearity of the spectral data. Hence the selection of the appropriate set of spectral variables is critical for the final quality of the ILS model, so ILS is almost always accompanied by variable selection. Selection can be made based on chemical and spectral knowledge about the analyte of interest and the interferents. In the absence of this information, empirical selection methods such as genetic algorithms based on some quality criteria can be used [20].

Note that both the collinearity problem and the requirement of having more samples than sensors can be solved by using regression techniques which can handle collinear data, such as factor-based methods such as PCR and PLS. These use linear combinations of all the variables and reduce the number of regressor variables. PCR or PLS are usually preferred instead of ILS, although they are mathematically more complex.

A last note refers to the design of the calibration set. Although we do not need to know either the spectra or the concentration of the interferences, we must be sure that the calibration samples contain the analytes and interferences which might contribute to the response of the unknown samples. In this way, the calculated regression coefficients can remove the contribution of the interferences in the predictions. If the instrumental measurement on the unknown sample contains signals from non-modelled interferences, biased predictions are likely to be obtained (as in the CLS model).

3.4.2 Principal Components Regression (PCR)

PCR and PLS (Chapter 4) are factor-based regression methods that remove some limitations of CLS and ILS. Both PCR and PLS express the correlated information in the many measured variables in a new coordinate system of a few 'latent' variables (also called factors) that are linear combinations of the original variables. Regression is then based on these latent variables instead of the original ones. Since the number of latent variables that are used is usually much less than the number of original variables, there are fewer regression coefficients to be estimated (one for each latent variable) and, hence, these methods do not require as many calibration samples as ILS. On the other hand, the new latent variables are orthogonal, so that the collinearity problem found in ILS is also solved. Moreover, they maintain the ILS property that only the concentration of the analyte of interest in the calibration samples must be known in order to estimate the model. There is no need to know all the constituents that give a response, as opposed to CLS. The focus of this section is PCR, and most of the ideas here are also valid for PLS, which will be discussed in Chapter 4.

3.4.2.1 Model Formulation

The model in PCR is, as for the ILS model, described in eqn (3.25), except that only a few new variables (the *latent variables*) t are used instead of the original measured responses r. Hence the PCR model is of the form

$$\begin{aligned} c_{i,k} &= \theta_{0,k} + \theta_{1,k} t_{i,1} + \ldots + \theta_{a,k} t_{i,a} + \ldots + \theta_{A,k} t_{iA} + g_{i,k} \\ &= \theta_{0,k} + \mathbf{t}_{\mathrm{A}}^{i\mathrm{T}} \boldsymbol{\theta}_{k,\mathrm{A}} + g_{i,k} \end{aligned} \quad (3.29)$$

where $\theta_{a,k}$ are the true model coefficients that must be estimated and $g_{i,k}$ is the error term. The value A (called the pseudo-rank of the calibration matrix) is the number of factors that are important for regression (and is usually much smaller than the number of measurements). The calculation of t and of the regression coefficients is commented on in the next section.

3.4.2.2 Calibration

PCR creates a quantitative model in a two-step process: (1) the so-called principal components analysis (PCA) *scores* (they are described just below), $\mathbf{T}_{\mathrm{A}}$, of the I calibration samples are calculated for A factors and then (2) the scores are regressed against the analyte concentration.

3.4.2.2.1 Obtaining the PCA Scores. In PCA, the matrix of measured instrumental responses of I calibration samples $\mathbf{R}$ ($I \times J$) (often column-centred or autoscaled) is decomposed into the product of two smaller matrices:

$$\mathbf{R} = \mathbf{T}_{\mathrm{A}} \mathbf{P}_{\mathrm{A}}^{\mathrm{T}} + \mathbf{E} \quad (3.30)$$

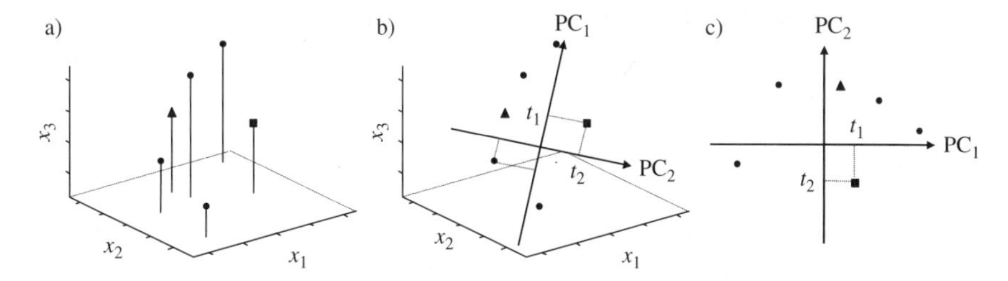

Figure 3.4 Principal components analysis of a 6 × 3 matrix: (a) the six samples in the original space of three measured variables; (b) the new axes (principal components PC_1 and PC_2) obtained from the SVD of the 6 × 3 matrix; (c) representation of the six samples in the space of the principal components. Note how the three original variables are correlated (the higher x_1 and x_2 are, the higher is x_3). Note also how by using only the coordinates (scores) of the samples on these two principal components, the relative position of the samples in the initial variable space is captured. This is possible because the original variables are correlated. Principal components regression (PCR) uses the scores on these two new variables (the two principal components) instead of the three originally measured variables.

where matrix $\mathbf{P}_A = [\mathbf{p}_{(1)}, \ldots, \mathbf{p}_{(A)}]$ $(J \times A)$ is related to the original set of variables and is called the matrix of **x**-*loadings*, matrix $\mathbf{T}_A = [\mathbf{t}_{(1)}, \ldots, \mathbf{t}_{(A)}]$ $(I \times A)$ is related to the calibration samples and is termed the matrix of *scores* and $\mathbf{E}$ $(I \times J)$ is the part of the data that is not modelled by the A factors. As mentioned above, A is the number of latent variables that are retained in the model. The loadings in $\mathbf{P}$ define a new coordinate system (a rotation of the original axis of the measured variables) and the scores $\mathbf{T}$ are the coordinates of the samples in this new coordinate system (see Figure 3.4 for a brief description).

Hence, both the ith row of $\mathbf{R}$ and the ith row of $\mathbf{T}_A$ describe the sample i: the ith row of $\mathbf{R}$ contains the measured values of the J original variables, whereas the ith row of $\mathbf{T}$ contains the coordinates of this sample in this new variable system. The number of columns in $\mathbf{T}$ is usually much smaller than the number of columns in $\mathbf{R}$ $(A << J)$, hence the scores are a compressed version (*i.e.* smaller number of variables) of the data in $\mathbf{R}$. Hence, by retaining only A factors, $\mathbf{R}_A = \mathbf{T}_A\mathbf{P}_A^T$ is an approximate reproduction of the original atomic spectra, $\mathbf{R}$, with a lower mathematical rank. It is assumed that $\mathbf{T}_A\mathbf{P}_A^T$ describes non-random sources of variation in $\mathbf{R}$ caused by changes in chemistry and is related to the constituents of interest, while $\mathbf{E}$ contains information which is not important for predicting concentration such as random noise.

The decomposition in eqn (3.30) is general for PCR, PLS and other regression methods. These methods differ in the criterion (and the algorithm) used for calculating $\mathbf{P}$ and, hence, they characterise the samples by different scores $\mathbf{T}$. In PCR, $\mathbf{T}$ and $\mathbf{P}$ are found from the PCA of the data matrix $\mathbf{R}$. Both the NIPALS algorithm [3] and the singular-value decomposition (SVD) (much used, see Appendix) of $\mathbf{R}$ can be used to obtain the $\mathbf{T}$ and $\mathbf{P}$ used in PCA/PCR. In PLS, other algorithms are used to obtain $\mathbf{T}$ and $\mathbf{P}$ (see Chapter 4).

3.4.2.2.2 Solving the Regression Model. Once the scores of the calibration standards, $\mathbf{T}_\mathrm{A}$, have been obtained by a PCA of the matrix of their atomic spectra, $\mathbf{R}$, the following system of equations is set up:

$$\mathbf{c}_k = \mathbf{T}_\mathrm{A}\boldsymbol{\theta}_{k,A} + \boldsymbol{\varepsilon}_k \tag{3.31}$$

where $\boldsymbol{\theta}_{k,A}$ is the vector of the regression coefficients and $\boldsymbol{\varepsilon}_k$ is a vector of errors. Note that this is similar to eqn (3.26), except that the scores $\mathbf{T}_\mathrm{A}$ are used instead of the original measured variables $\mathbf{R}$. Compared with the ILS model, eqn (3.31) has fewer regression coefficients to be estimated (only A coefficients, one for each column of $\mathbf{T}$), so the number of calibration samples does not need to be as large as for ILS.

The OLS solution for $\boldsymbol{\theta}_{k,A}$ has the same form as the ILS solution but with $\mathbf{T}_\mathrm{A}$ and $\mathbf{q}_{k,A}$ instead of $\mathbf{R}$ and $\mathbf{b}_k$:

$$\mathbf{q}_{k,A} = \mathbf{T}_\mathrm{A}{}^{+}\mathbf{c}_k \tag{3.32}$$

An additional advantage of PCR is that the score vectors are orthogonal (PCA provides orthogonal scores), so that the collinearity problems found in ILS are avoided and the resulting parameter estimates $q_{k,a}$ are stable. Note also that this model also maintains the form described in eqn (3.3). This can be seen by starting with the original ILS model [eqn (3.26)] and multiplying by $\mathbf{P}_\mathrm{A}\mathbf{P}_\mathrm{A}{}^\mathrm{T} = \mathbf{I}$:

$$\mathbf{c}_k = \mathbf{R}\mathbf{P}_\mathrm{A}\mathbf{P}_\mathrm{A}{}^\mathrm{T}\boldsymbol{\beta}_k + \boldsymbol{\varepsilon}_k \tag{3.33}$$

which gives eqn (3.3) by changing $\mathbf{T}_\mathrm{A} = \mathbf{R}\mathbf{P}_\mathrm{A}$ and $\boldsymbol{\theta}_{k,A} = \mathbf{P}_\mathrm{A}{}^\mathrm{T}\boldsymbol{\beta}_k$. Hence the coefficients $\mathbf{b}_k$ in eqn (3.3) can be calculated from the estimated $\mathbf{q}_{k,A}$ as

$$\mathbf{b}_{k,\mathrm{A,PCR}} = \mathbf{P}_\mathrm{A}\mathbf{q}_{k,\mathrm{A}} \tag{3.34}$$

where $\mathbf{P}_\mathrm{A}$ only contains the loadings for the A significant factors.

3.4.2.3 Prediction

The concentration of analyte k in an unknown sample can be predicted in two equivalent ways:

(a) By using the coefficients given by eqn (3.34):

$$c_k = \mathbf{r}^\mathrm{T}\mathbf{b}_{k,\mathrm{A,PCR}} \tag{3.35}$$

(b) By obtaining the scores corresponding to the unknown sample as $\mathbf{t}_\mathrm{A}{}^\mathrm{T} = \mathbf{r}^\mathrm{T}\mathbf{P}_\mathrm{A}$ and using the coefficients from eqn (3.32):

$$c_k = \mathbf{t}_\mathrm{A}{}^\mathrm{T}\mathbf{q}_{k,\mathrm{A}} \tag{3.36}$$

3.4.2.4 Selection of the Optimal Number of Factors in PCR

Unlike CLS or ILS, which calculate only one model, several PCR models can be calculated by varying the number of factors. In other words, we must decide into how many latent variables (columns of **T**) we will 'compress' the original matrix **R**. Compressing **R** too much (*i.e.* using a too small number of factors *A*) will introduce systematic error since the few new latent variables are not able to describe the main variations in the measured data; this is termed *underfitting* and it will also be commented on in Chapter 4. On the other hand, an unnecessarily large number of factors *A* (low data compression rate) will include factors that model noise, thus leading to overfitting, increased collinearity and bad predictions.

Hence the optimal number of factors (also termed *dimensionality* or *complexity* of the PCR model), which will give the best compromise between modelling enough systematic data and not modelling too much noise, must be found. There are different possibilities to select the optimal number of factors, among which the usual one consists in building models with an increasing number of factors and measuring their predictive ability using samples of known concentrations [4,5,21,22]. Usually, *A* corresponds to the first local minimum of the plot (predictive ability versus number of factors). Statistical tests for determining whether an additional factor is significant have also been described. These procedures are described in Chapter 4 for the PLS model, since they are the same as for PCR.

3.4.2.5 Considerations About PCR

PCR, being a factor-based model, has the best features of the ILS method and is generally more accurate and robust. First, PCR retains the advantage of ILS over CLS that only the concentration of the analyte of interest must be known in advance to calculate the model. Second, it overcomes the limitations of ILS regarding collinearity, variable selection and minimum number of samples. The key for all these advantages is the data compression step based on PCA. PCR can use all the spectral variables without being affected by collinearity, since all spectral variables are reduced to only a few scores (the columns in $\mathbf{T}_A$) which are orthogonal. Moreover, the number of calibration samples does not need to be as high as in ILS because the number of coefficients to be estimated is low (one for each latent variable entering the model). As a consequence, variable selection is not required in PCR for mathematical reasons. Finally, since the principal components with the smallest explained spectral variance (those which would produce the largest variance in the estimated coefficients) have been deleted [these are included in matrix **E** in eqn (3.30)], more reliable estimates of the model coefficients are obtained, noise effects are excluded (because they are relegated to unused factors) and hence a good predictive model can be obtained to calibrate atomic spectroscopic instruments.

There is still room for improvement in the PCR method. While the advantage of PCR over ILS is that the regression is performed on scores instead of on the original measured variables, the limitation is that the PCA factors are calculated independently of any knowledge of the concentration of the analyte of interest. Hence the scores that are used for regression, although they are related to the main variations in the spectral x-data, are not necessarily correlated with the property of interest. Imagine a situation in which the analyte of interest produces only a very small (but sufficient) variation in the spectra of the sample, while other constituents in the sample make the spectra vary much more. The principal components will describe the largest variations in the spectra and hence the scores will not be correlated with the concentration of the analyte of interest, unless we include the last principal components in the model to account for those small variations. Hence the usual top-down method can introduce irrelevant factors that degrade the predictive ability of the model. This is improved in the PLS method (see Chapter 4), in which the factors are not decided based on the x-variation only, but are calculated so that the sample scores on those factors are correlated as much as possible with the property of interest. Another point is that we commented that PCR does not need variable selection for mathematical requirements. However, the inclusion of non-informative features or sensors that contribute non-linearities can degrade the predictive performance of the model [5] and, hence, variable selection is also often used in order to improve the prediction ability of the PCR model.

3.5 Examples of Practical Applications

Only a small number of applications combining multiple linear regression and atomic spectrometry have been published. Most of them have been covered in two reviews [23,24], which readers are strongly encouraged to consult. Since most applications of MLR models have been studied together with the use of PLS models, the practical examples will be referred to in Chapter 4, in order to avoid unnecessary repetitions here.

3.6 Appendix

The singular-value decomposition (SVD) [25] decomposes a matrix **X** as a product of three matrices:

$$\mathbf{X} = \mathbf{UDP}^{\mathrm{T}} \tag{3.37}$$

where the column vectors of $\mathbf{U}(I \times R)$ and $\mathbf{P}(J \times R)$ are orthonormal and $\mathbf{D}(R \times R)$ is diagonal which contains (in the diagonal) the singular values of **X** in order $\sigma_1 \geq \sigma_2 \geq \ldots \geq \sigma_A \geq 0$. The rank of **X** is R, but the three matrices can be partitioned to include only the relevant A columns, to obtain $\mathbf{U}_A$, $\mathbf{P}_A$ and $\mathbf{D}_A = \mathrm{diag}(\sigma_1, \ldots, \sigma_A)$. Then, the rebuilt matrix

$$\mathbf{X}_A = \mathbf{U}_A\mathbf{D}_A\mathbf{P}_A{}^{\mathrm{T}} \tag{3.38}$$

is an updated version of **X** without the irrelevant variation (usually noise). This decomposition is also related to the scores **T** and loadings **P** obtained from PCA of $\mathbf{X}^T\mathbf{X}$ since the matrix of loadings **P** is the same in both cases and the matrix of scores is given by $\mathbf{T}=\mathbf{UD}$. Hence the columns in **T** are ordered in decreasing order of explained variance so the first factors (and thus a reduced number of variables) express the most important information in the data with no significant loss.

References

1. E. Sanchez and B. R. Kowalski, Tensorial calibration: I. First-order calibration, *J. Chemom.*, 2(4), 1988, 247–263.
2. K. S. Booksh and B. R. Kowalski, Theory of analytical chemistry, *Anal. Chem.*, 66, 1994, 782A–791A.
3. H. Martens and T. Næs, *Multivariate Calibration*, Wiley, New York, 1991.
4. B. R. Kowalski and M. B. Seasholtz, Recent developments in multivariate calibration, *J. Chemom.*, 5(3), 1991, 129–145.
5. E. V. Thomas, A primer on multivariate calibration, *Anal. Chem.*, 66, 1994, 795A–804A.
6. D. M. Haaland and E. V. Thomas, Partial least-squares methods for spectral analyses. 1. Relation to other quantitative calibration methods and the extraction of qualitative information, *Anal. Chem.*, 60(11), 1988, 1193–1202.
7. C. R. Rao and H. Toutenburg, *Linear Models: Least Squares and Alternatives*, Springer, New York, 1999.
8. S. Sekulic, M. B. Seasholtz, Z. Wang, B. R. Kowalski, S. E. Lee and B. R. Holt, Nonlinear multivariate calibration methods in analytical chemistry, *Anal. Chem.*, 65, 1993, 835A–845A.
9. A. Ben-Israel and T. N. E. Greville, *Generalized Inverses. Theory and Applications*, 2nd edn., Springer, New York, 2003.
10. J. Militký and M. Meloun, Use of the mean quadratic error of prediction for the construction of biased linear models, *Anal. Chim. Acta*, 227(2), 1993, 267–271.
11. D. A. Belsley, E. Kuh and R. E. Welsch, Regression Diagnostics: Identifying Influential Data and Sources of Collinearity, Wiley, New York, 1980.
12. P. F. Velleman and R. E. Welsch, Efficient computing of regression diagnostics, *Am. Stat.*, 35, 1981, 234–242.
13. D. W. Marquardt, Generalized inverses, ridge regression, biased linear estimation and nonlinear estimation, *Technometrics*, 12, 1970, 591–612.
14. M. Otto and W. Wegscheider, Spectrophotometric multicomponent analysis applied to trace metal determinations, *Anal. Chem.*, 57(1), 1985, 63–69.
15. A. Lorber, Error propagation and figures of merit for quantification by solving matrix equations, *Anal. Chem.*, 58(6), 1986, 1167–1172.
16. R. Manne, Analysis of two partial-least-squares algorithms for multivariate calibration, *Chemom. Intell. Lab. Syst.*, 2(1–3), 1987, 187–197.

17. R. Marbach and H. M. Heise, On the efficiency of algorithms for multivariate linear calibration used in analytical spectroscopy, *Trends Anal. Chem.*, 11(8), 1992, 270–275.
18. T. Næs and H. Martens, Principal component regression in NIR analysis: viewpoints, background details and selection of components, *J. Chemom.*, 2(2), 1988, 155–167.
19. H. Martens, T. Karstang and T. Næs, Improved selectivity in spectroscopy by multivariate calibration, *J. Chemom.*, 1(4), 1987, 201–219.
20. R. Leardi, Genetic algorithms in chemometrics and chemistry: a review, *J. Chemom.*, 15(7), 2001, 559–569.
21. R. Marbach and H. M. Heise, Calibration modeling by partial least-squares and principal component regression and its optimisation using an improved leverage correction for prediction testing, *Chemom. Intell. Lab. Syst.*, 9(1), 1990, 45–63.
22. P. Geladi and B. R. Kowalski, Partial least-squares regression: a tutorial, *Anal. Chim. Acta*, 185, 1986, 1–17.
23. J. M. Andrade, M. J. Cal-Prieto, M. P. Gómez-Carracedo, A. Carlosena and D. Prada, A tutorial on multivariate calibration in atomic spectrometry techniques, *J. Anal. At. Spectrom.*, 23, 2008, 15–28.
24. G. Marco, Improving the analytical performances of inductively coupled plasma optical emission spectrometry by multivariate analysis techniques, *Ann. Chim. (Rome)*, 94(1–2), 2004, 1–15.
25. G. W. Stewart, On the early history of the singular value decomposition, *SIAM Rev.*, 35, 1993, 551–566.

CHAPTER 4

Partial Least-Squares Regression

JOSE MANUEL ANDRADE-GARDA,[a] RICARD BOQUÉ-MARTÍ,[b] JOAN FERRÉ-BALDRICH[b] AND ALATZNE CARLOSENA-ZUBIETA[a]

[a] Department of Analytical Chemistry, University of A Coruña, A Coruña, Spain; [b] Department of Analytical and Organic Chemistry, University Rovira i Virgili, Tarragona, Spain

4.1 A Graphical Approach to the Basic Partial Least-squares Algorithm

This chapter discusses partial least-squares (PLS) regression. The objective is to introduce newcomers to the most common procedures of this multivariate regression technique, explaining its advantages and limitations. As in the other chapters of the book, we will not focus extensively on the mathematics behind the method. Instead, we will present it graphically in an intuitive way, sufficient to understand the basis of the algorithm. Although the plots offer a simplified view of the method, they are good tools to visualise it even for those without a mathematical background. In this respect, it is worth stressing again that although spectroscopists act mainly as end users of the multivariate techniques, they should know the capabilities and limitations of the calibration techniques, in the same way as they master what is behind (for instance) the extraction procedure linked to a slurry sampling methodology. Advanced references will be given for readers who wish to seek a deeper understanding of this method.

RSC Analytical Spectroscopy Monographs No. 10
Basic Chemometric Techniques in Atomic Spectroscopy
Edited by Jose Manuel Andrade-Garda

Published by the Royal Society of Chemistry, www.rsc.org

Throughout this chapter, the terms '*instrumental response*', '*independent variables*' or '*predictors*' (this last term is the preferred one) denote the atomic spectra, whereas '*dependent*', '*predictand*' or '*predicted variable*' (the second term is preferred) refer to concentration(s) of the analyte(s).

PLS regression was first developed by Hermann Wold in 1975 in the econometrics and social sciences arenas [1,2]. The classical regression methods suffered from severe problems when many collinear variables and reduced numbers of samples were available. After working unsatisfactorily with maximum likelihood methods, he proposed the more empirical PLS approach, based on local least-squares fits. Later, his son Svante Wold popularised PLS in chemistry and it became one of the most commonly used regression techniques in chemometrics. H. Wold developed the PLS algorithm on an intuitive, empirical basis without a deep statistical corpus sustaining it. The theoretical background has been developed steadily since then, although some mathematicians are still reluctant to accept it and stress its drawbacks [3,4]. Nevertheless, molecular spectroscopists and industrial scientists/technicians, and also many other scientists in fields such as economy, marketing and structure–activity relationship, constantly report useful predictive models. More importantly, PLS can solve many practical problems which are not easy to treat by other means. The original algorithm was modified slightly by Svante Wold to suit data from science and technology better [5].

As with the regression methods in the previous chapter, PLS will handle a set of spectral variables (the absorbances recorded while the atomic peak is registered, *e.g.* every 0.01 s) in order to predict a given property (concentration of the analyte) of the sample. This corresponds to typical situations in atomic spectrometry such as graphite furnace measurements or the widely applied flow-injection analysis–atomic absorption spectrometry (FIA–AAS) coupling (which was discussed in Chapter 1). To generalise the discussions, in all those cases the overall signal registered by the detector is assumed to be constituted by several overlapping peaks or absorption bands, one corresponding to the analyte and the others corresponding to interferences (if present) or different atomisation processes (see Figure 4.1 for a general example). It follows that atomic spectrometric signals can be treated exactly as any other molecular spectroscopic signal, for which multivariate regression has shown long-standing success.

In PLS, the spectral measurements (atomic absorbances or intensities registered at different times or wavelengths) constitute a set of independent variables (or, better, predictors) which, in general, is called the **X**-block. The variable that has to be predicted is the dependent variable (or predictand) and is called the **Y**-block. Actually, PLS can be used to predict not only one y-variable, but several, hence the term 'block'. This was, indeed, the problem that H. Wold addressed: how to use a set of variables to predict the behaviour of several others. To be as general as possible, we will consider that situation here. Obvious simplifications will be made to consider the prediction of only one dependent **y** variable. Although the prediction of several analytes is not common in atomic spectroscopy so far, it has sometimes been applied in molecular spectroscopy. Potential applications of this specific PLS ability may be the

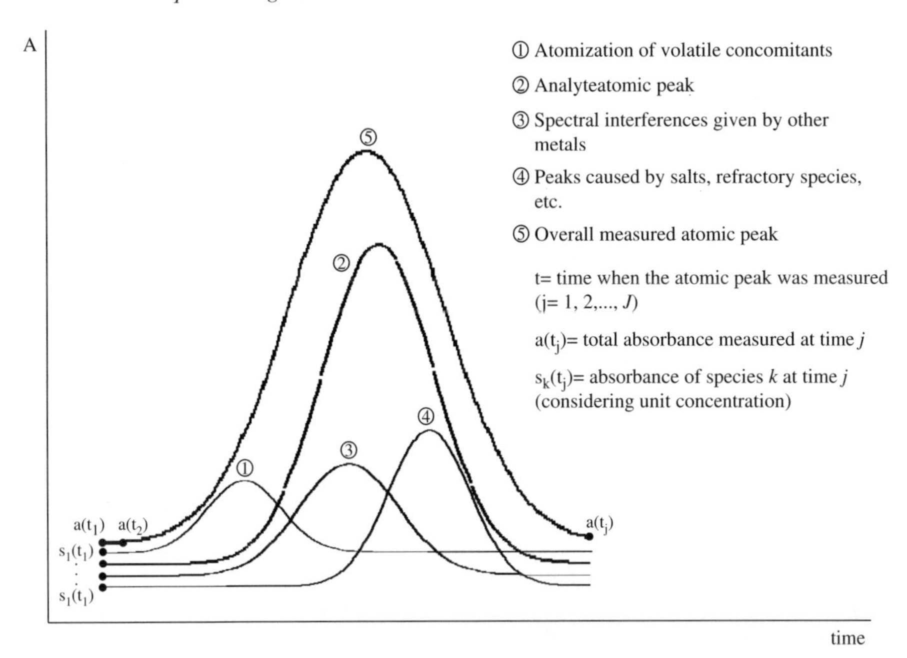

Figure 4.1 General representation of the overall atomic signal registered on the detector as composed of different overlapping signals. Reproduced from Ref. [12], with permission from the Royal Society of Chemistry.

quantification of several metals from complex data matrices measured by inductively coupled plasma mass spectrometry (ICP-MS) (involving many intensities from different mass/charge ratios), ICP-MS–MS measurements, isotope studies or 'simultaneous' FIA–AAS measurement of several analytes which yield complex signals.

In the previous chapter, it was commented that the ordinary least-squares approach applied to multivariate data (multivariate linear regression, MLR) suffered from serious uncertainty problems when the independent variables were collinear. Principal components regression (PCR) could solve the collinearity problem and provide additional benefits of factor-based regression methods, such as noise filtering. Recall that the PCR 'compressed' the original **X**-block (*e.g.* matrix of absorbances) into a new block of scores **T**, containing fewer variables (the so-called factors, latent variables or principal components) and then regression was performed between **T** and the property of interest (the **y**-variable). It was also explained that PCR, however, does not take into account the values **y** we want to predict to define the factors. Hence the factors **T** used for regression may not be the most correlated with the property of interest. A more advanced approach is to decompose the original **X**-block into some abstract factors **T** which explain the maximum information on the predictors while being, at the same time, correlated as much as possible with the **Y** (the concentrations of the analytes). These abstract factors should, as in PCR, also solve the multicollinearity problems.

Multicollinearity problems can be solved from the very beginning by considering the principal components (PCs) of the two data sets (**X** and **Y**) instead of the original absorbances. Recall that the PCs are independent of each other (orthogonal) and that a reduced number of them is commonly adequate to describe most of the information (variability) of the initial data sets. Both properties [orthogonality and high percentage of explained variability (variance) in the data] will be used here. For practical reasons (discussed in Section 4.3.3), we first mean centre the data. Hence, in Figure 4.2 $\bar{\mathbf{x}}^T$ and $\bar{\mathbf{y}}^T$ are the vectors containing the averages of each column. The superscript T (^{T}) means the transpose of the matrix or vector. Also in the figure, **T** and **U** are the PCA scores of the **X** (absorbances, intensities, *m*/*z* ratios, *etc.*) and **Y** (concentration) matrices; and **P** and **Q** are their loadings, respectively. **E** and **F** are the error matrices, which are the differences between the original data and the data accounted for in **T** and **U**. The more factors are included in **T** and **U**, the lower are the values in **E** and **F**. Up to this point, this plot would be a common PCA decomposition (the one PCR uses) but, as was noted in the previous chapter, PCR might not yield satisfactory solutions, because **T** may not be optimally correlated with the property of interest. Therefore, our problem is to define **T** and **U** in such a way that the maximum of the information of **X** and **Y** is extracted and, at the same time, related to each other.

How can one relate **T**, **U**, **P** and **Q** in such a way? First, our previous knowledge of the analytical technique suggests that these blocks of data, which represent two different aspects of the same true materials (solutions, slurries,

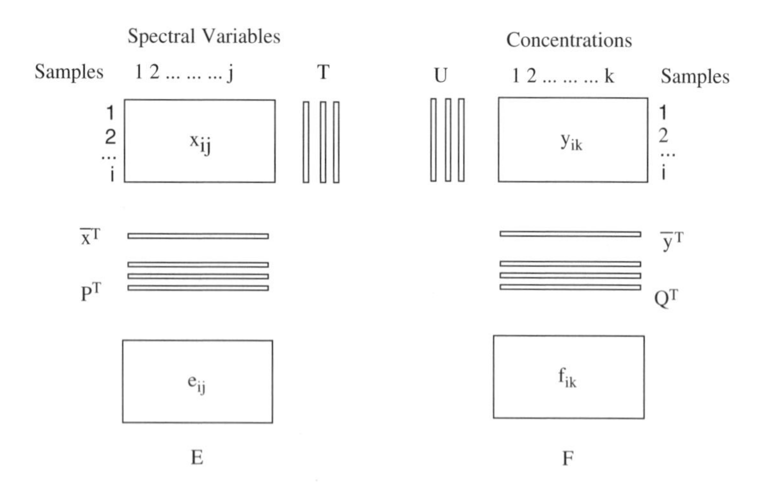

Figure 4.2 Starting point: original data and how a traditional PCA would decompose them. Since this is not an optimal solution, we need to define the scores and loadings in some other way that not only describe the main information present on **X** and **Y**, but at the same time relate them.

etc.) must be related (we do not know how, but they must!). The algorithm developed by H. Wold (called 'nonlinear iterative partial least squares', NIPALS; sometimes it is also termed 'noniterative partial least squares') started from this idea and was formulated as presented below. The following ideas have roots in works by Geladi (and co-workers) and Otto [6–8]. We considered seven major steps:

Step 1: Let us take any **y**, for instance the *y*-variable with the largest variance. This is not mandatory since the procedure will iterate until convergence (i.e., a solution is used as the starting point to obtain a new improved solution and so on). Of course, if there is only one *y*-variable (*e.g.* the concentrations of an analyte), you select that one. Grossly speaking, this **y** will act as a first, preliminary, 'scores vector' for the **Y**-block **u**. Use this scores vector to calculate the 'loadings' of **X** (recall that data = scores × loadings). Since these are not the true loadings, we changed the name of the matrix from **P** to **W** (*i.e.* from 'loadings' to 'loading weights' for the **X**-block, which is a more correct name); see Figure 4.3. It is worth noting that in the following explanations 'vector' will denote a 'column vector', as it is commonly stated, for example **x**. Hence a horizontal vector, or 'row', will be denoted by its transpose, for example $\mathbf{x}^T$. Mathematically:

$$\text{Let } \mathbf{u}_1 = \text{any } \mathbf{y} \tag{4.1}$$

where the subscript 1 denotes that these calculations are for the first factor (or latent variable). Next, calculate the weights of the **X**-block:

$$\mathbf{w}_1{}^T = \mathbf{u}_1{}^T\mathbf{X}/\mathbf{u}_1{}^T\mathbf{u}_1 \tag{4.2}$$

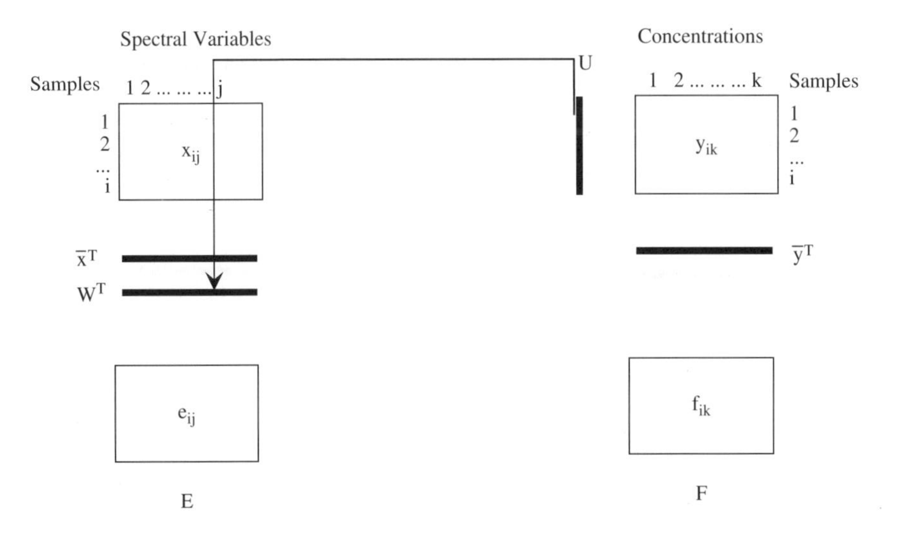

Figure 4.3 Step 1 of the NIPALS algorithm: mean centring, starting point and first calculations.

and scale them to length one:

$$\mathbf{w}_1{}^{T} = \mathbf{w}_1{}^{T}/||\mathbf{w}_1|| \tag{4.3}$$

where $||\mathbf{w}_1|| = (\mathbf{w}_1{}^{T}\mathbf{w}_1)^{\frac{1}{2}}$ is the module of column vector $\mathbf{w}_1$. This scaling makes all weights lie in the 0–1 range.

Step 2: Now we can define the **X**-scores using the original absorbances (**X**) and these just-calculated '**X**-loadings' (strictly, the **X**-weights); see Figure 4.4.

$$\mathbf{t}_1 = \mathbf{X}\mathbf{w}_1 \tag{4.4}$$

Step 3: The **X**-scores are used to calculate the **Y**-loadings; see Figure 4.5.

$$\mathbf{q}_1{}^{T} = \mathbf{t}_1{}^{T}\mathbf{Y}/\mathbf{t}_1{}^{T}\mathbf{t}_1 \tag{4.5}$$

Now, since data=scores × loadings, we can recalculate the first **Y**-scores vector; see Figure 4.6.

$$\mathbf{u}_{1,\mathrm{new}} = \mathbf{Y}\mathbf{q}_1/\mathbf{q}_1{}^{T}\mathbf{q}_1 \tag{4.6}$$

The new scores vector, $\mathbf{u}_{1,\mathrm{new}}$ ('white column' in Figure 4.6) will be a better estimation than our preliminary estimation ('black column' in Figure 4.6). Next, this newly calculated **u** vector is used as the input to start again the process above. After several iterations (the procedure is very fast), all vectors will converge to a stable solution (convergence is usually set by the software according to a rounding error) and we can finish the calculation of the first factor.

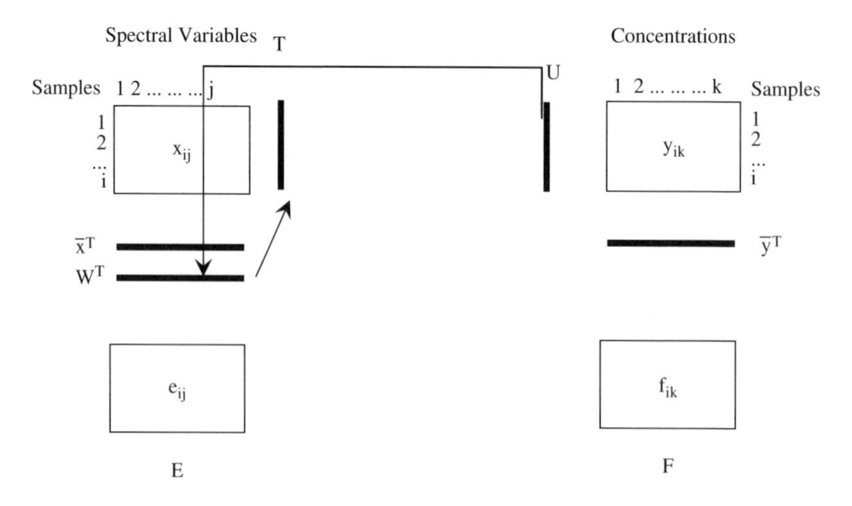

Figure 4.4 Step 2 of the NIPALS algorithm: scores for the **X**-block, taking into account the information in the concentrations (**Y**-block).

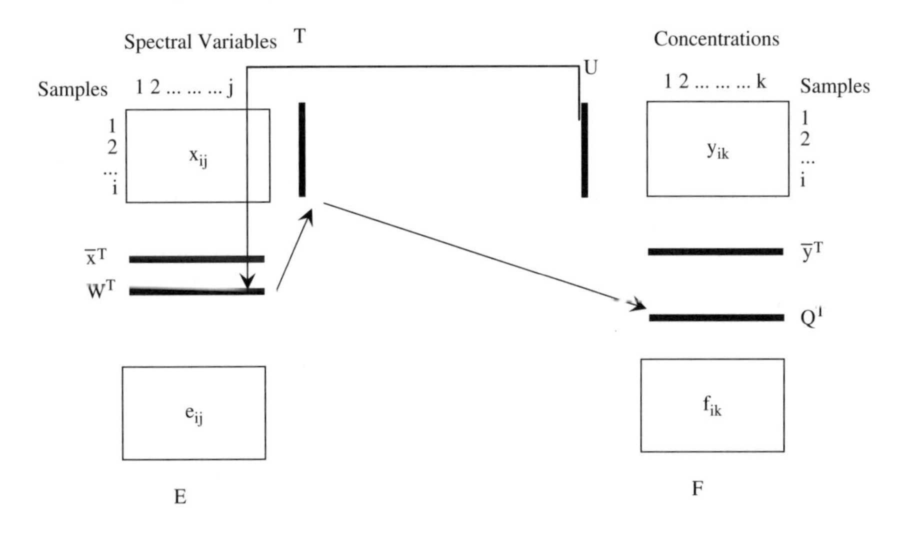

Figure 4.5 Calculation of the **Y**-loadings taking into account the information on the **X**-block.

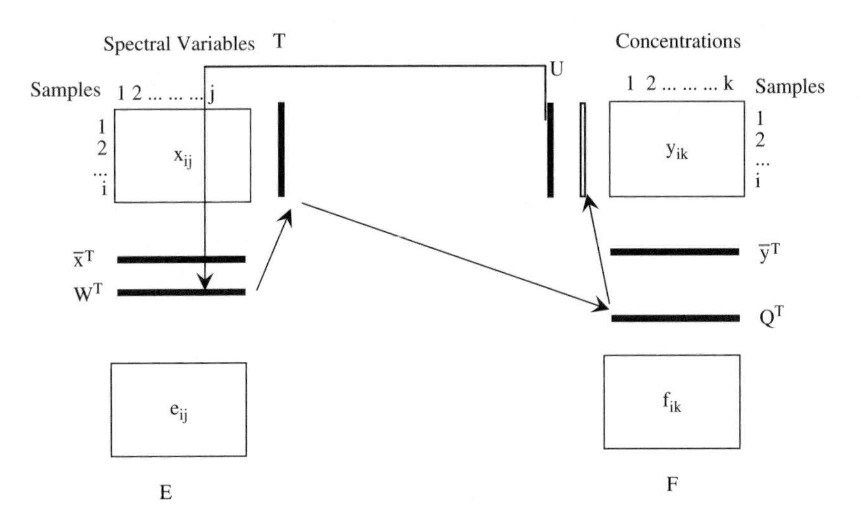

Figure 4.6 Recalculation of the first **Y**-scores vector, once the loop was closed.

Although these factors are fairly similar to those described in Chapter 3, there are some relevant differences. In effect, in PCA (and hence in PCR), the factors explain most of the variance in the **X**-domain regardless of whether such variance is or is not related to the analyte, whereas in PLS they are calculated so that they explain not only variance in the spectral variables but also variance which is related to the property of interest. A typical example might be a series of graphite furnace measurements where a baseline shift appears due to some variation in the temperature of the furnace because of a drift in the power supply. The variance associated with the baseline shift will generate a factor that should not be considered to predict the concentration of the analyte.

Step 4: Nevertheless, so far we 'only' extracted information from the **X**- and **Y**-blocks. Now a regression step is needed to obtain a predictive model. This is achieved by establishing an 'inner' relationship between **u** (the scores representing the concentrations of the analytes we want to predict) and **t** (the scores representing the spectral absorbances) that we just calculated for the **X**- and **Y**-blocks. The simplest one is an ordinary regression (note that the 'regression coefficient' is just a scalar because we are relating two vectors):

$$\hat{\mathbf{u}}_1 = b_1\mathbf{t}_1 \tag{4.7}$$

so that the inner regression coefficient is calculated as

$$b = \mathbf{u}_1{}^{\mathrm{T}}\mathbf{t}_1/(\mathbf{t}_1{}^{\mathrm{T}}\mathbf{t}_1) \tag{4.8}$$

where the hat (^) means predicted or estimated and the subscript 1 refers to the factor being considered. Figure 4.7 shows how this is performed. Experience demonstrates that this simple linear relationship is sufficient in most situations. Further developments considered a more complex inner relationship (*e.g.* a second-order polynomial) in order to take account of strong nonlinearities in the data set. This was done by using splines or local regression [9–11]. Nevertheless, polynomial PLS have not, in general, outperformed linear PLS unless severe nonlinearities exist (which is seldom the case) [12–17]. Hence the discussions presented in this chapter will be restricted to the common 'linear' PLS algorithm.

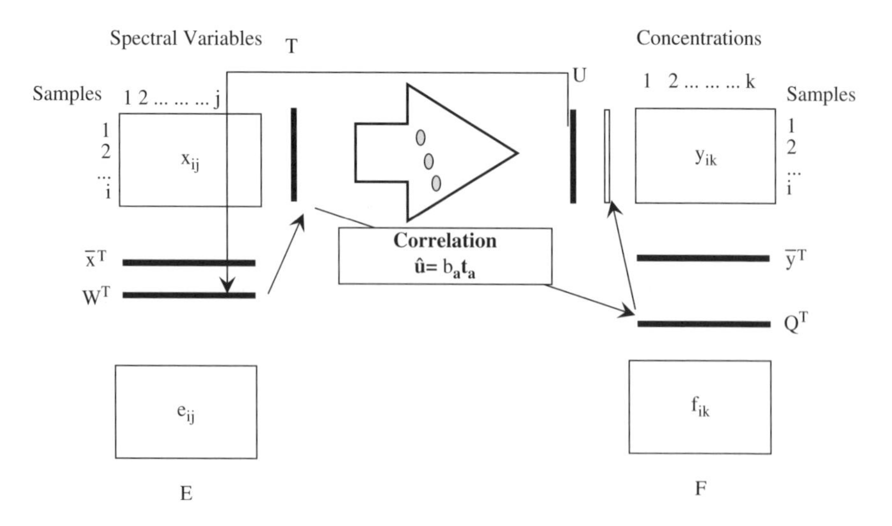

Figure 4.7 Setting the PLS model by calculating the regression coefficients between **t** and **u**.

Step 5: Finally, we can calculate the loadings of the **X**-block (recall that in step 1 we resourced to **X**-block 'loading weights') as

$$\mathbf{p}_1{}^{\mathrm{T}} = \mathbf{t}_1{}^{\mathrm{T}}\mathbf{X}/\mathbf{t}_1{}^{\mathrm{T}}\mathbf{t}_1 \tag{4.9}$$

Step 6: In order to extract the second factor (or latent variable), the information linked to the first factor has to be subtracted from the original data and a sort of 'residual' matrices are obtained for the **X**- and **Y**-blocks as

$$\mathbf{E} = \mathbf{X} - \mathbf{t}_1\mathbf{p}_1{}^{\mathrm{T}} \tag{4.10a}$$

$$\mathbf{F} = \mathbf{Y} - b\mathbf{t}_1\mathbf{q}_1{}^{\mathrm{T}} \tag{4.10b}$$

Now **E** and **F** do not contain the information explained by the first factor. These matrices are used as the new **X** and **Y** in Step 1. The overall process is repeated until the desired number of factors indicated by the user has been extracted. For each factor (latent variable), **p**, **w** and **q** vectors are obtained, which can be gathered into matrices **P**, **W** and **Q**. When only a *y*-variable is predicted, the algorithm does not iterate for estimating **u**.

Step 7: Note that from eqn (4.8), a coefficient was calculated to relate the scores of the first factor to the property of interest. Similar coefficients are required for each of the *A* factors included in the PLS model. Hence the prediction of the concentration of the analyte of interest in an unknown sample will go through first calculating the scores of that sample in the model and then applying the estimated *b*-coefficients.

It is also possible to derive a vector of regression coefficients that applies directly to the (preprocessed) measured data. This vector can be calculated as

$$\mathbf{b} = \mathbf{W}(\mathbf{P}^{\mathrm{T}}\mathbf{W})^{-1}\mathbf{q} \tag{4.11a}$$

with $\mathbf{q} = (\mathbf{T}^{\mathrm{T}}\mathbf{T})^{-1}\mathbf{T}^{\mathrm{T}} \cdot \mathbf{y}$, when only one analyte is to be predicted (PLS1) and as

$$\mathbf{B} = \mathbf{W}(\mathbf{P}^{\mathrm{T}}\mathbf{W})^{-1}\mathbf{Q}^{\mathrm{T}} \tag{4.11b}$$

for the PLS2 case, with $\mathbf{Q} = (\mathbf{T}^{\mathrm{T}}\mathbf{T})^{-1}\mathbf{T}^{\mathrm{T}} \cdot \mathbf{Y}$. Here the superscript ($^{-1}$) denotes the inverse of the matrix. From this matrix (vector if only an analyte is predicted) of regression coefficients, the concentration of the analyte(s) in a new sample can be predicted as

$$\hat{y} = \mathbf{x}^{\mathrm{T}}\mathbf{b} \tag{4.12a}$$

for one analyte and as

$$\hat{\mathbf{Y}} = \mathbf{x}^{\mathrm{T}}\mathbf{B} \tag{4.12b}$$

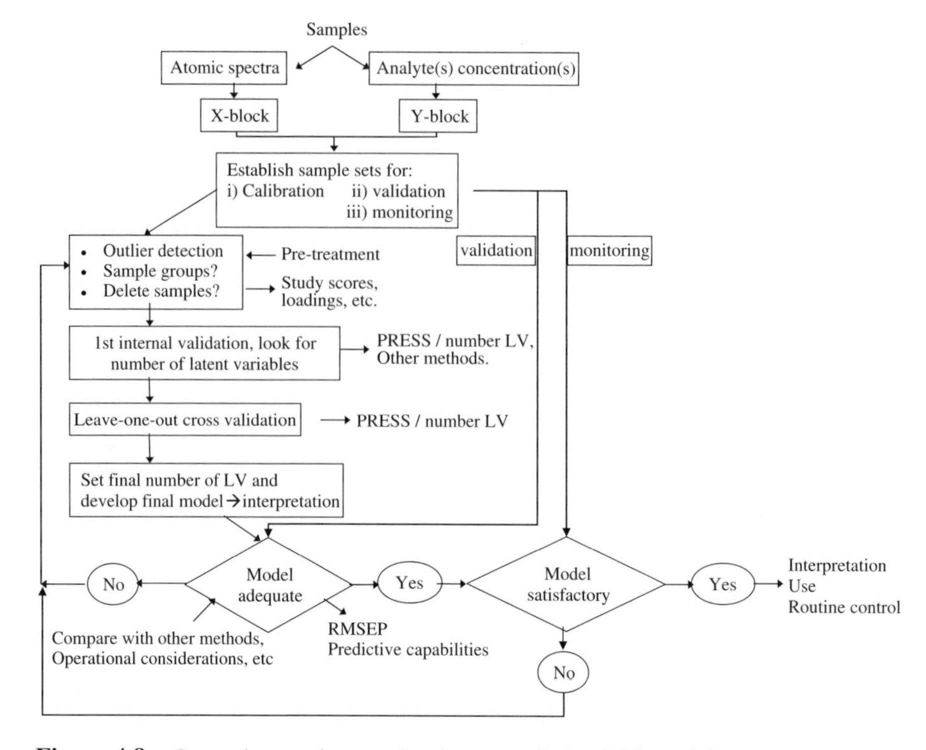

Figure 4.8 General procedure to develop a predictive PLS model.

for the PLS2 algorithm, where **x** is the vector with the preprocessed experimental absorbances (intensities, *etc.*) of the new sample.

Dimensionality reduction comes from the fact that only A latent variables (fewer than the original number of x-variables) are considered to develop the model and, thus, matrices **W**, **P** and **T** contain a reduced number of rows or columns.

The main steps required to obtain a PLS model are schematised in Figure 4.8 and considered in the following sections.

Once the basis of the PLS algorithm has been presented, it is easier to understand the advantages of this multivariate regression technique over the simpler ones MLR and PCR. Some of these advantages are already obtained in PCR. They can be summarised as follows:

1. Collinearity among the x-variables (*e.g.* absorbances at consecutive times of the atomic peaks) is not a problem. The latent variables calculated in PLS, like the PCs, resume the most relevant information of the whole data set by taking linear combinations of the x-variables. The scores of the **X**-block are orthogonal (the factors are independent of each other) and the corresponding weights are orthonormal (their maximum magnitude is 1). As for PCR, this means that the information explained by

one factor (*i.e.* the underlying phenomenon causing some feature on the atomic peak) is different from the phenomenon involved in other factor(s). In contrast, the scores and loadings associated with the **Y**-block do not hold this relationship [5]. This is not relevant in applications where only an analyte has to be predicted and, therefore, we will not go into further details.

2. PLS is very efficient in using only the information in **X** that is related to the analyte. Hence it will lead to more parsimonious models than PCR, that is, in general, PLS will require fewer factors than PCR to obtain a model with a similar predictive performance.
3. Since a reduced number of factors is considered, there are no matrix inversion problems related to the size of the matrices being inverted.
4. PLS models lead to relatively simple but very informative graphs showing the relationships among samples, among variables and between samples and variables. Chemical/physical interpretations of these graphs (although this is not always simple) may provide interesting insights into the system under study that were not evident by observing the measured **X**-data alone. Unfortunately, this important advantage has not always been fully exploited when PLS models are reported. Because PLS is based on factors, several simple diagnostic statistics can be defined and diagnostic plots can help to evaluate whether outliers are influencing the model, whether some variables are more relevant than others to define the regression or whether a curved distribution of samples is visualised, among others.
5. If severe nonlinearities might be present, the linear inner relation can be modified to a quadratic or cubic one. This strong nonlinear situation might arise whenever problems occur on the detector or monochromator, malfunction of the automatic sampler in ETAAS, strong influence of the concomitants on the signal, when the linear range for the analyte is too short or when LIBS, LA–ICP-MS measurements or isotope dilution are carried out (see Chapter 1 for more possibilities).
6. PLS converges fairly fast and is broadly implemented in commercial software.
7. A very important advantage of PLS is that it supports errors in both the **X**- and **Y**-blocks. Questions on whether the **y**-values (the concentrations of the analyte in the standards) can be wrong may appear exotic but the real fact is that nowadays we have such powerful instruments, so precise and so sensitive that the statement that 'there are no errors in the concentrations' can be discussed. As Brereton pointed out [18], the standards are, most of the time, prepared by weighing and/or diluting and the quality of the volumetric hardware (pipettes, flasks, *etc.*) has not improved as dramatically as the electronics, accessories and, in general, instruments. Therefore, the concentration of the analyte may have a non-negligible uncertainty. This is of special concern for trace and ultra-trace analyses. Thus, MLR cannot cope with such uncertainties, but PLS can because it performs a sort of 'averaging' (*i.e.* factors) which can remove most

random noise. More detailed and advanced discussions on this topic can be found in DiFoggio's paper [19], which considered different aspects of error propagation and PLS feasibility. Although the last part of the paper is difficult for beginners, the first pages can be followed easily and it shows several important hints to keep in mind. Discussions were centred on molecular spectrometry, but are of general application. In Chapter 1, several discussions were given on the uncertainty of the standards, mainly for isotope dilution analysis, and some more details can be seen in a paper by Laborda *et al.* [20] regarding a FIA–HPLC–MS-ICP system.

8. PLS can be applied to predict various analytes simultaneously (this is known as PLS-2 block modelling, which is the algorithm described in the previous section), although this is only recommended when they are correlated. Otherwise it is advisable to develop a PLS model for each analyte (*i.e.* a single **y**-vector per model).

These relevant issues will be considered in more detail in the following sections. There, the main steps to develop a PLS model will be discussed with a practical example.

4.2 Sample Sets

The quality of a model depends on the quality of the samples used to calculate it (or, to say it using the univariate approach, the quality of any traditional univariate calibration cannot be better than the quality of the standards employed to measure the analyte). Although this statement is trivial, the discussion on how many samples and which samples are required to develop a good predictive model is still open, so only general comments will be given. Below, we consider that the quality of the measurement device fits the purpose of the analytical problem.

This problem is familiar to atomic spectroscopists because we rarely can reproduce the matrix of the original samples (sometimes this is completely impossible; see Chapter 1). How can we know in advance how many concomitants are in our sample, and what their atomic profiles are?' The answer is that, simply, we cannot anticipate all this information.

Further, in atomic spectrometry we must face the serious problem that the behaviour (atomisation/excitation characteristics) of the analyte in the calibration samples should be the same as in the future unknown samples where the analyte of interest has to be quantified, otherwise peak displacement and changes of the peak shape may cause serious bias in the predictions. Fortunately, many atomic techniques analyse aqueous extracts or acid solutions of the (pretreated) samples and current working procedures match the amount of acids in the calibration and treated samples, so the matrices become rather similar. Current practices in method development involve studying potential interferents. The analyte is fixed at some average concentration (sometimes studies are made at different concentrations) and the effects of a wide number of potential interferents are tested. They include major cations, anions and other metals. Disappointingly, matching concomitants is not so straightforward and some doubts remain about the 'similarity' of the calibration and unknown samples. The golden rule here is that standards and future samples should be treated in exactly the same manner. In this sense, multivariate calibration is a step within the analytical process and this should be considered when calculating figures of merit, bias, *etc.*

Atomic spectroscopists can, therefore, take advantage of the information derived from their studies on the interference of the concomitants and design a working space where the samples should be contained (several methodologies were presented in Chapter 2 to obtain such designs). In a classical univariate approach, most analysts would restrict the applicability of the analytical procedure to situations where the interferents become limited to narrow, predefined ranges (where they do not interfere with the analyte signal). Although this has been an effective solution in many studies, we suggest profiting from the advantages of the multivariate regression models, particularly of PLS, to open up new possibilities.

Two issues can be recalled here: (i) the PLS algorithm will extract and use, from the overall bulk of raw absorbances, only the spectral information that is related to the concentration of the analyte; and (ii) PLS benefits from the use of as many standards as possible and as different as possible in order to model the characteristics that determine the – hidden – relationship between the spectra and the analyte. Thus the calibration standards for PLS regression should be designed in such a way that they embrace not only the (linear) range of the analyte but also the usual ranges of the interfering concomitants. It is possible, therefore, to include calibration samples that resemble the different samples that may arrive at the laboratory. This is especially important in those situations where the ranges of the concomitants cannot be under the control of the analyst but they vary within some 'reasonably common' ranges. Typical examples include environmental analyses, industrial control of raw materials, *etc.*

Then, the particular experimental setup to prepare a set of calibration samples can be deployed following any of the procedures explained in Chapter 2 (experimental design and optimisation). We found it very useful to set a Plackett–Burman design at each level of the analyte concentration but, of course, other possibilities exist. The important point here is to obtain a sufficiently large number of calibration samples so that any (reasonable) combination of concomitants and analyte that might appear in the problem samples has been considered previously in the calibration. This is the general rule that '*you should model first what you want to predict afterwards*' [18].

As an example, Figure 4.9 shows the overall set of atomic peaks of the calibration samples stacked on the same plot. This corresponds to a study where antimony had to be determined by ETAAS in natural waters (drinking water, wells, springs, *etc.*) [12]. An extensive study of potential interferents [21] revealed that several ions may interfere in the atomisation process. A PLS multivariate study was carried out as an alternative to classical approaches and standard solutions were prepared with various concentration of Sb and the

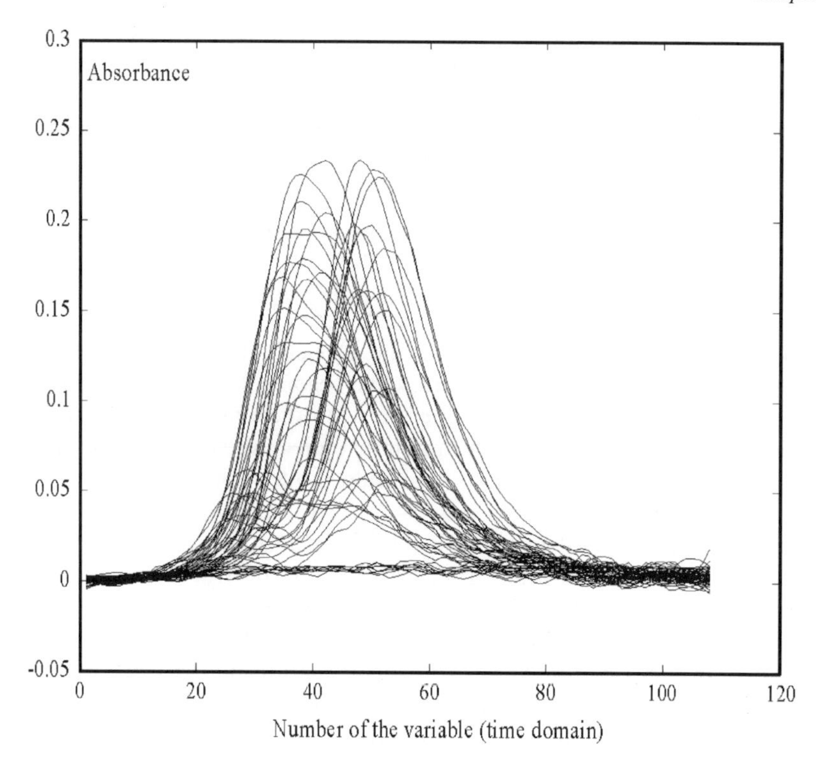

Figure 4.9 General appearance of the overall set of atomic peaks of the calibration standards used to determine Sb in natural waters.

concentrations of the seven ions affecting the atomic signal (Ca^{2+}, Fe^{3+}, Na^{+}, Mg^{2+}, Cl^{-}, SO_4^{2-} and PO_4^{3-}).

Nevertheless, obtaining a calibration model with a good fit for the calibration samples is not an indication that the model will be useful for predicting future samples. The more factors are considered in the PLS model, the better is the fit, although, probably, the prediction of new samples is worse. Hence another set of known samples, the *validation set*, is required to test the models. This can be a small group of samples prepared within the overall domain of the calibration samples (remember that extrapolation is not allowed in regression models, not even in the univariate models), with different concentrations of the analyte (we want to test that the model quantifies it correctly) and different concentrations of the concomitants. Then, if the predictions are good, we can be reasonably sure that the model can avoid the undesirable influence of the concomitants.

4.3 Data Pretreatment

Data pretreatment or data preprocessing is the mathematical manipulation of the data prior to the main analysis. Preprocessing is very important in multivariate regression to remove or reduce irrelevant sources of variation (either random or systematic) [22] so that the regression step can focus better on the important variability that must be modelled. Unfortunately, there is no golden rule of thumb ('*the rule is that there is no rule*'). A particular treatment can improve a particular model for a particular property and particular **X**-data, but prejudice another model for different types of data. Hence each case needs particular studies and different trials are essential to find the most appropriate data pretreatment (this usually involves an iterative process between the analyst and the preprocessing options). The best selection will be the one leading to the best prediction capabilities. There are several well-known situations for which certain established treatments commonly lead to improved results. This topic has been developed widely for molecular spectrometry. In the atomic spectrometry field, preprocessing has not been considered in great detail and still needs to be studied further.

4.3.1 Baseline Correction

The baseline of the atomic peak can steadily increase its absorbance because of instrumental malfunction, drift, a sample constituent giving rise to proportional interferences or an atomisation problem. The general appearance is that the atomic peaks are 'stacked', but this arrangement is not related to the concentration of the analyte. Baseline correction is a general practice in atomic spectrometry to measure accurately the height or the total area under the peak. In multivariate analysis, the procedure is the same in order to start all atomic peaks in the same absorbance position, in general at zero scale. The most basic baseline correction is '*offset correction*', in which a constant value is subtracted to all absorbances that define the atomic peak (Figure 4.10a).

4.3.2 Smoothing

Signal smoothing is also frequent in atomic spectrometry because random fluctuations add noise to the atomic peak profile (see, *e.g.*, Figure 4.10b). Since noise is random and does not belong to the true signal of the analyte, we would like to filter it. A note of caution is needed here. Smoothing reduces the noise and the signal; too much smoothing and the signal is reduced in intensity and resolution, too little smoothing and noise remains. In Brereton's words [23], a key role of chemometrics is to obtain as informative a signal as possible after removing the noise.

A variety of smoothing methods can be found in introductory textbooks [*e.g.* 8,22,23]. The two classical, simple forms of smoothing are the 'moving-average window' smoothing and the Savitzky–Golay filter. They are implemented in most commercial software packages managing atomic data. Both techniques consider that the random noise can be averaged out (or, at the very least, its magnitude reduced significantly) by averaging a number of consecutive signal points. How this average is done defines the different smoothing procedures.

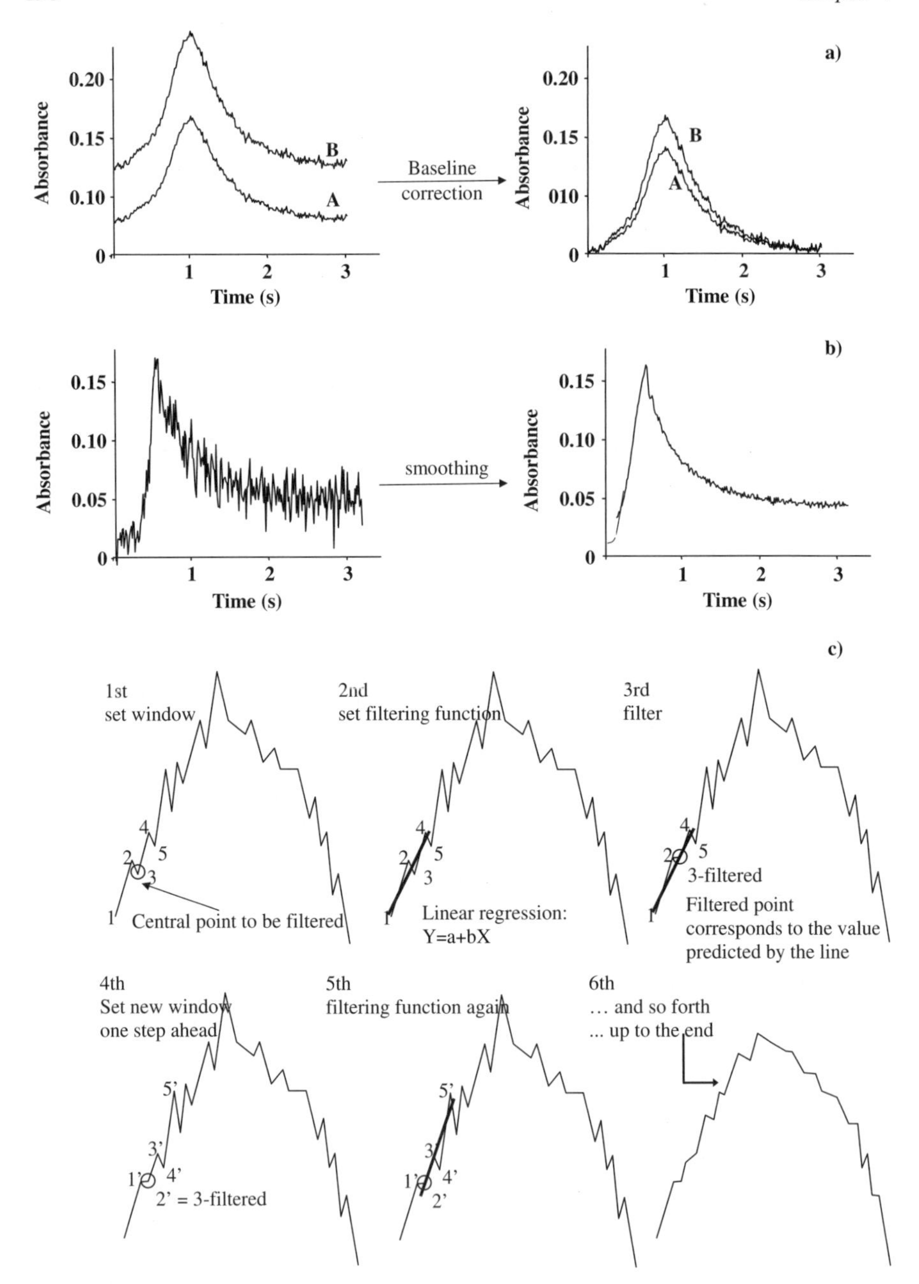

Figure 4.10 Graphical representation of (a) baseline correction, (b) smoothing and (c) simplified view of a Savitzky–Golay filter (moving window, five points per window, linear interpolation).

The classical *moving-average window* considers z consecutive values (the '*window*') in such a way that the value in the middle is replaced by the average of the z values considered in the window. Here, z is an odd number that can range anywhere from 3 to 5, 7, 11, *etc*. Once the absorbance at the middle of the window has been smoothed, the window is moved one position (for example, to the right) and the new absorbance in the middle of the new window is filtered, and so forth until the window has passed over all the atomic peak (see Figure 4.10c). In general, the wider is z, the more pronounced the smoothing and the larger the (possible) distortion of the peak. Hence the length z has to be tested for each particular situation in order to achieve a trade-off between noise reduction and peak distortion [24].

If z is expressed as $(2d+1)$, the moving-average window will not filter the first d values of the peak or the last d values (no possibility of moving the window), so the two extremes of the atomic peak cannot be filtered. Further, moving-average windows assume that the behaviour of the absorbances within the window follows a straight line (this is implicit in the use of the average), but it is intuitive that frequently even small parts of the atomic peak are often best approximated by curves. Considering this, Abraham Savitzky and Marcel Golay published in 1964 [25] a filtering method based on local polynomial regression. Their paper is one of the most widely cited publications in the journal *Analytical Chemistry* and is classed by that journal as one of its '10 seminal papers', saying *'it can be argued that the dawn of the computer-controlled analytical instrument can be traced to this article'* [26]. Savitzky and Golay's original paper contained several typographical errors that were corrected by Steinier, Termonia and Deltour [27].

The Savitzky–Golay method works like the moving-average window but performs a local polynomial regression (of degree n) on the z points included in the window. This means that each time the window is moved, a new fit is performed to the points into the window. The value of the polynomial in the centre of the window will substitute the noisy value. The degree of the polynomial is fixed before running the filter and most often a polynomial of degree 2 (quadratic) or 3 (cubic) provides sufficiently satisfactory results. The main advantage of this filter is that it preserves the features of the peak such as relative maxima, minima and width, which are usually 'flattened' by other averaging techniques such as the moving-average window.

As a rule of thumb, Otto [8] suggested three main guidelines to select the most appropriate filter: (i) around 95% of the noise is filtered in just one application, so single smoothing is often sufficient, (ii) the filter width should correspond to the full width at half-maximum of a peak and (iii) the filter width should be small if the height rather than the area is used.

4.3.3 Mean Centring and Autoscaling

PLS is based on extracting factors (latent variables) that, like the PCs – see Chapter 3 – share a common origin (the zero value of the scales of the factors/PCs). How to set this origin is not trivial, mostly because neither PCA (PCR) nor PLS is invariant to changes in the scales of the variables. The simplest possibility is to set the origin of the factors as the origin of the experimental variables (the origin of all time–absorbance variables is the null value), but sometimes this is not optimal, mainly when the variables do not have the same origins or share a common scale (*e.g.* zero for pH is not the same as zero for an absorbance) [28].

Another option is mean centring, which removes the part of the data that is common to all the samples and sets the origin of the latent variables at the centre of the multivariate point swarm (*i.e.* the origin of all factors is at the average values of all variables for all samples). Mean centring is done by subtracting from each column of the data matrix the average of that column (variable). Then, the resulting data matrix, from which the PLS model will be developed, will have positive and negative values (see Figure 4.11a). In regression, the model (for mean-centred data) does not have an intercept. In factor-based methods, mean centring makes the model focus on the main variations in the data instead of focussing on describing the raw data themselves. Without mean centring, the first latent variable would go from the origin of coordinates of the original variables to the centre of the data, *i.e.* the first factor would describe the mean of the data; the following latent variables would describe what is not described by the first latent variables, *i.e.* what is not 'the mean'. For this reason, often the loadings of the first latent variable for the raw data resemble the mean of the data set and the second latent variable of raw data resembles the first latent variable for mean-centred data. In that case, mean-centring makes the interpretation of the model easier.

Autoscaling implies that each column is mean-centred and also divided by its standard deviation, very useful for situations where the variables are measured on different scales. Too different scales mean that the variances are too different and this would bias the model – factors – towards the variables with the highest variances, even though they might not necessarily be the most relevant ones to solve the problem at hand. Autoscaling is not always advantageous in spectral studies. The reason is that it converts all variables to the same variance (unity variance, Figure 4.11b). This results in the most informative variables describing the main signal being given the same importance as the variables that, mainly, describe noise. This makes it more difficult for the regression algorithm to find the important systematic variation in the data and random variability will enter into the most important factors. This will reduce the prediction ability of the model. Ideally, noise should be restricted to those factors that are not used in the calibration model. Nevertheless, autoscaling would be useful whenever the spectroscopist decides that a spectral variable has no more significance or relevance than others [23].

It has to be stressed that the selection of the most adequate scaling mode (or not to scale at all) has to be done on a trial-and-error basis.

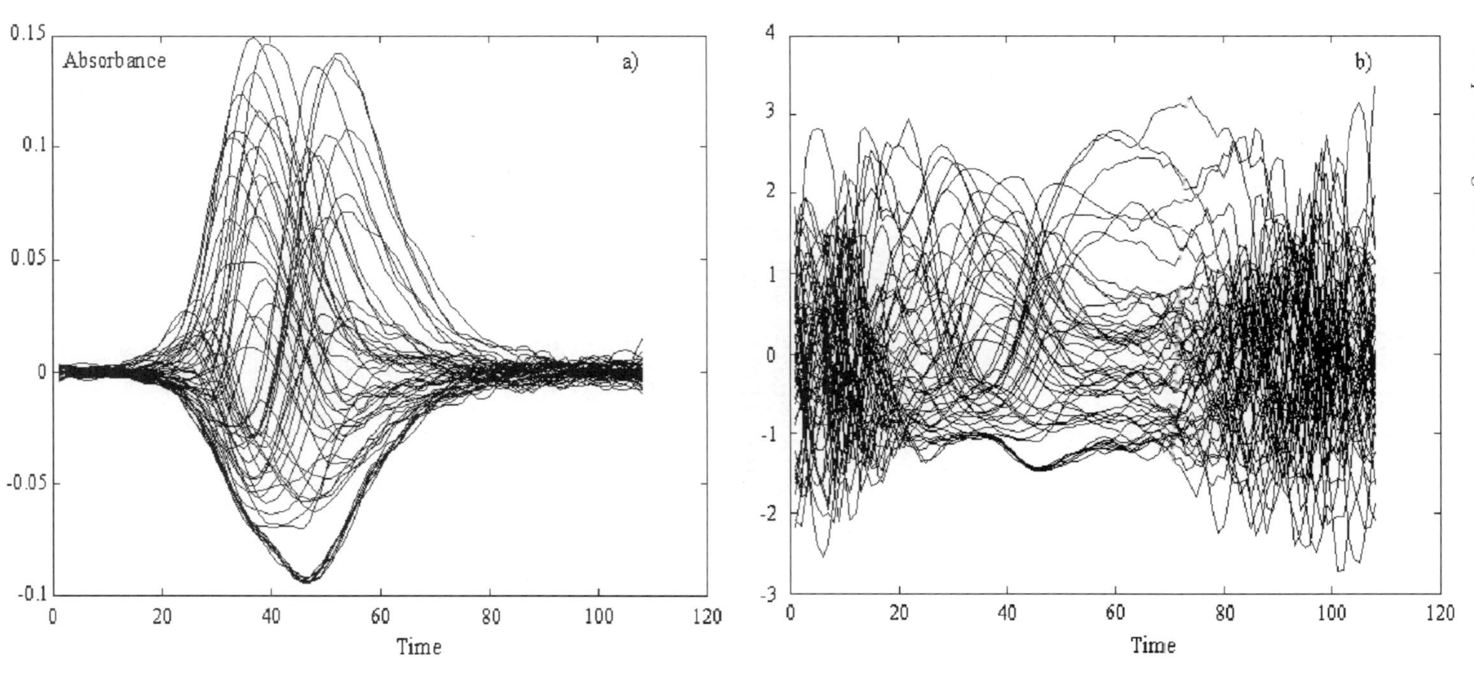

Figure 4.11 Two typical pretreatments to the original spectral data shown in Figure 4.9: (a) mean centring and (b) autoscaling.

4.3.4 Derivatives

First- and second-order derivatives are commonly applied by spectroscopists to tackle constant offsets (*e.g*. drift in the baseline; for which the first derivative is sufficient) or proportional drift and/or scatter effects (which need application of the second derivative). Further, as is known, inflection points (also termed '*spectral shoulders*') in the spectral peaks suggest that several species yield overlapped spectra. The place where a first-derivative plot intercepts the abscissa indicates where an inflection point occurs in the original spectra. On the other hand, the second derivative will show a minimum whose position denotes the inflection point on the original peak.

First- and second-order derivatives are often performed by the Savitzky–Golay algorithm, for which some special coefficients are used. The advantage of using the Savitzky–Golay algorithm over 'crude' derivatives (based on simply calculating increments of signal divided by increments of x-variables) is that the Savitzky–Golay algorithm, in addition, performs a smoothing. This is relevant because derivatives (see next paragraph) increase the noise in the preprocessed data. Often this is done automatically, but some commercial software requires fixing the window. Here, we refer to the same discussion as for the size of the windows when smoothing was discussed above. The common appearance of first- and second-order derivatives is presented in Figure 4.12.

Three main drawbacks can be associated with the use of derivatives: (i) the noise can be substantially increased [we need original signals with a very high signal-to-noise (S/N) ratio] and, as a consequence, (ii) the models can be unstable or not very appropriate to new unknowns, due to the presence of different random error distributions or different S/N ratios (even though the calibration may appear very good), and finally, (iii) the models can be difficult to interpret because the loadings do not have a direct relationship with the spectral characteristics (their maxima and/or minima are difficult to relate to the original spectra).

4.4 Dimensionality of the Model

Dimensionality (or complexity) of the regression model refers to the number of factors included in the regression model. This is probably one of the most critical parameters that have to be optimised in order to obtain good predictions. Unfortunately, there is no unique criterion on how to set it. In this introductory book, only the most common procedure is explained in detail. The reader should be aware, however, that it is just *a* procedure, not *the* procedure. We will also mention two other recent possibilities that, in our experience, yield good results and are intuitive.

Before going into the details on how to set the dimensionality, we first recommend taking a look at the sample sets to verify that no clear outlying samples will immediately influence (negatively) the models. One can start by examining all the spectra overlaid in order to see rapidly whether some are different. A histogram of the concentration values may help to assess their

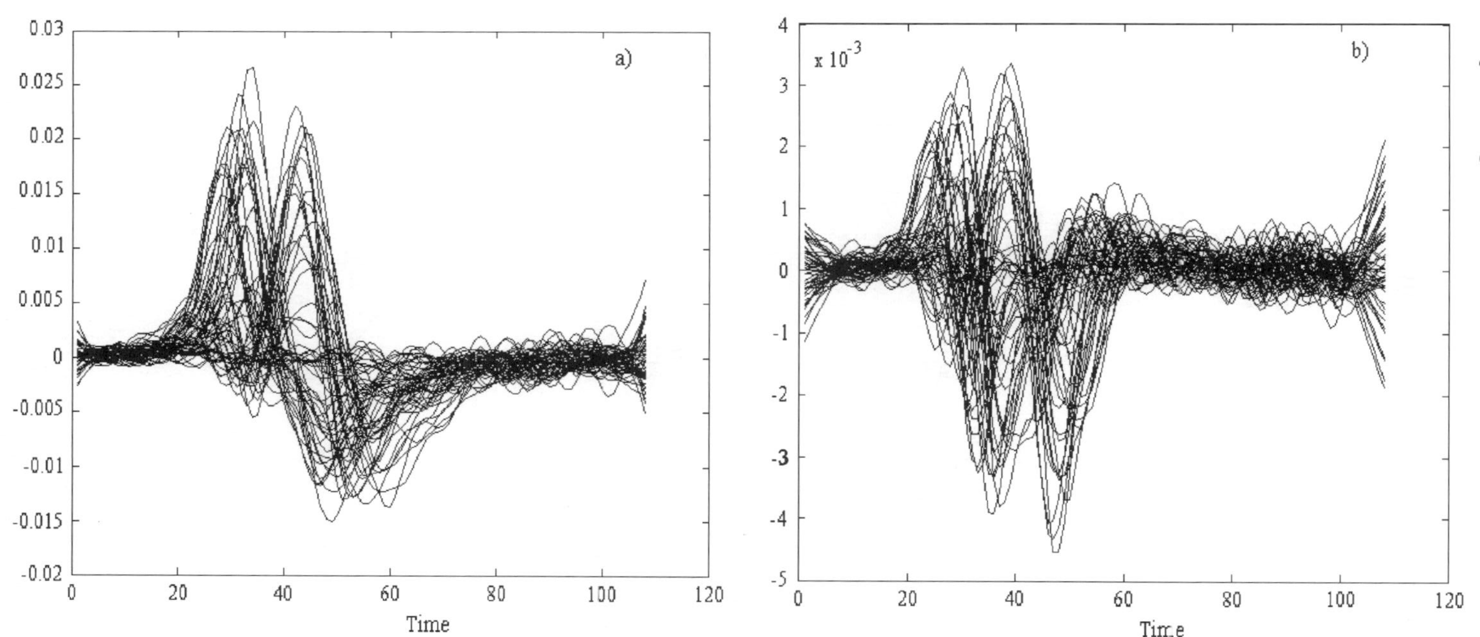

Figure 4.12 Two derivatives applied to the original spectral data shown in Figure 4.9 as a pretreatment step: (a) first- and (b) second-order derivatives.

distribution and whether one (or more) is clearly different from the main bulk of standards (typographical errors are not infrequent!). After spectral visualisation, it is useful to combine the calibration and validation samples temporarily and perform a preliminary PLS model, no matter what the true dimensionality. The aim is to plot the scores of the samples and verify that no sample(s) is(are) clearly different from the main bulk (see Figure 4.13). These outlying samples may have a large effect on the model. However, this does not mean that the effect is necessarily bad. A sample may be an outlier because it has unique characteristics, such as a unique combination of interferents or

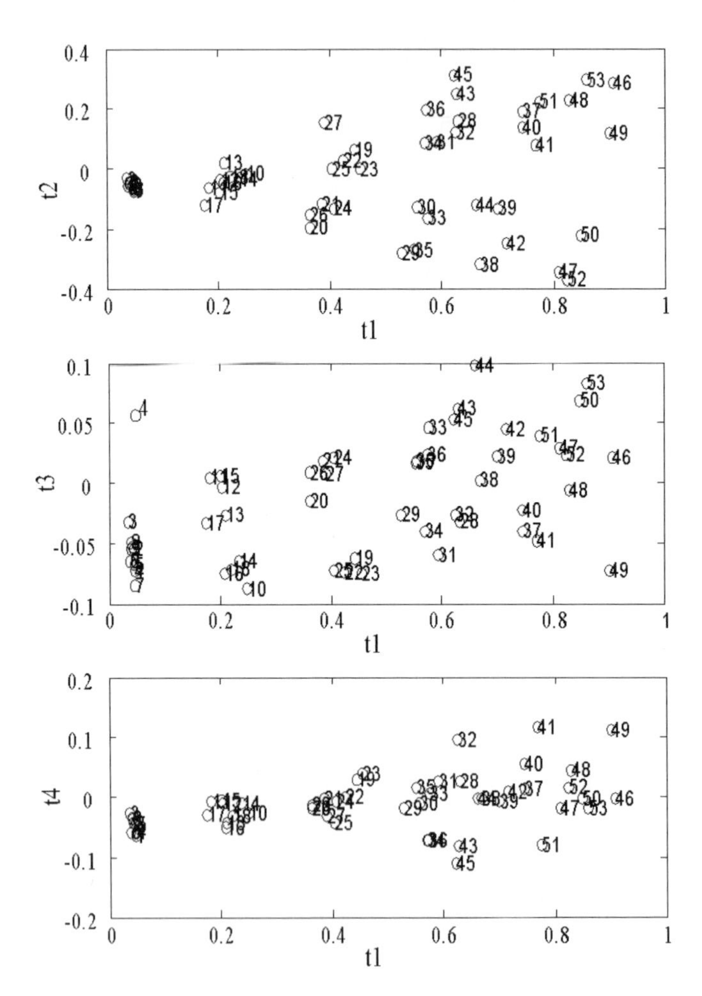

Figure 4.13 Preliminary PLS model developed to seek out outlying samples using the scores plots. The scores of the most important factor (**t**1) are always plotted on the abscissa. Sample 4 has a different behaviour because the tail of the atomic peak had a strange elevation and this was detected on the third factor. Note that the samples show some grouping because of the experimental design used to prepare the calibration set.

extreme values of the properties being modelled. Sometimes, these unique characteristics may be desirable (for example, a relatively high value of the constituent of interest will allow the model to expand the calibration range) and detecting the outlier simply warns us of the presence of this particular sample and even that it would be desirable to gather more samples similar to this one. On the other hand, of course, the sample can be an outlier with a bad effect on the model, such as those samples that do not belong to the population of samples we are trying to model or those samples that result from experimental or instrumental errors. These latter samples should be deleted since they will have an undesirable effect on the model. Note also that the detection of outliers may depend on the number of factors considered in the model. Hence setting the optimum dimensionality and looking for outliers are two intertwined processes that the user has to perform iteratively.

As stated above, searching for the optimum number of factors (number of latent variables or dimensionality) is a critical step in model development. A too low number of factors will not model adequately the behaviour of the spectra and their relationships with the analyte; hence prediction will not be satisfactory because we will use less information than we should – this is called *underfitting*. The opposite situation is termed *overfitting*: using too many factors to describe the system. In this case, calibration will be excellent (the more factors we include in the model, the better the fit of the model to the calibration samples will be), but the generalisation capabilities can be poor and, again, the predictions will be highly unsatisfactory. In this case we included too much information (*e.g.* noise) in the model that is particular to the calibration samples only. Although underfitting is not frequent (we usually can see clearly when the predictions of the calibration samples are not good, so we do not feel confident on the predictions obtained for the validation samples), overfitting is more likely to occur. Underfitting can be associated with bias (the average difference between the predicted and the true values of the analyte in the calibration samples), whereas overfitting can be associated with an increase in the variance in the predictions [29]. It follows that decreasing the bias of a model can be only achieved at the cost of increasing the variance on the predictions (and *vice versa*)! Hence adequate dimensionality of the model has to be established, fairly often, as a compromise between bias and variance [29,30] or, in other words, between fitting ability (model error) and prediction ability (represented by the overall estimation error of unknowns, through the use of a validation set) [24].

The problem is complex (as is any trade-off balance) and it can be approached graphically by the well-known Taguchi's loss function, widely used in quality control, slightly modified to account for Faber's discussions [30]. Thus, Figure 4.14 shows that, in general, the overall error decreases sharply when the first factors are introduced into the model. At this point, the lower the number of factors, the larger is the bias and the lower the explained variance. When more factors are included in the model, more spectral variance is used to relate the spectra to the concentration of the standards. Accordingly, the bias decreases but, at the same time, the variance in the predictions

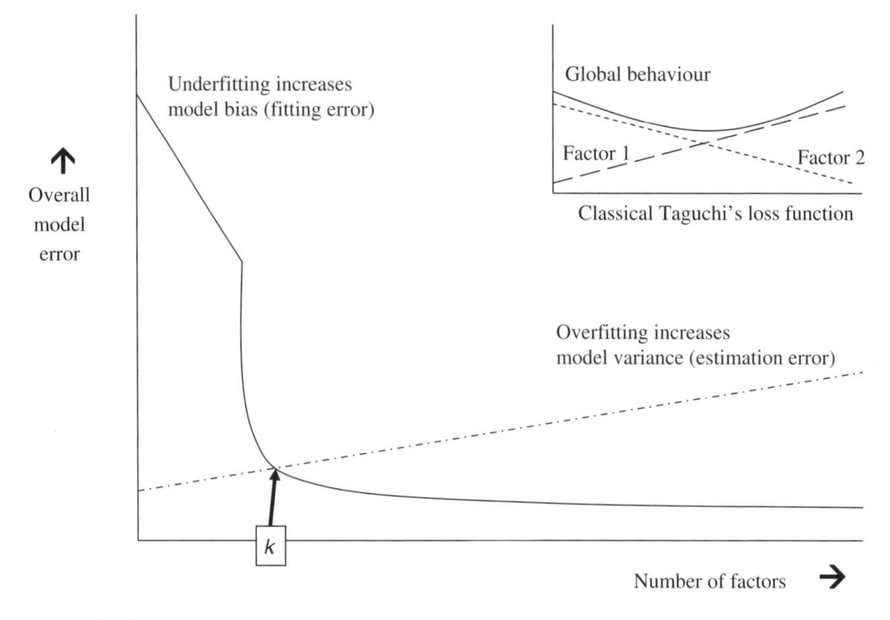

Figure 4.14 Taguchi's loss function to show the trade-off between bias (underfitting) and variance (overfitting); see Ref. [30] for more mathematical details. Here, *k* would be the optimum dimensionality of the PLS model.

increases smoothly. More details about bias and figures of merit are explained in Section 4.6.

Another important concept is parsimony. This means that if you have to select among several models that perform more or less equally, the preferred model is the one that has fewer factors, because the fewer factors are used, the less sensitive its predictions will be to non-relevant spectral phenomena. Hence the PLS model will still be useful for future unknowns even when slight differences appear (*e.g.* some smooth instrumental drift, slightly higher noise, *etc.*). Such a model is said to be more parsimonious than the others.

A way to start evaluating how a model performs when different factors are considered is to evaluate how much of the information (*i.e.* variance) in the **X**- and **Y**-blocks is explained. We expect that whichever the particular number of factors is the optimum, no more relevant information would enter the model after such a value. Almost any software will calculate the amount of variance explained by the model. A typical output will appear as in Table 4.1. There, it is clear that not all information in **X** is useful to predict the concentration of Sb in the standards, probably because of the interfering phenomena caused by the concomitants. It is worth noting that 'only' around 68% of the information in **X** is related to around 98% of the information in **Y** ([Sb]). This type of table is not always so clear and a fairly important number of factors may be required to model a large percentage of information in **X** and, more importantly, in **Y**. As a first gross approach, one can say that the 'optimal' dimensionality should be

Table 4.1 Amount of information explained by PLS models with a different number of factors. Original data correspond to Figure 4.9, mean centred.

Number of factors	*Variance explained in **X**-block (%)*		*Variance explained in **Y**-block (%)*	
	Factor	*Accumulated*	*Factor*	*Accumulated*
1	68.42	68.42	97.96	97.96
2	25.13	93.55	0.41	98.38
3	4.53	98.08	0.70	99.08
4	1.29	99.37	0.07	99.15
5	0.14	99.52	0.12	99.27
6	0.06	99.57	0.20	99.47
7	0.04	99.61	0.10	99.57
8	0.04	99.65	0.06	99.64
9	0.04	99.69	0.03	99.67
10	0.04	99.73	0.03	99.70
11	0.03	99.75	0.04	99.75
12	0.03	99.78	0.04	99.79
13	0.02	99.80	0.05	99.83
14	0.03	99.82	0.02	99.85

anything between two and five factors; considering more than six factors would in principle include only noise in the model.

How can we fine-tune the model? Well, remember that we have prepared a different set of samples to validate the model. You can use this set to evaluate which model predicts best. If you have a sufficient number of samples, it would be even better to prepare a small '*testing*' or '*control*' set (different from the calibration and validation sets and without a too large number of samples) to visualise which model seems best and, then, proceed with the validation set. This strategy has to be applied when developing artificial neural networks (see Chapter 5).

4.4.1 Cross-validation

Often the number of samples for calibration is limited and it is not possible to split the data into a calibration set and a validation set containing representative samples that are representative enough for calibration and for validation. As we want a satisfactory model that predicts future samples well, we should include as many different samples in the calibration set as possible. This leads us to the severe problem that we do not have samples for the validation set. Such a problem could be solved if we were able to perform calibration with the whole set of samples and validation as well (without predicting the same samples that we have used to calculate the model). There are different options but, roughly speaking, most of them can be classified under the generic term '*cross-validation*'. More advanced discussions can be found elsewhere [31–33].

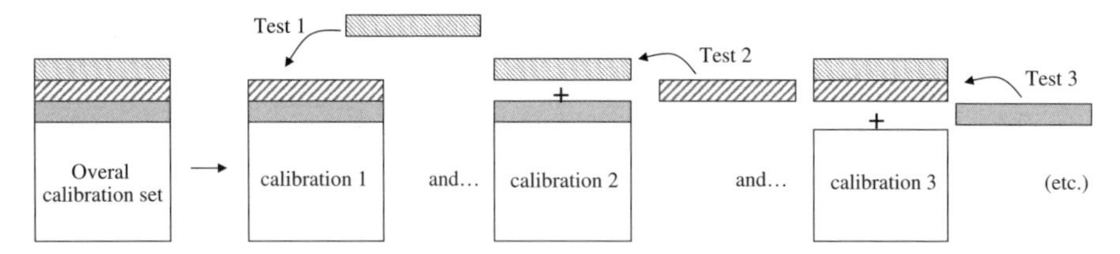

Figure 4.15 Schedule of the leave-one-out cross-validation scheme. Any cross-validation procedure will perform in the same way although considering more samples at each validation segment.

Cross-validation works by temporarily extracting some samples (the group is usually called a *segment*) from the calibration set, developing the model with the remaining ones and then predicting those that had been left out. Since those samples are known, a prediction error can be calculated and so the 'best' predictive model can be searched for. Of course, we should not base our decision on a unique segment but repeat the procedure several times. How many? In other words, how large should the segment be? Wiklund *et al.* [34] proposed seven segments as a good option. The extreme situation is *leave-one-out cross-validation* (LOOCV), in which the number of segments is equal to the number of calibration samples. In such a case, only a sample is deleted and the process is repeated as many times as there are samples (see Figure 4.15). LOOCV has become a common standard practice even though it has been reported that it tends to select unnecessarily large model dimensionalities [5,35,36]. The latter problem is of most relevance for practical purposes and it can be overcome using an external validation set (if available, of course) [37] in order to verify the model proposed by LOOCV by testing the neighbouring models to that suggested by LOOCV [38].

In any case, the cross-validation process is repeated a number of times and the squared prediction errors are summed. This leads to a statistic [predicted residual sum of squares (PRESS), the sum of the squared errors] that varies as a function of model dimensionality. Typically a graph (PRESS plot) is used to draw conclusions. The best number of components is the one that minimises the overall prediction error (see Figure 4.16). Sometimes it is possible (depending on the software you can handle) to visualise in detail how the samples behaved in the LOOCV process and, thus, detect if some sample can be considered an outlier (see Figure 4.16a). Although Figure 4.16b is close to an ideal situation because the first minimum is very well defined, two different situations frequently occur:

1. A plateau is obtained instead of a minimum. In this case, we should be careful and spend time testing the model (*e.g.* different cross-validation schemes) and, more importantly, validating it with new samples (parsimony is key here).
2. Neither a minimum nor a plateau is seen, but a random behaviour. This suggests problems with either the model or the data (too different samples, *etc.*) [22].

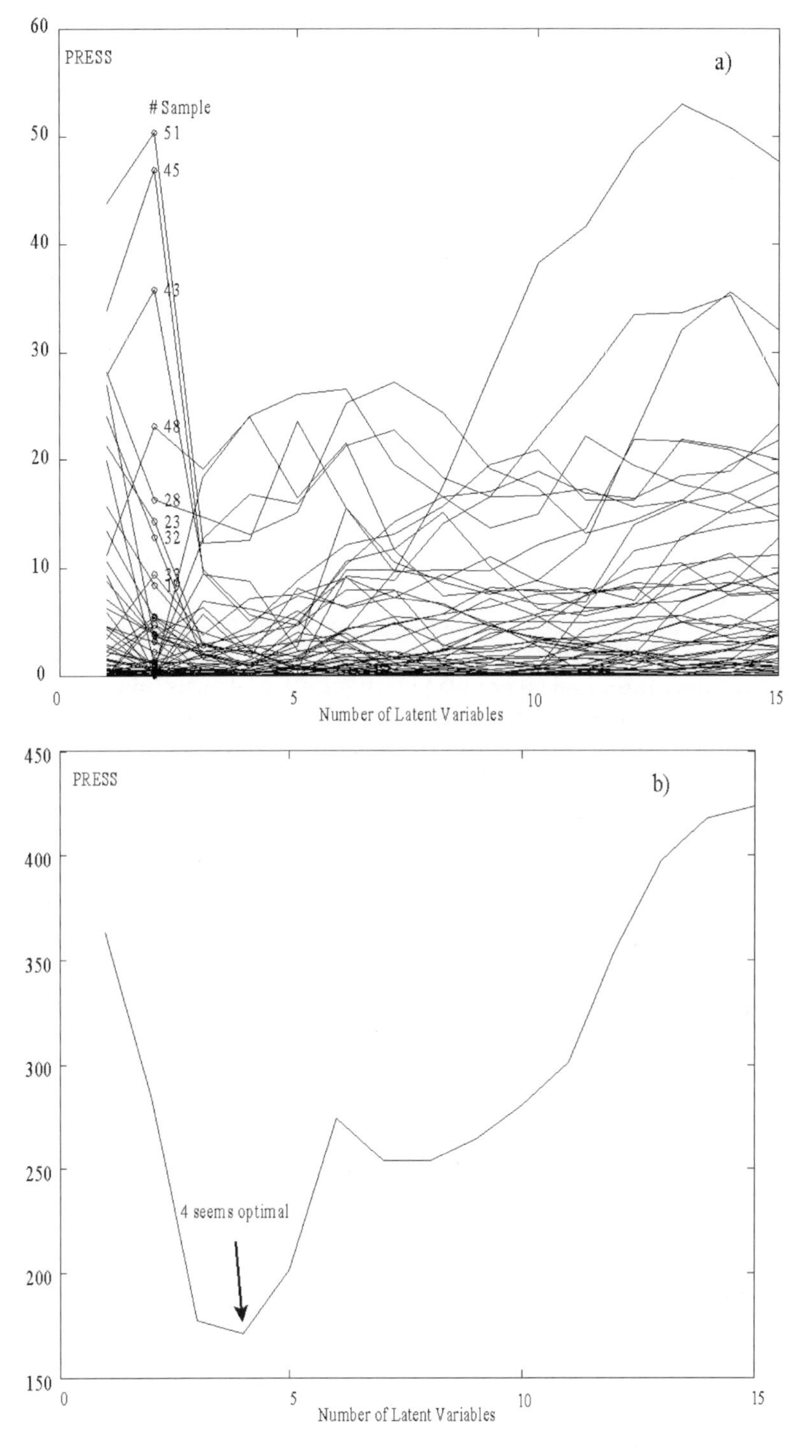

Figure 4.16 Typical example of a PRESS plot to select the model dimensionality (data shown in Figure 4.9, mean centred): (a) overall PRESS and (b) behaviour of each sample for each number of factors considered in the model. Four factors seem to be the optimal choice here as the minimum is clearly defined.

There are many other options to select the dimensionality of PLS models. Unfortunately, we should not expect them to yield the same results and, therefore, some consensus should be reached. The problem can be more difficult whenever calibration sets cannot be deployed according to an experimental design (*e.g.* in industrial applications; see [38] for a worked example).

We will only introduce here a classical and two recent alternatives because of their potential usefulness. These discussions may be avoided if the reader prefers to focus only on the criterion given above.

4.4.2 Other Approaches to Select the Dimensionality of the PLS Model

A simple and classical method is Wold's criterion [39], which resembles the well-known *F*-test, defined as the ratio between two successive values of PRESS (obtained by cross-validation). The optimum dimensionality is set as the number of factors for which the ratio does not exceeds unity (at that moment the residual error for a model containing A components becomes larger than that for a model with only $A - 1$ components). The adjusted Wold's criterion limits the upper ratio to 0.90 or 0.95 [35]. Figure 4.17 depicts how this criterion behaves when applied to the calibration data set of the working example developed to determine Sb in natural waters. This plot shows that the third pair (formed by the third and fourth factors) yields a PRESS ratio that is slightly lower than one, so probably the best number of factors to be included in the model would be three or four.

The randomisation test proposed by Wiklund *et al.* [34] assesses the statistical significance of each individual component that enters the model. This had been studied previously, *e.g.* using a *t*- or *F*-test (for instance, Wold's criterion seen above), but they are all based on unrealistic assumptions about the data, *e.g.* the absence of spectral noise; see [34] for more advanced explanations and examples. A pragmatic data-driven approach is therefore called for and it has been studied in some detail recently [34,40]. We have included it here because it is simple, fairly intuitive and fast and it seems promising for many applications.

The so-called *randomisation test* is a data-driven approach and therefore ideally suited for avoiding unrealistic assumptions. For an excellent description of this methodology, see van der Voet [41,42]. The rationale behind the randomisation test in the context of regression modelling is to evaluate how good models are for which the relation between the **X**-block and the **Y**-block is (artificially set to) random. Randomisation amounts to permuting indices. For that reason, the randomisation test is often referred to as a permutation test (sometimes also termed 'Y-scrambling'). Clearly, 'scrambling' the elements of vector **y**, while keeping the corresponding rows in **X** fixed, destroys any relationship that might exist between the **X**- and **Y**-variables. Randomisation therefore yields PLS regression models that should reflect the absence of a real association between the **X**- and **Y**-variables – in other words, purely random

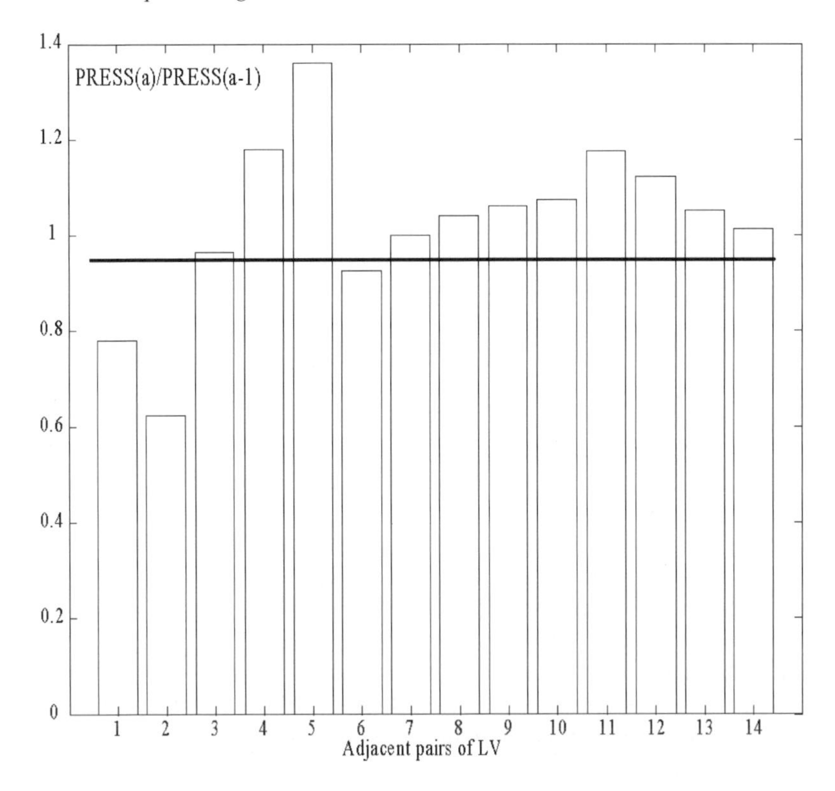

Figure 4.17 Selection of the number of latent variables to be included in the PLS model according to the modified Wold's criterion.

models. For each of these random models, a test statistic is calculated. Wiklund *et al.* [34] opted for the covariance between the **X**-variables score vector (**t**) and the vector of *y*-values because it is a natural measure of association. Geometrically, it is the inner product of the **t**-score vector and the **y**-vector. Clearly, the value of a test statistic obtained after randomisation should be indistinguishable from a chance fluctuation. These values will therefore be referred to as 'noise values'. Repeating this calculation a number of times (say 1000) generates a histogram for the null-hypothesis distribution, *i.e.* the distribution that holds when the component under scrutiny is due to chance – the null-hypothesis (H_0). Finally, the risk of over-fitting (alpha error) is calculated as the percentage of 'noise values' that exceeds the statistic obtained for the original data – the value under test. The (only) difference from a conventional statistical test is that the alpha error follows from a data-driven histogram of 'noise values' instead of a theoretical distribution that is tabulated, *e.g.* *t* or *F*.

PoLiSh–PLS is an entirely different approach applicable only to signals, (e.g. atomic spectra) not to non-ordered variables. PoLiSh combines two important ideas: (i) to perform an inner smoothing of the loading weights vectors obtained at each iteration step of the NIPALS procedure (or any other matrix decomposition algorithm being used) in order to remove noise from the essential components, which remains to accumulate in the later (noisy) ones and (ii) to measure the structure of the PLS regression vectors (the **b**-coefficients outlined in eqn 4.11) using the Durbin–Watson criterion to select the optimal PLS dimensionality based on the signal-to-noise ratio of the regression vector. The first idea would allow a better differentiation between important (commonly, the low-numbered) and noisy (generally, the high-numbered) components [43]. The second idea uses an objective criterion to determine whether the **b**-coefficients (calculated for a given number of components) are structured or not (if they can be considered random, the corresponding component should be discarded). Briefly, the Durbin–Watson test is generally applied to test whether a series of residuals from a regression are random; similarly, it can also be applied to test if random behaviour appears on the loadings and weights when successive components are included into the model. Experience [35,36,43] has shown that the best results are obtained when the **b**-coefficients are used.

Another straightforward approach consists in verifying whether two consecutive **b**-vectors are strongly correlated and, therefore, that the second one does not contain new information. Closeness can be easily quantified by the angle between two consecutive vectors and this parameter responds to a classical way to compare vectors. This approach can be denoted the 'Angles criterion'. PoLiSh will, accordingly, function in three steps: (1) Savitzky–Golay smoothing of the weights vectors obtained it each iteration step of the algorithm to develop a PLS model; (2) application of the Durbin–Watson test to the **b**-vectors; and (3) calculation of the angles between successive **b**-vectors. Step two will lead to a graph (Durbin–Watson statistic versus dimensionality) where a sharp increase indicates an increase in randomness (*i.e.* the component adds significant noise to the model); see Figure 4.18. Step 3 will also yield a plot where the first angle giving a minimum corresponds to the optimal dimensionality. PoLiSh was recently compared with Monte Carlo cross-validation and the criteria explained above, with good results [35,38].

4.5 Diagnostics

As mentioned above, PLS can render very useful plots to diagnose whether the model seems adequate or not. The diagnostic plots that one can use depend somewhat on the commercial software available, but probably the following sections take account of the most common and typical ones.

4.5.1 *t vs t* Plots

Although gross outliers should be deleted in advance (see the initial paragraphs in Section 4.4), it is likely to occur that some sample with some particular characteristic (spectrum, noise, *etc.*) shows a suspicious or different behaviour. Two plots inspecting the space of the spectra and the relationship of the **X**-space with the **Y**-space will be useful here. The **X**-spectral space is searched for by the '*t–t scores plot*'. Here the scores of the second, third, fourth, *etc.*,

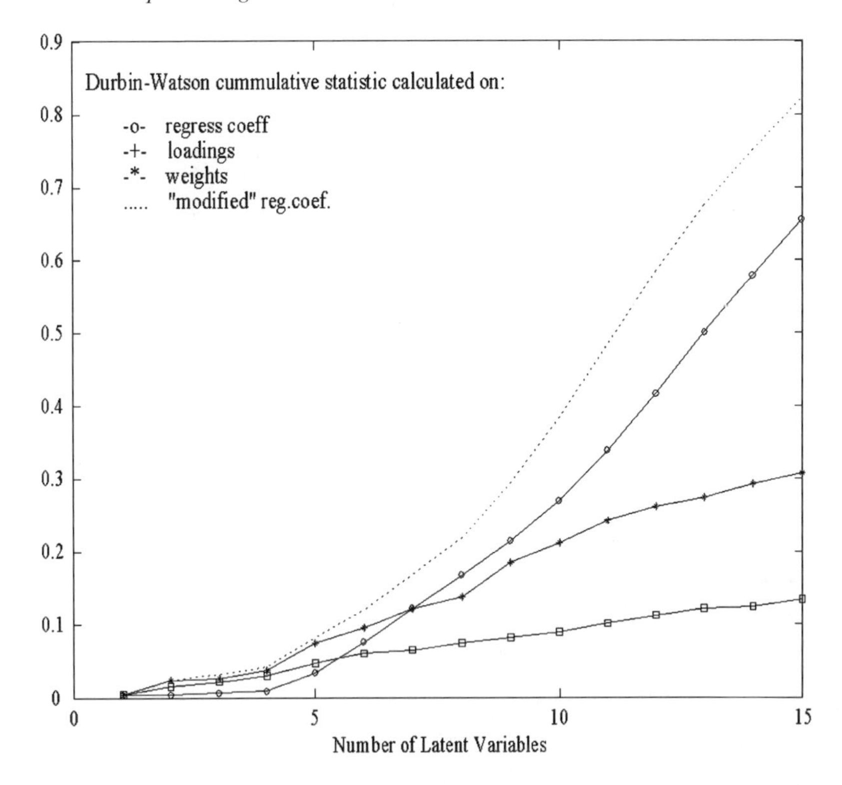

Figure 4.18 PoLiSh strategy to set the number of latent variables to be included in the PLS model. Application of the Durbin–Watson criterion to any series suggests that after latent variable 5, random noise is introduced into the model. This should be the maximum dimensionality to consider in the model (original data in Figure 4.9, mean centred).

factors are plotted against the first one so that the most relevant distribution of the samples in the spectral domain is visualised. Figure 4.19 shows how the samples used to develop a four-factor model using our worked example distribute fairly homogeneously within the working space. The particular shape shown here is caused by the experimental designs (mean-centred data). No sample is at a particularly suspicious position.

4.5.2 *t vs u* Plots

The relationship amongst the **X**- (here, atomic spectra visualised in Figure 4.9) and **Y**-blocks (here, [Sb]), can be evaluated by the '*t–u plot*'. This plot shows how the different factors (latent variables) account for the relation between **X** and **Y**. Hence we can inspect whether a sample stands out because of an anomalous position in the relation. Also, we can visualise whether there is a fairly good straight relationship between the blocks. Whenever the samples follow a clear curved shape, we can reasonably assume that the model has not

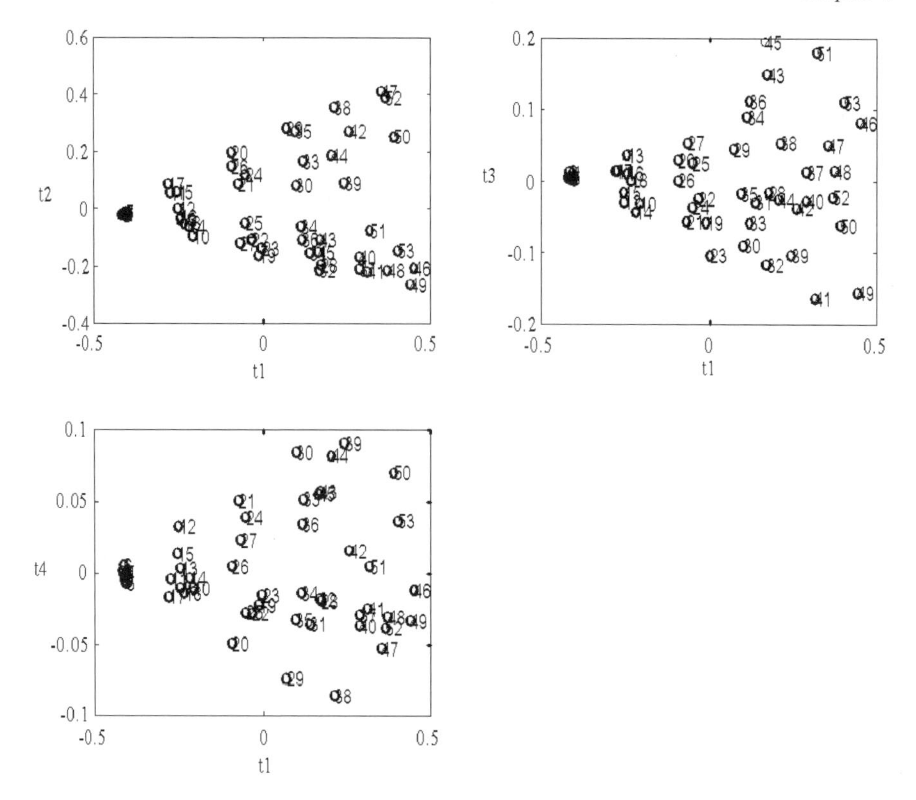

Figure 4.19 Diagnostic plots: '*t vs t scores*' to inspect for anomalous samples on the spectral space (**X**-block).

modelled some information properly or that the model should not be of first order (a straight-line functional relationship) but of a higher order. Possible solutions to this problem might be either to use a nonlinear PLS inner relationship (*e.g.* a second-order polynomial as the inner relationship) or to increase the **X**-matrix with squared terms (see. [5] for a detailed example). Figure 4.20 reveals that no sample is outside the general working space and that the four factors considered in the model hold a fairly acceptable linear (straight) relationship with the concentration of Sb in the standards, even when the combinations of concomitants in the standards may affect the atomic peaks. The first factor has, obviously, an outstanding relevance in the model. According to Table 4.1, this factor extracts 'only' around 68% of the spectral variance that is linearly related to about 98% of the variance in the Sb concentration in the solutions. The linear relationship is not so evident for the fourth factor, but this is a common situation and its mission in the model should be understood as some kind of modelling of subtle spectral characteristics. It is clear that those latent variables that have the samples following a linear (straight-line) pattern will be better predictor latent variables than those where the samples are widely dispersed [44,45].

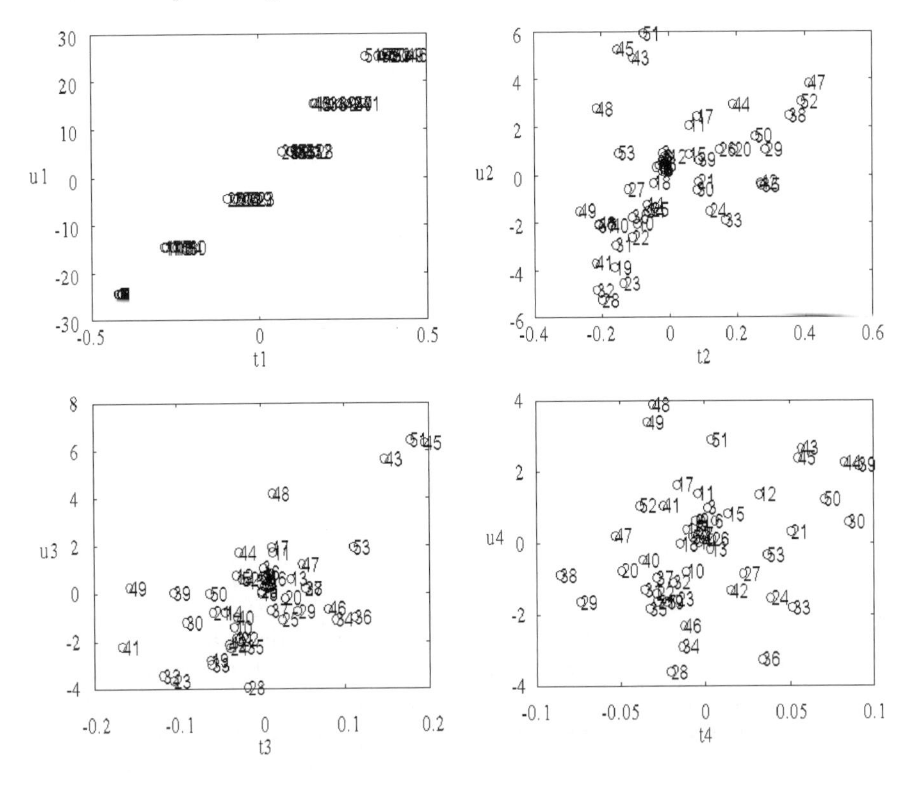

Figure 4.20 Diagnostic plots: '*t vs u scores*' to inspect for anomalous samples and strong non-linearities on the relationship between the spectral space (**X**-block) and the property of interest (**Y**-block).

4.5.3 The T^2, *h* and *Q* statistics

In order to find outliers in the calibration set, it is advantageous to extract some more information from the spectra that constitute the **X** data matrix for PLS regression. It is particularly relevant to ascertain whether a given sample is different from the average of the other calibration samples. Here, 'average' is a slightly ambiguous term because we can evaluate how similar our sample is to either the average of the most relevant spectral characteristics included in the regression model (*i.e.* the spectral characteristics explained by the latent variables or factors), and/or the average of the discarded information (*i.e.* the spectral characteristics – variance – not included in the regression model); see Figure 4.21 for a schematic representation of both concepts.

The first issue can be addressed using two very similar and mathematically related statistics:

1. The T^2 test or Hotelling's T^2 test, (named after Harold Hotelling, who developed a test to generalise Student's *t*-test [46]). This test is used to verify, in a multivariate sense, whether a sample belongs to a distribution

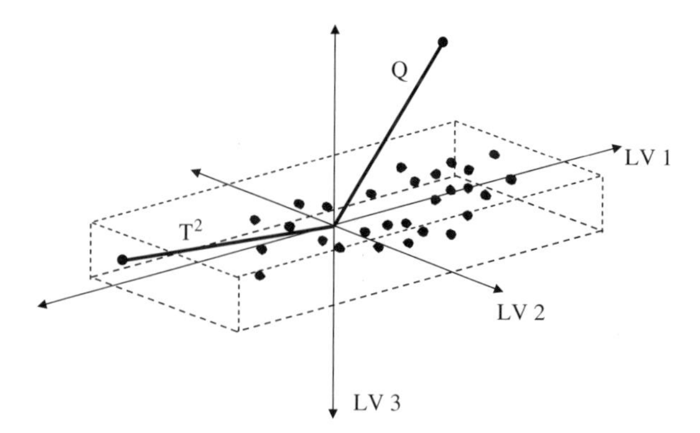

Figure 4.21 Graphical representation of the T^2 and Q statistics. The sample on the left uses the T^2 statistic to assess whether it has similar patterns to the average ones included within the PLS model. The point on the right uses the Q statistic to verify whether that sample has some unusual amount of variation (residual) compared with the residuals of the other samples (differences are magnified to simplify visualisation).

whose average is given by the average vector of the calibration samples. From a chemical standpoint, this is equivalent to testing whether the main spectral characteristics of a given sample are similar to the average values shown by the calibration samples – but only those that have been taken into account by the first latent variables. This means that we test whether the sample is close to the average value within the multivariate space defined by the PLS latent variables (this can be termed the model '*hyper-box*'); see Figure 4.22 for a graphical example. In this control chart, none of the calibration samples pass the limit, although the increasing pattern of the statistic suggests that the last samples have some spectral characteristics which are slightly different from those of the first samples. Recalling that the samples are ordered according to the concentration of Sb, it is seen that the spectra of some solutions containing some of the highest Sb concentrations became a bit more affected by the presence of large amounts of concomitants.

Its original formulation (considering that the data had been centred previously, which is the general case) is $T_i^2 = I \cdot \mathbf{x}_i^{\mathrm{T}} \mathbf{C}^{-1} \mathbf{x}_i$, where I is the total number of samples ($i = 1, \ldots, I$), $\mathbf{x}_i$ is the spectrum (column vector) of J variables for sample i and $\mathbf{C}$ is a $(J \times J)$ covariance matrix. In PLS, calculations are made on the normalised scores, so $T_i^2 = \mathbf{t}_i \boldsymbol{\lambda}^{-1} \mathbf{t}_i^{\mathrm{T}} = \mathbf{x}_i \mathbf{P} \boldsymbol{\lambda}^{-1} \mathbf{P}^{\mathrm{T}} \mathbf{x}_i^{\mathrm{T}}$; now, $\boldsymbol{\lambda}^{-1}$ is the diagonal matrix containing the inverse of the eigenvalues associated with the A factors used into the model, $\mathbf{t}$ is the ath-factor-scores vector of sample i and $\mathbf{P}$ is the loadings matrix [47,48]. This test is calculated for each sample in the calibration set (see Figure 4.22), although it can also be applied to new samples in order to prove that they are similar to the calibration ones. A 95% confidence interval upper limit can be calculated for the T^2 statistic and, thus, the software which

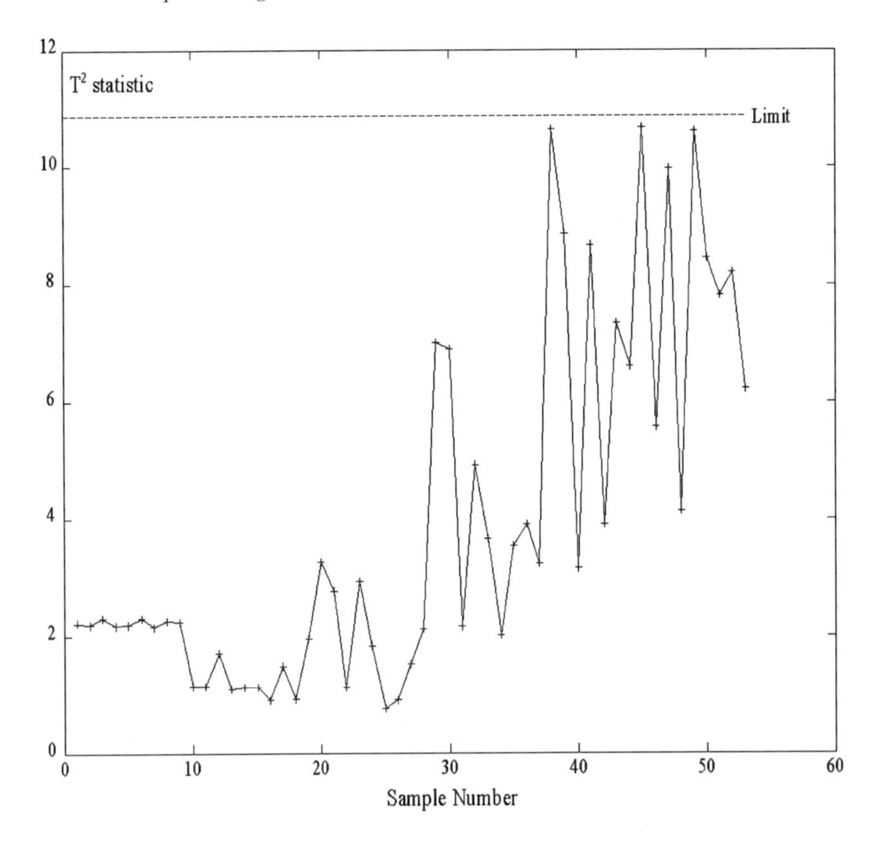

Figure 4.22 Example of a T^2 control chart to test for outlying samples on the calibration set. Four factors were used to develop the model (original spectra from Figure 4.9, mean centred).

implements this test will allow for a simple decision-making. The derivation of such a limit is not simple and will not be included here. Interested readers are encouraged to consult Ref. [47].

The h-statistic, or leverage, appears naturally when the predictions are calculated from the model. The leverage is calculated on the scores subspace and, thus, $h_i = 1/I + \left[\sum_{a=1}^{A} t_{ia}^2 / \left(\sum_{i=1}^{I} t_{ia}^2\right)\right]$ [23], where the letters have the same meaning as above. The leverage can be interpreted geometrically throughout its relationship with the Mahalanobis distance. The Mahalanobis distance (named after Prasanta Chandra Mahalanobis) is a measure of the distance of a data point (sample) to the average of a multivariate data set. The Mahalanobis distance, D^2, and the leverage are related as $h_i = 1/I + D^2/(I-1)$ [49], where I is the number of data points in the calibration set. Also, $D^2 = \mathbf{x}_i^{\mathrm{T}} \mathbf{C}^{-1} \mathbf{x}_i$, where $\mathbf{x}$, i and $\mathbf{C}$ have the same meanings as above. Since the influence that a sample has on the model depends on the relative position of the sample in the multivariate space, the leverage flags those samples with a considerable influence on the regression vectors and their associated estimated concentrations. It can be computed for each sample in the calibration and validation sets and plotted on a graph (similar to Figure 4.22) to make decisions. As a rule of thumb, a leverage larger than $2A/I$ (A and I were defined above) or $3A/I$ flags a sample as having a large influence in the model [22,47]. Note, however, that the leverage alone does not indicate whether this influence is good or bad. For example, samples with the highest or lowest concentrations will have a large leverage but they are not necessarily bad. The leverage only underlines which samples require a more dedicated inspection because they have the largest influence in the model. Of course, a sample with a high or low concentration of the analyte can do so, although not necessarily in a bad manner. The other diagnostic tools will help you to make a decision.

Additional, complementary, information can be obtained by determining whether a sample has some non-systematic variation that has not been explained by the model. This can be performed by comparing this unexplained variation with the non-systematic variation in the other samples. This is done by comparing the multivariate residuals of the samples so that large residuals will denote those samples containing unexpected spectral features compared with the other samples [24]. The residuals can be calculated straightforwardly as $\mathbf{E} = \mathbf{X} - \mathbf{T}\mathbf{P}^{\mathrm{T}}$, where $\mathbf{X}$ is the original (centred) data matrix and $\mathbf{T}$ and $\mathbf{P}$ the scores and loadings of the $\mathbf{X}$ matrix considering only A-latent variables [44]. As for the two previous statistics, a 95% confidence interval upper limit can be calculated *ad hoc* [47]. Figure 4.23 shows what a typical Q chart looks like. Here, sample 44 seems rather different and the other diagnostics were inspected to decide about its deletion from the spectral data (which would involve the development of a new model). Since that sample was not highlighted by any other statistic it was decided to keep it into the model. The same can be argued for the other samples that became slightly above the limit.

4.5.4 Studentised Concentration Residuals

The Studentised concentration residuals are frequently called just 'Studentised residuals', but one has to be aware that the residuals we will use now should not be confounded with those in the previous section. Now we are concerned with studying whether some sample cannot be predicted properly. Hence 'residual' is interpreted in the usual way that atomic spectroscopists do; *i.e.* the difference between the true and the predicted concentrations of the analyte ($\hat{y} - y_{\text{reference}}$), where the hat (ˆ) means predicted or estimated. Of course, the simplest possibility (also used in classical univariate regression) is to plot the residuals against the number of the sample or the concentration and, often, the outlying samples will be obvious from the graph.

A better approach to improve visualisation and decision-making is to standardise the residuals. A very simple approach is just to subtract the mean of the overall set of residuals and divide by their standard deviation. Brereton suggested that a simple plot of these standardised residuals is sufficient in

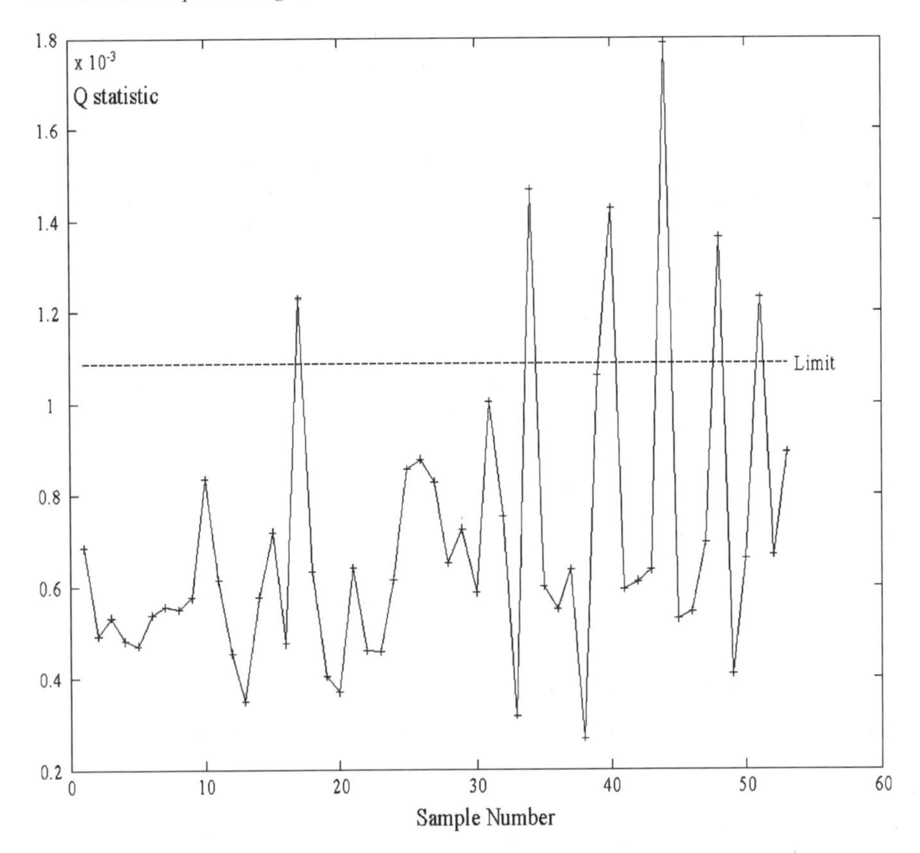

Figure 4.23 Example of a Q control chart to evaluate the presence of outlying samples on the calibration set. Four factors were used to develop the model (original spectra from Figure 4.9, mean centred).

most circumstances [23], and Esbensen *et al.* [45] recommended making a normal-probability plot (this simple transformation corresponds to the typical z-score in univariate studies).

Most common in PLS modelling are the so-called 'Studentised residuals', which are used to tackle the (possible) variation in the measurement errors when the samples are different from the average. A straightforward way to consider this is to include the leverage in the calculations as it measures the difference amongst the samples and the average of the overall set. Thus, for each sample, the Studentised residual can be calculated easily using the equation Studentised residual$=(\hat{y}-y_{\text{reference}})/[s(1-h)^{1/2}]$, where s is the standard deviation of all residuals from the calibration set (using $I-A-1$ as the degrees of freedom) and h is the leverage defined above. Samples for which the predictions show Studentised residuals larger than ± 3 should be inspected carefully.

The '*Studentised residuals vs leverage*' plot is a two-dimensional graph commonly employed to detect suspicious samples. Its interpretation is straightforward as samples with extreme leverages and/or extreme Studentised residuals are visualised easily. Some general rules can be derived according to the location of the samples on the plot (see Figure 4.24a) [22] (consider only absolute values). Region 1 denotes low leverages and low residuals and none of the samples here harms the model. Region 2 (*e.g.* sample 79) characterises samples with high residuals but whose spectra are similar to the average of the calibration set (low leverage); these samples should be inspected for the concentrations of the analyte as there might be typographical errors, wrong calculations and so on. Another possibility is that the sample has a unique y-value. Region 3 comprises large leverages (important differences from the average spectrum) although the sample is well modelled (low residuals). Samples here are influential but not necessarily 'bad outliers' in the sense that they do not have a bad influence in the model. They typically correspond to samples with the lowest and/or highest concentrations of analyte. Samples in region 4 (*e.g.* sample 78) have not been modelled properly by the PLS regression and, further, they are different from the average. Probably they are outliers and should be discarded from the model. Going back to the example that we are dealing with, Figure 4.24b shows that there are no suspicious samples so probably the warnings for sample 44 in the previous section are not severe and, hence, it can remain on the model.

4.5.5 'Predicted vs Reference' Plot

This is, by far, the most often used plot to assess regression models. It displays the y-values predicted by the model (on either the calibration or the validation set) against the reference ones. Good models should yield a 45° line (that is, perfect predictions), so we would like models whose predictions are close to such ideal behaviour. Samples with large departures from the line denote a bias in the predictions (as in classical univariate calibration), particularly when they are near the extremes of the working range. In addition, a homogeneous distribution of the samples across the working range is to be preferred, without isolated samples (with too low or too high concentrations) as they are influential and prone to bias the model if something is wrong with them.

The *root mean square error of calibration* (RMSEC) is a measure of the average difference between the predicted and the actual values and, thus, gives an overall view of the fitting ability of the model (how good the model is for predicting the same samples that were used to calculate the model). It is defined as $\text{RMSEC}=[\Sigma(\hat{y} - y_{\text{reference}})^2/(I - A - 1)]^{1/2}$, where I and A have the usual meanings and the 1 takes account of the mean centring of the data). As detailed in Section 4.4, it decreases continuously as the number of factors included in the PLS model increases. This means that it is not a good diagnostic to select the best predictive PLS model (it leads to severely overfitted models). The LOOCV error is a more appropriate diagnostic as it is more related to the predictive capabilities of the model; when cross-validation is employed, the average error is usually termed RMSEC-CV or LOOCV-RMSEC. In general, the RMSEC is lower than the LOOCV-RMSEC (see Figure 4.25). In all cases, their units are the same as the original concentrations (here, ng ml^{-1}) and can be clearly

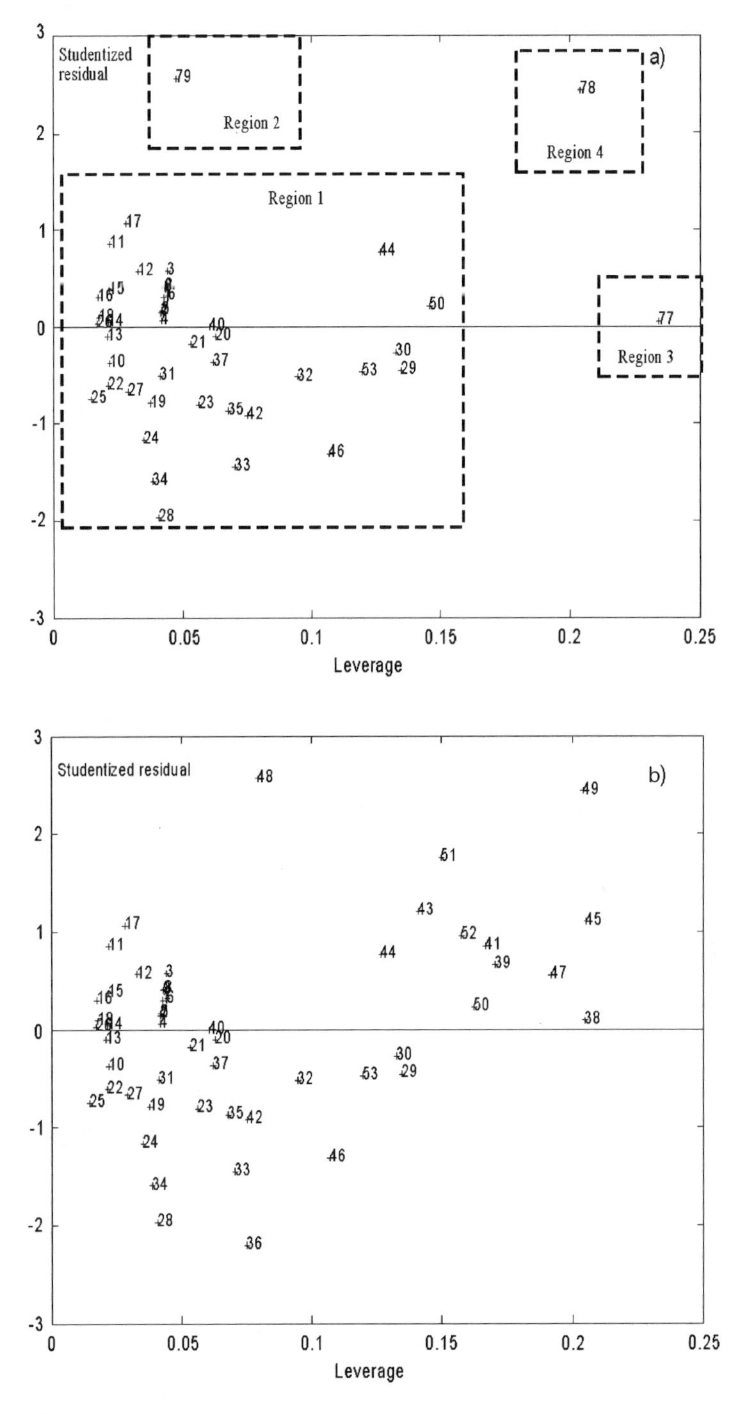

Figure 4.24 'Studentised residuals *vs* leverage' plot to detect outlier samples on the calibration set: (a) general rules displayed on a hypothetical case and (b) diagnostics for the data from the worked example (Figure 4.9, mean centred, four factors in the PLS model).

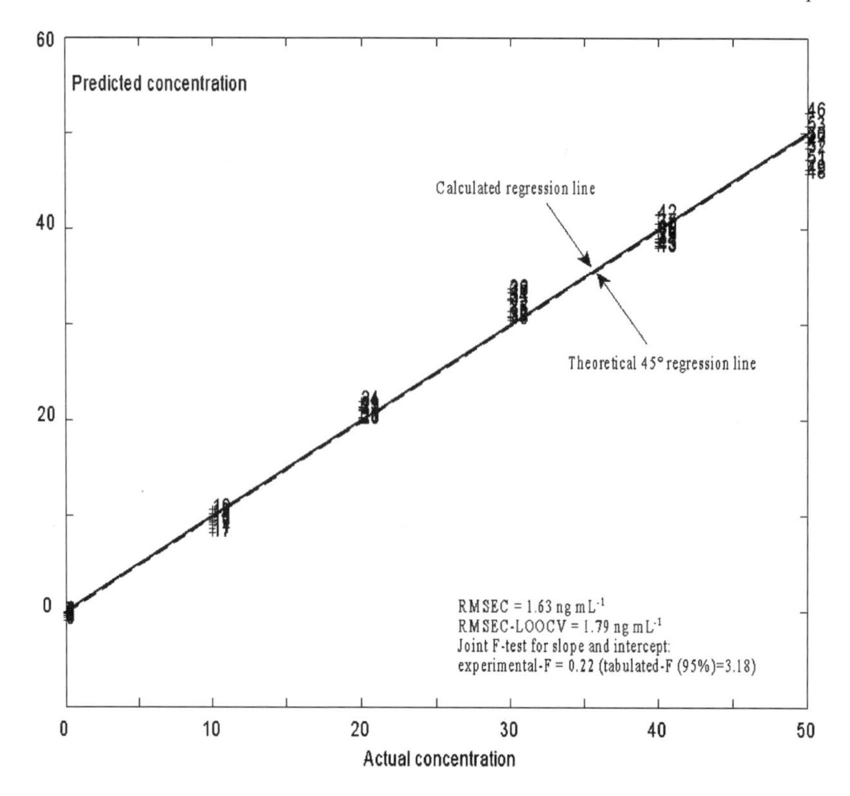

Figure 4.25 Predicted *vs* reference concentration plot and several related statistics to evaluate the adequacy of the PLS model (original data, Figure 4.9, mean centred, four factors in the PLS model).

influenced by outlying predictions (*e.g.* typesetting errors, as discussed in the previous section).

In order to test whether the model yields biased predictions, a regression line can be calculated using as 'dependent' variable the predicted concentrations given by the PLS models and as 'independent' variable the reference values. In order to test simultaneously whether the intercept and the slope are zero and one, respectively (*i.e.* the 45° regression line), the joint F-test should be used as

$$F = \frac{a^2 + 2(1-b)\bar{x} + (1-b)^2\left(\sum_{i=1}^{n} x_i^2/I\right)}{2s_{y/x}^2/I} \quad (4.13)$$

with 2 and $I-2$ degrees of freedom for the numerator and the denominator, respectively; I is the number of data points; a is the intercept; b is the slope and $s_{y/x}$ is standard error of the regression. If a large extrapolation for the intercept is required (*e.g.* the working range is 100–200 μg ml^{-1}), the test might yield wrong conclusions because any slight deviation from 1 in the slope leads to intercepts far from zero. In this case, application of the classical univariate test to asses that the slope statistically equals unity is advisable [t-test=$(1-b)/s_b$, where s_b is the standard error of the slope]. Another different t-test will be studied to assess the bias in Section 4.6.

Figure 4.25 shows an example (real data shown in Figure 4.9, mean centred, four factors in the PLS model) where outliers are not present and a good experimental F-test (much lower than the tabulated F, 95% confidence), showing that the prediction model is not biased. The unique concern is that the predictions were slightly spread at highest concentrations (50 ng ml^{-1}). Here it was decided to limit the applicability of the method to water samples with less than 30–40 ng ml^{-1}, which is high enough for the water samples we studied and it includes the World Health Organization (WHO) maximum level. This boundary corresponds to the concept of 'working linear range' in classical calibration.

4.5.6 Validation

Validation is the key step to accept or reject a regression model. It is worth noting that *sensu stricto* there are no perfect models. A model might be appropriate (satisfactory) to solve for a particular problem but totally useless to address another one. The difference is in what is called fit-for-purpose. When developing a predictive model, one should be aware of the needs of the client or, at least, on the degree of agreement that should be imposed on it. Do you need to report the concentration of Sb within ±3 or ±1 ng ml^{-1} of the 'true' values? In the former case, the model for the case study we are dealing with seems excellent at the calibration stages, but not very useful for the latter case.

Once the most promising model has been selected, we have to apply it to a set of new samples (*i.e.* not included on the calibration stages) but whose 'true' values are known, for example, because they have also been analysed by other method. The use of quotation marks above stresses that the reference values will (unavoidably) have some associated error and, therefore, it might be too optimistic to consider them strictly as 'true'. Hence you have to be aware that even 'wrong' predictions might not indicate that your model is wrong. A similar discussion was presented in Chapter 1 for isotope ratio measurements. There, it was shown that the concentrations of the analytes in the calibration standards may have a high degree of uncertainty. This is a difficult problem and recent advances have been published [50–53], although some more work is still needed because when the errors in the reference values are too high (not too rare in industrial applications of molecular spectroscopy), the equations yield negative corrections that cannot be accounted for.

The new samples constitute what is called the *validation set* and, of course, the more validation samples you have, the more representative the conclusions will be. The RMSEP (root mean square error of prediction) statistic measures how well the model predicts new samples. It is calculated as

$RMSEP = [\Sigma(\hat{y} - y_{reference})^2/I_t]^{1/2}$ (I_t = number of validation samples), with concentration units. The behaviour of RMSEP is depicted at Figure 4.14, *i.e.* it is high whenever a too low or too high number of latent variables are included in the model and it decreases more or less sharply at the vicinity of the optimum number of factors. Nevertheless, you should not use the validation set to select the model because, then, you would need another true validation set. As was explained in previous sections, if you have many samples (which is seldom the case) you can develop three sample sets: one for calibration, one for fine-tuning the model and, finally, another one for a true validation.

As the validation set is used to test the model under 'real circumstances', it may occur that one or several samples are different from the calibration samples. When we use experimental designs to set the working space, this is not frequent, but it may well happen in quality control of raw materials in production, environmental samples, *etc.* Therefore, it is advisable to apply the Hotelling's T^2 or the leverage diagnostics to the validation set. This would prevent wrong predictions from biasing the RMSEP and, therefore, the final decision on the acceptance of the PLS model.

We can summarise some other ideas for evaluating the predictive ability of the PLS model. First, you can compare the average error (RMSEP) with the concentration levels of the standards (in calibration) and evaluate whether you (or your client) can accept the magnitude of this error (fit-for-purpose). Then, it is interesting to calculate the so-called '*ratio of prediction to deviation*', which is just RPD=SD/SEP, where, SD is the standard deviation of the concentrations of the validation samples and SEP is the bias-corrected standard error of prediction (for SEP, see Section 4.6 for more details). As a rule of thumb, an RPD ratio lower than 3 suggests the model has poor predictive capabilities [54].

4.5.7 Chemical Interpretation of the Model

An important advantage of PLS over other multivariate regression methods (*e.g.* MLR or artificial neural networks) is that a chemical interpretation of the main parameters is possible. This cannot be taken for granted, of course, because we are using abstract factors after all, but derivation of a general chemical explanation should be attempted. Often the first factors can be related to some chemical (spectral) meaning and, undoubtedly, this gives more confidence on what is going on within the model. Often, MLR regression coefficients are purely mathematical numbers and their interpretation has, in general, little meaning, whereas drawings of the PCR loadings will focus on the variance of the spectra, not necessarily relevant to predicting the analyte. Artificial neural networks (to be explained in the next chapter) are particularly 'recalcitrant' to any interpretation, although some attempts have been made in recent years to address this issue (more explanations will be given in next chapter).

Therefore, we will consider here two graphical representations that are relatively common. They rely on comparing the general shapes of the main loading vectors and regression coefficients with the original spectra (without scaling). In this way, it is possible to detect similarities and discover relevant associations. Here, it is as important to get a broad overall view as to concentrate on some particular details relevant for the model (*e.g.* the maximum loadings and/or regression coefficients –in an absolute sense). When this is attempted for our example, Figures 4.26 and 4.27 are obtained.

In Figure 4.26, the loadings of the first latent variable can be roughly identified with the average spectral profile, whereas the loadings of the second, third and fourth latent variables present a first-derivative shape with maxima and minima just where the atomic peaks have maxima or inflection points. This suggests that the model uses several factors to address the undesirable effects

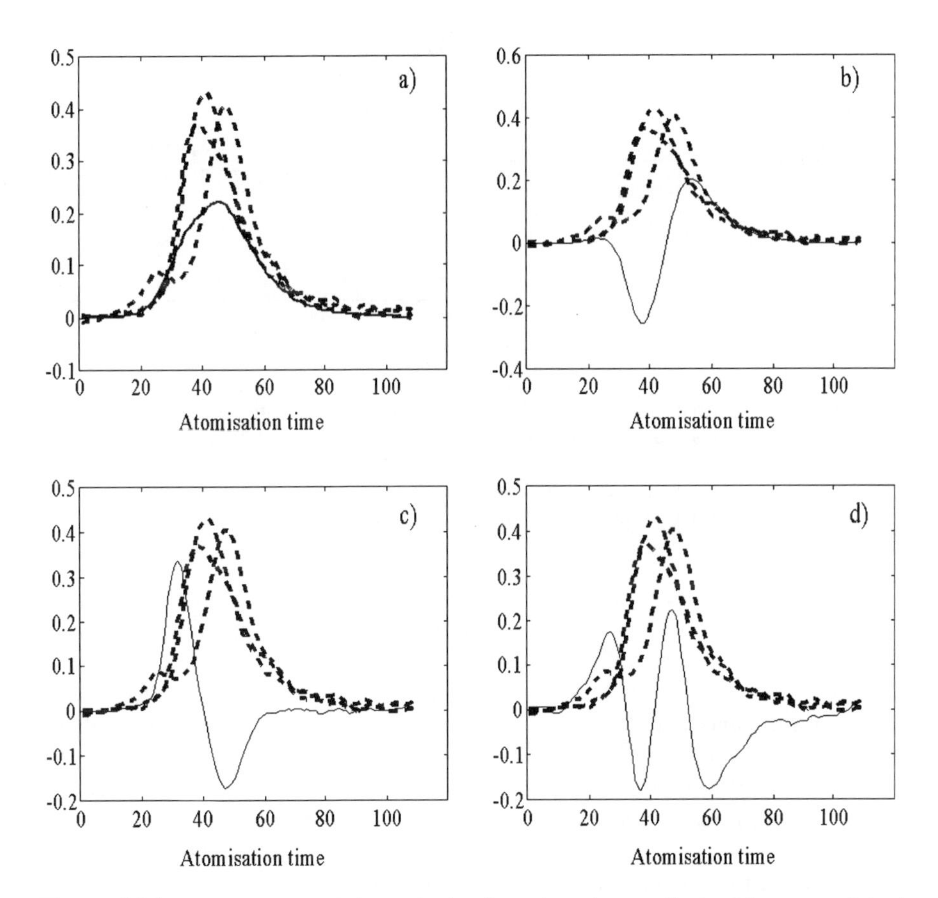

Figure 4.26 Comparison of the main loadings (continuous line) with some original spectra from Figure 4.9 (dotted lines) for the main latent variables: (a) first, (b) second, (c) third and (d) fourth. Note that largest positive loadings correspond to spectral inflection points ('shoulders') whereas minima correspond to several maxima. Adapted from Ref. [12], with permission from the Royal Society of Chemistry.

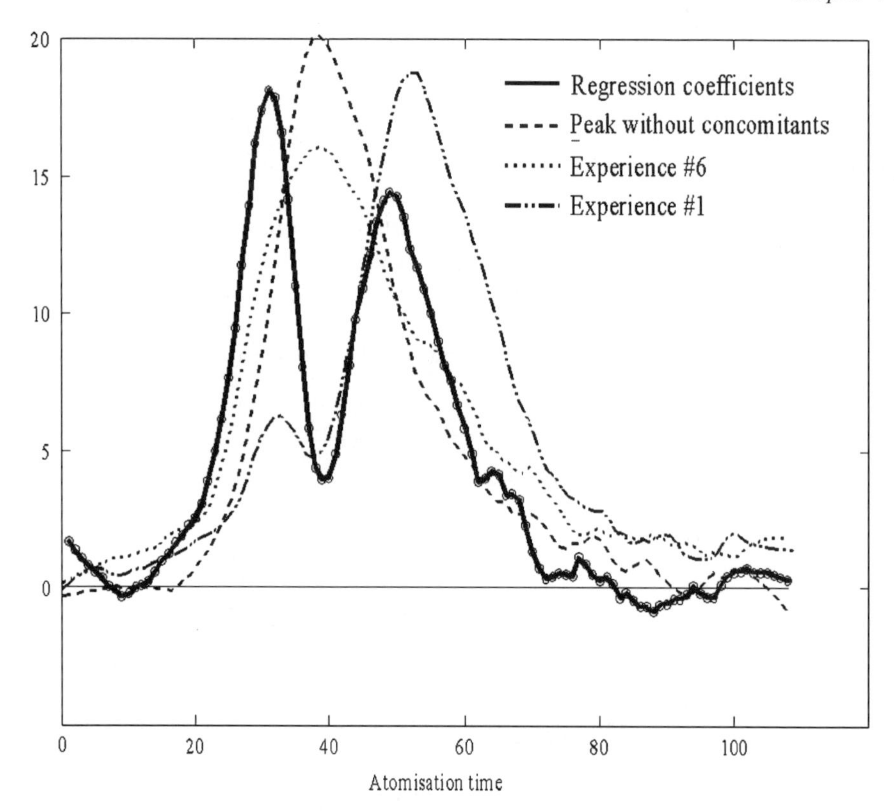

Figure 4.27 Comparison of the regression coefficient vector with some original spectra from Figure 4.9 (in particular, experiments 1 and 6 from an experimental design). The largest positive regression coefficients correspond to spectral inflection points ('shoulders') rather than to spectral maxima. Adapted from Ref. [12], with permission from the Royal Society of Chemistry.

that some interferences or concomitants might be causing on those regions (atomisation times). This appears clearer when the final regression coefficients are considered (Figure 4.27) as the model has positive implications for those variables defining the secondary (frontal) peak and, surprisingly, those variables after the maxima of the atomic peaks (which sometimes became displaced to the right as some experiments of the experimental designs induced displacements of the atomic peaks; see [12] for more details). It is noteworthy that this location also coincides with a shoulder on the non-displaced peaks. Note also the relatively scarce importance of the regression coefficients located in the region of the main maximum of the atomic peaks.

The above yields the overall conclusion that the PLS regression model considered that the inflection points present in some atomic spectra are more relevant to predict the concentration of Sb in the aqueous samples than the typical atomic peak maxima. This is because the maxima become much more affected by the effects induced by the concomitants than the shoulders themselves.

4.6 Multivariate Figures of Merit

As in traditional methods that use univariate calibrations, the description of a method of analysis that uses multivariate calibration must also include the corresponding estimated figures of merit, including accuracy (trueness and precision), selectivity, sensitivity, linearity, limit of detection (LOD), limit of quantification (LOQ) and robustness. In this chapter, only the most common figures of merit are described. For a more extensive review, see [55]. Also, for a practical calculation of figures of merit in an atomic spectroscopic application, see [12].

A note of caution is needed here. The figures of merit presented in this section refer to the multivariate calibration model. This multivariate model, built with standards, is then applied to future real samples. If standards and real samples match, as should be the case in most applications, the calibration model is the essential step of the overall analytical procedure. However, if real samples require additional steps (basically preprocessing steps such as extractions, preconcentrations, *etc.*) different from those of the standards, then the calibration model is just one more step in the whole procedure. If the previous steps are not the same, this means that the figures of merit calculated for the model do not refer to the whole analytical procedure and, therefore, other approaches should be undertaken to calculate them [56].

4.6.1 Accuracy (Trueness and Precision)

The ISO Guide 3534-1 [57] defines accuracy as '*the closeness of agreement between a test result and the accepted reference value*', with a note stating that '*the term accuracy, when applied to a set of test results, involves a combination of random components and a common systematic error or bias component*'. Accuracy is expressed, then, as two components: *trueness* and *precision*. Trueness is defined as '*the closeness of agreement between the average value obtained from a large set of test results and an accepted reference value*' and it is normally expressed in terms of bias. Finally, precision is defined as '*the closeness of agreement between independent test results obtained under stipulated conditions*'.

In multivariate calibration, *accuracy* reports the closeness of agreement between the reference value and the value found by the calibration model and is generally expressed as the root mean square error of prediction (RMSEP, as described in section 4.5.6) for a set of validation samples:

$$\mathrm{RMSEP} = \sqrt{\frac{\sum_{i=1}^{It} (\hat{y} - y_{\mathrm{reference}})^2}{I_t}} \tag{4.14}$$

where I_t is the number of validation samples. RMSEP expresses the average error to be expected with the future predictions when the calibration model is applied to unknown samples.

Bias represents the average difference between predicted and measured y-values for the I_t samples in the validation set:

$$\text{bias} = \frac{\sum_{i=1}^{It} (\hat{y} - y_{\text{reference}})}{I_t} \quad (4.15)$$

Note that this is an average bias for all validation samples, which does not hold for each individual prediction. As an example, just assume that a model is built for your data with zero PLS components, so you will use the average of the concentrations for the calibration set to predict future samples. Then you will underestimate high values and overestimate low values, simply because you predict with the average all the time. You will obtain biased predictions, but the average bias, as expressed in eqn (4.15), would be fairly small, even not statistically significant. Underfitting is, as already discussed in Section 4.4, the main component of bias (but not the only one). Bearing these considerations in mind, bias is still a useful statistic for certain applications.

Finally, the precision of the multivariate calibration method can be estimated from the standard error of prediction (SEP) (also called *standard error of performance*), corrected for bias:

$$\text{SEP} = \sqrt{\frac{\sum_{i=1}^{It} (\hat{y} - y_{\text{reference}} - \text{bias})^2}{I_t - 1}} \quad (4.16)$$

It can be shown [58] that the relationship between bias and SEP is

$$\text{RMSEP}^2 = \text{bias}^2 + \text{SEP}^2 \quad (4.17)$$

which is consistent with the ISO definition of accuracy, *i.e.* a combination of trueness and precision. When bias is small, then RMSEP $\approx$ SEP.

The general recommendation for reporting accuracy for a multivariate calibration method would be as follows:

1. Calculate bias and check for its statistical significance. A simple way to perform this is through a Student's t-test [59]:

$$t_{\text{bias}} = \frac{|\text{bias}| \times \sqrt{I_t}}{\text{SEP}} \quad (4.18)$$

If t_{bias} is greater than the critical t value at a given confidence level (*i.e.* 95%) and $(I_t - 1)$ degrees of freedom, then there is evidence that the bias is significant. Another way to check for bias has been described in Section 4.5.5

2. Report RMSEP, as a global measure of accuracy, when bias is not significant.
3. Report bias and SEP, together with RMSEP, when the bias is significant. Correct future results for significant bias.

4.6.2 Sample-specific Standard Error of Prediction

The summary statistics for accuracy described in the previous section are just average statistics for the whole set of samples. They are important, because they allow monitoring of changes when the calibration model is optimised (*i.e.* a different data pretreatment or optimal number of factors is used). However, they do not provide an indication of the uncertainty for individual predicted concentrations. *Uncertainty* is defined as '*a parameter, associated with the result of a measurement, which characterises the dispersion of the values that could reasonably be attributed to the measurand*' [60]. Therefore, uncertainty gives an idea of the quality of the result since it provides the range of values in which the analyst believes that the 'true concentration' of the analyte is situated. Its estimation is a requirement for analytical laboratories [61] and is especially important when analytical results have to be compared with an established legal threshold.

There are two basic ways of estimating uncertainty in multivariate calibration, namely, error propagation and resampling strategies, such as jack-knife or bootstrap. We will focus on error propagation because it leads to closed-form expressions, which give a fundamental insight into the statistical properties of the calibration model and are highly convenient for estimating other figures of merit, such as detection and quantification limits.

A general mathematical expression has been derived to estimate standard errors $s(\hat{y})$ in the predicted concentrations, based on the Errors in Variables (EIV) theory [62], which, after some simplifications, reduces to the following equation [63]:

$$s(\hat{y}) = \sqrt{\mathrm{RMSEC}^2 \times \left(1 + h_{\mathrm{pre}}\right) - s_{\mathrm{ref}}{}^2} \tag{4.19}$$

where h_{pre} is the leverage of the sample being predicted, which quantifies the distance from the prediction sample to the training samples in the A-dimensional space (A being the number of PLS factors). For a centred model it is calculated as $h_{\mathrm{pre}} = 1/I + \mathbf{t}_{\mathrm{pre}}^{\mathrm{T}}(\mathbf{T}^{\mathrm{T}}\mathbf{T})\mathbf{t}_{\mathrm{pre}}$, where $\mathbf{t}_{\mathrm{pre}}$ is the $(A \times 1)$ score vector of the predicted sample and $\mathbf{T}$ the $(I \times A)$ matrix of scores for the calibration set. s_{ref} is an estimate of the precision of the reference method (or value). The subtraction of this term in eqn (4.19) corrects for the finite precision of the reference method (value). This means that predictions obtained by the multivariate PLS model can be more precise than a reference method [50,64]. When the precision of the reference method (value) is small or it is not available, eqn (4.19) simplifies to

$$s(\hat{y}) = \mathrm{RMSEC} \times \sqrt{1 + h_{\mathrm{pre}}} \tag{4.20}$$

Equation (4.20) was proposed by Höskuldsson [65] many years ago and has been adopted by the American Society for Testing and Materials (ASTM) [59]. It generalises the univariate expression to the multivariate context and concisely describes the error propagated from three uncertainty sources to the standard error of the predicted concentration: calibration concentration errors, errors in calibration instrumental signals and errors in test sample signals. Equations (4.19) and (4.20) assume that calibrations standards are representative of the test or future samples. However, if the test or future (real) sample presents uncalibrated components or spectral artefacts, the residuals will be abnormally large. In this case, the sample should be classified as an outlier and the analyte concentration cannot be predicted by the current model. This constitutes the basis of the excellent outlier detection capabilities of first-order multivariate methodologies.

As in univariate calibration, prediction intervals (PIs) can be constructed from the above estimated standard error of prediction, by means of a Student's t-statistic, as:

$$\mathrm{PI}(\hat{y}) = \hat{y} \pm t_{\nu,1-\alpha/2} \times \mathrm{RMSEC} \times \sqrt{1 + h_{\mathrm{pre}}} \tag{4.21}$$

It has been suggested that the number of degrees of freedom used in the calculation of RMSEC and $t_{\nu,1-\alpha/2}$ is determined by the approach of pseudo-degrees of freedom proposed by Van der Voet [66].

Other estimates of the standard error of prediction have been proposed. A fairly popular and user-friendly one, although with limited value for some data, is contained in the Unscrambler software (CAMO ASA, Oslo, Norway).

Finally, analyte concentrations predicted by multivariate calibration methods should be reported together with the corresponding estimate of the standard error of prediction.

4.6.3 Sensitivity

In multivariate calibration, not all the recorded signal is used for prediction, only the part that can be uniquely assigned to the analyte. This part, called the net analyte signal (NAS), is calculated for a given sample as $\mathbf{r}^* = \hat{y}_i \mathbf{s}^*$, where $\hat{y}_i$ is the predicted concentration for the ith sample and $\mathbf{s}^*$ is the net sensitivity vector [67]. Since the PLS model is calculated on mean-centred data, $\hat{y}_i$ is the mean-centred prediction and $\mathbf{r}^*$ is the NAS for the mean-centred spectrum. The net sensitivity characterises the model and is calculated as $\mathbf{s}^* = \mathbf{b}/||\mathbf{b}||^2$, where $\mathbf{b}$ is the vector of regression coefficients and $||\cdot||$ indicates the Euclidean norm. Using the NAS, the PLS model can be represented like a usual univariate calibration model, *i.e.* as a scatter plot of the concentrations of standards and their signal, $||\mathbf{r}^*||$ (see Figure 4.28).

Hence the sensitivity of the method is the inverse of the slope of the regression line (which is also the norm of the net sensitivity vector):

$$\mathbf{s} = ||\mathbf{s}^*|| = 1/||\mathbf{b}|| \tag{4.22}$$

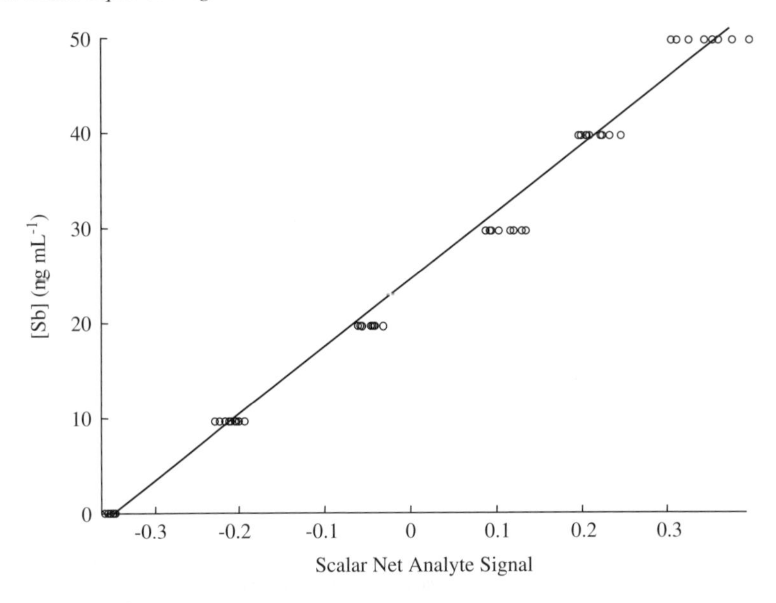

Figure 4.28 Pseudo-univariate representation of the PLS model implemented for the case study (four latent variables and mean-centred data). Reproduced from Ref. [12], with permission from the Royal Society of Chemistry.

4.6.4 Selectivity

In multivariate calibration, selectivity is commonly used to measure the amount of signal that cannot be used for prediction because of the overlap between the signal of the analyte and the signal of the interferences [68,69]. For inverse models, such as PLS, selectivity is usually calculated for each calibration sample as

$$\xi = ||\mathbf{r}^*||/||\mathbf{r}|| \tag{4.23}$$

where $\mathbf{r}^*$ and $\mathbf{r}$ are the NAS and the spectrum of the sample, respectively. Note that samples with the same amount of analyte but different amounts of interferences will have the same $||\mathbf{r}^*||$ but different $||\mathbf{r}||$ and hence a different selectivity value. This poses a problem for defining a unique selectivity value that characterises the PLS model. However, advantage can be taken of the experimental design of the calibration standards and calculate a global measure of selectivity. A plot of the norm of the NAS (mean-centred data) *versus* the norm of the measured spectra will follow a linear trend (see Figure 4.29 for the case study employed in this chapter), the slope of which is a global measure of selectivity that represents the model [12].

4.6.5 Limit of Detection

The limit of detection is defined as '*the true net concentration or amount of the analyte in the material to be analysed which will lead, with probability $(1-\beta)$, to*

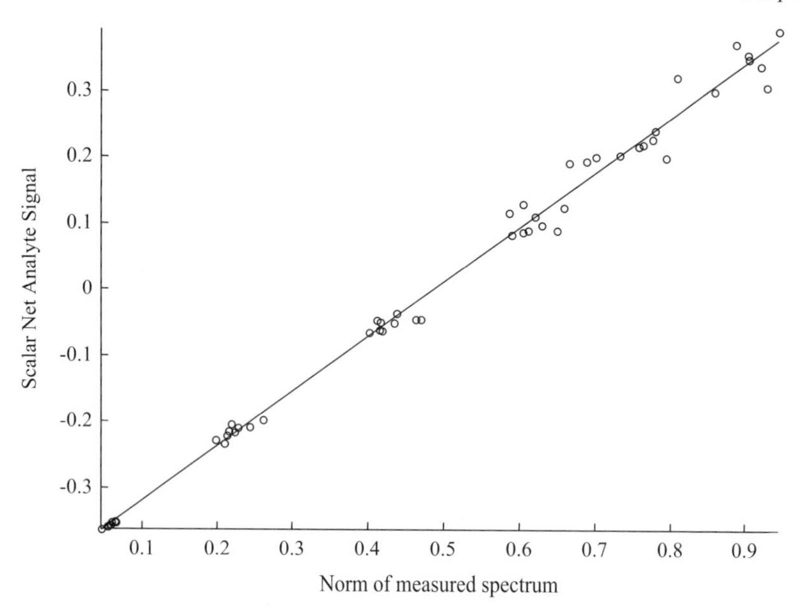

Figure 4.29 Plot of the norm of the NAS (mean-centred data) *versus* the norm of the measured spectra. The slope is a measure of the selectivity of the PLS model, whose value (here 0.83) indicates that approximately 83% of the measured spectra in the case study is used for prediction and that 17% of the measured signal is lost due to presence of the interferences.

the conclusion that the concentration or amount of the analyte in the analysed material is larger than that in the blank material' [70,71]. In more recent documents, ISO and IUPAC recommend that detection limits (minimum detectable amounts) should be derived from the theory of hypothesis testing and taking into account the probabilities of false positives (α) and false negatives (β). The concept of detection limit is illustrated in Figure 4.30.

The limit of detection is the analyte concentration that leads to a *correct* positive detection decision with sufficiently high probability ($1-\beta$). The detection decision amounts to comparing the prediction $\hat{y}$ with the critical level (Lc). This level is estimated in such a way that it allows for a positive detection decision with probability α when, in reality, the analyte is absent. The critical level is only determined by the distribution of the prediction under the null hypothesis (H_0: not present). By contrast, the limit of detection is also determined by the distribution of $\hat{y}$ under the alternative hypothesis (H_1: present at certain level).

The limit of detection of a multivariate PLS calibration method can be estimated from eqn (4.24) [12], which is based on an approximate expression for the sample-specific standard error of prediction, as developed by Höskuldsson [65]:

$$\mathrm{LOD} = \Delta(\alpha, \beta, \nu) \times \mathrm{RMSEC} \times \sqrt{1 + h_0} \tag{4.24}$$

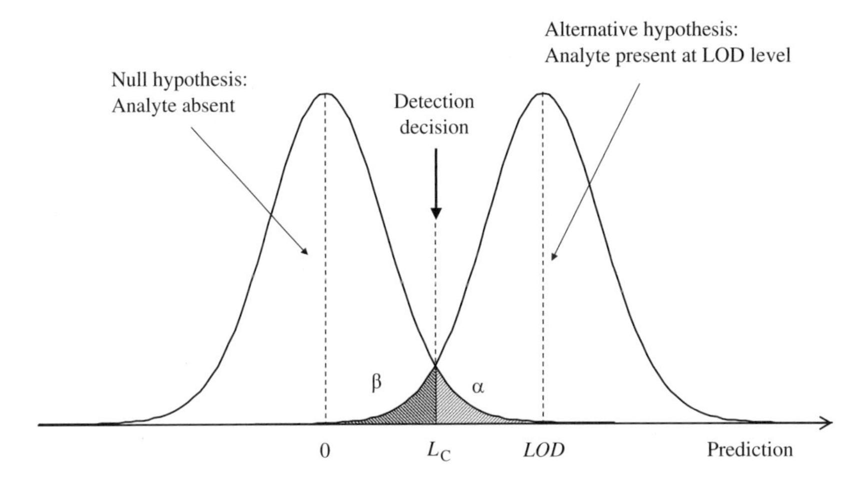

Figure 4.30 The concept of the detection limit.

The root mean squared error of calibration (RMSEC) has been defined above. The leverage, h_0, quantifies the distance of the predicted sample (at zero concentration level) to the mean of the calibration set in the *A*-dimensional space. For a centred model (spectra were mean centred before developing the model), it is calculated as $h_0 = 1/I + \mathbf{t}_0^{\mathrm{T}}(\mathbf{T}^{\mathrm{T}}\mathbf{T})^{-1}\mathbf{t}_0$, where $\mathbf{t}_0$ is the $(A \times 1)$ score vector of the predicted sample and $\mathbf{T}$ the $(I \times A)$ matrix of scores for the calibration set. h_0 can be estimated as an average value of the leverages of a set of validation samples having zero concentration of the analyte. Finally, the term $\Delta(\alpha,\beta,v)$ is a statistical parameter that takes into account the α and β probabilities of falsely stating the presence/absence of analyte, respectively, as recommended elsewhere [70,71]. When the number of degrees of freedom is high ($v \geq 25$), as is usually the case in multivariate calibration models and $\alpha = \beta$, then $\Delta(\alpha,\beta,v)$ can be safely approached to $2t_{1-\alpha,v}$.

A different approach, suggested by Ortiz *et al.* [72], is to perform standard univariate regression using a 'surrogate' signal variable, obtained from the sample multivariate signal and directly related to the concentration of the analyte. In the case of multivariate PLS calibration, the surrogate variable is the concentration of analyte in the calibration samples predicted by the optimal PLS model. From the univariate regression line (signal *versus* analyte concentration) it is then possible to derive figures of merit, such as the limit of detection, among others.

4.6.6 Limit of Quantification

The limit of quantification, LOQ, is a figure of merit that expresses the ability of a measurement process to *quantify* adequately an analyte and it is defined as '*the lowest amount or concentration of analyte that can be determined with an acceptable level of precision and accuracy*' [73]. In practice, the quantification limit is generally expressed as the concentration that can be determined with a specified relative standard deviation (RSD) (usually 10%). Thus:

$$\mathrm{LOQ} = k_{\mathrm{Q}} s_{\mathrm{Q}} \tag{4.25}$$

where $k_{\mathrm{Q}} = 10$ if the required RSD = 10% and s_{Q} is the sample-specific standard error of prediction at the level of the limit of quantification. To estimate s_{Q}, a number of independent determinations (at least seven) must be carried out on a sample which is known to contain the analyte at a concentration close to the LOQ. Usually, the LOQ estimation is carried out as part of the study to determine the working range of the analytical method and it is common practice to fix the LOQ as the lowest standard concentration of the calibration range.

4.7 Examples of Practical Applications

Notwithstanding PLS multivariate regression has not been broadly applied in atomic spectroscopy, it is by far the most common multivariate regression technique employed so far. In the following we present an extensive review of those applications combining multivariate regression and atomic absorption spectrometry published during the period 1990–2008. It is worth noting that, despite their relatively small number, they are more frequent for complex atomic measurements where information is difficult to extract (ICP and LIBS) and where analyte atomisation can be accompanied by complex interferents (*e.g.* ETAAS).

4.7.1 Flame and Electrothermal Atomic Spectrometry

Regarding FAAS, only a study by Flores *et al.* [74] employed PLS to quantify Cd in marine and river sediments measured by direct solid sampling FAAS.

There are three pioneering studies that, in our opinion, are very important for interested readers. Baxter and Öhman [75] reported that background correction in ETAAS has several problems that lead to incompletely corrected interferences and that such a problem could be overcome by PLS. In order to do so, they also applied the generalised standard additions method (GSAM) to develop the calibration set and then the final model. Among the different studies, they combined signal averaging of the replicates and signal smoothing before implementing the PLS model.

ETAAS was also combined with PLS by Baxter *et al.* [76] to determine As in marine sediments. They demonstrated that the classical standard additions method does not correct for spectral interferences (their main problem) because of mutual interactions between the two analytes of interest (As and Al). PLS-2 block was applied to quantify simultaneously both elements (PLS-2 block is the notation for the PLS regression when several analytes are predicted simultaneously; it is not widely applied, except when the analytes are correlated). Surprisingly, PLS has not attracted much attention among ETAAS practitioners.

Since the work of Baxter *et al.* [75,76] around 1990, we have not found many more recent applications and it was not until 2003 that Felipe-Sotelo *et al.* [77] presented another application. They considered a problem where a major element (Fe) caused spectral and chemical interferences on a minor element (Cr), which had to be quantified in natural waters. They demonstrated that linear PLS handled (eventual) nonlinearities since polynomial PLS and locally weighted regression (nonlinear models) did not outperform its results. Further, it was found that linear PLS was able to model three typical effects which currently occur in ETAAS: peak shift, peak enhancement (depletion) and random noise.

The situation where several major concomitants affect the atomic signal of the trace element(s) to be measured (whose concentrations are several orders of magnitude lower) is more complex. PLS was recently demonstrated to give good results [12] when a proper experimental design was developed to quantify Sb in waters by ETAAS. Instead of the traditional approach where only an experimental design is deployed to establish the calibration (and validation) set, a saturated experimental design considering the concomitants was deployed at each of the concentration levels considered for the analyte. Once again, polynomial PLS performed worse than or equal than linear PLS, demonstrating that linear models are good enough. Further, multivariate figures of merit were calculated following IUPAC and ISO guidelines. In these two later papers, the authors studied graphically the factor loadings and regression coefficients to gain an insight into how the models behaved.

More recently, Felipe-Sotelo *et al.* [78] modelled complex interfering effects on Sb when soil, sediments and fly ash samples were analysed by ultrasonic slurry sampling–ETAAS (USS–ETAAS). Sometimes, spectral and chemical interferences cannot be totally resolved in slurries using chemical modifiers, ashing programmes, *etc.*, because of the absence of a sample pretreatment step to eliminate/reduce the sample matrix. Hence the molecular absorption signal is so high and structured that background correctors cannot be totally effective. In addition, alternative wavelengths may not be a solution due to their low sensitivity when trace levels are measured. To circumvent these problems, the authors employed current PLS, second-order polynomial PLS and artificial neural networks (ANNs; see Chapter 5) to develop predictive models on experimentally designed calibration sets. Validation with five certified reference materials (CRMs) showed that the main limitations of the models were related to the USS–ETAAS technique; *i.e.* the mass/volume ratio and the low content of analyte in some solid matrices (which forced the introduction of too much sample matrix into the atomiser). Both PLS and ANN gave good results since they could handle severe problems such as peak displacement, peak enhancement/depletion and peak tailing. Nevertheless, PLS was preferred because the loadings and correlation coefficients could be interpreted chemically. Moreover, this approach allowed a reduction in the laboratory workload to optimise the analytical procedure by around 50%.

4.7.2 Inductively Coupled Plasma

Inductively coupled plasma (ICP) needs careful extraction of the relevant information to obtain satisfactory models and several chemometric studies were found. Indeed, many ICP spectroscopists have applied multivariate regression and other multivariate methods to a number of problems. An excellent review has recently been published on chemometric modelling and applications of inductively coupled plasma optical emission spectrometry (ICP-OES) [79].

Several papers can be traced back to the early 1990s: Glick *et al.* [80] compensated for spectral and stray light interferences in an ICP-OES-photodiode array spectrometer; Ivaldi *et al.* [81] extracted ICP-OES information using least-squares regression; Danzer and co-workers used PCR and multi-line PCR and PLS on ICP-OES spectra [82,83]. Two other applications are those from Van Veen *et al.* [84], applying Kalman filtering to ICP-OES spectra and reviewing several procedures to perform background correction and multicomponent analysis [85]. Sadler and Littlejohn [86] applied PLS to detect uncorrected additive interferences. Venth *et al.* [87] compared PLS and canonical correlation analysis to solve isobaric and polyatomic ion interferences in Mo–Zr alloys measured by ICP-MS. Pimentel *et al.* [88] applied PLS and PCR to measure simultaneously five metals (Mn, Mo, Cr, Ni and Fe) in steel samples, using a low-resolution ICP with diode-array detection.

Rupprecht and Probst [89] corrected ICP-MS spectral and non-spectral interferences by different multivariate regression methods. They studied multivariate ordinary least-squares (MLR), PCR and PLSR and compared them with ordinary least-squares regression. Further, they tested different data pretreatments; namely mean centring, autoscaling, scaling 0 to 1 and internal standardisation. The best model was developed using PLS, mean-centred data (also internal standardisation would be fine) and variable selection according to the regression coefficients (studied in a previous model).

Moberg *et al.* [90] used ICP-MS to determine Cd in fly ash and metal alloys; since severe spectral overlaps arose, multivariate regression outperformed other univariate approaches. They also studied whether day-to-day ICP-MS recalibration could be avoided so that the calibration set could be constructed on several runs.

Haaland *et al.* [91] developed a so-called multi-window classical least-squares method for ICP-OES measurements [charge-couple device (CCD) detector arrays]. Essentially, it consisted in performing a classical least-squares regression in each of the spectral windows which were measured and combining the concentration predictions (for a given analyte). The methodology was compared with PLS and it proved superior and capable of handling interferences from several concomitants.

Griffiths *et al.* [92] quantified Pt, Pd and Rh in autocatalyst digests by ICP with a CCD detector array. They compared univariate techniques (pure standards, pure standards with interelement correction factors and matrix-matched standards) and PLS, the latter being superior in general, although less effective at low concentrations due to spectral noise. They also studied the effect of using the gross signal or background-corrected signals, being more successful the former option. In subsequent work, Griffiths *et al.* [93] studied how to reduce the ICP-OES (segmented-array CCD detector) raw variables (5684 wavelengths per spectrum). This application holds many similarities with classical molecular spectrometry, from where they selected two advanced algorithms, applied in three steps: (i) application of an Uninformative Variable Elimination PLS algorithm (UVE-PLS), which identifies variables with close-to-zero regression coefficients, (ii) application of an Informative Variable Degradation-PLS, which ranked variables using a ratio calculated as the regression coefficient divided by its estimated standard error, and (iii) selection of the variables according to that ratio. Interestingly, they had to autoscale data instead of using the more frequent mean-centring pretreatment.

Other multivariate methods have been applied to ICP spectra for quantitative measurements. As examples, they include multicomponent spectral fitting (which is incorporated in several commercial instrument software) [81]; matrix projection, which avoids measurement of background species [94,95]; generalised standard additions [96]; and Bayesian analysis [97].

4.7.3 Laser-induced Breakdown Spectrometry

Laser-induced breakdown spectrometry (LIBS) also required multivariate regression to solve complex problems. Here, the first paper was found date back to 2000 and it was related to jewellery studies. Amador-Hernández *et al.* [98] showed that PLS models were highly satisfactory for measuring Au and Ag in Au–Ag–Cu alloys. This study was further completed by Jurado-López and Luque de Castro [99], who compared the 'hybridised' LIBS–PLS technique with analytical scanning microscopy, ICP-OES, FAAS and LIBS. Although the accuracy of LIBS–PLS did not outperform the official procedure, several characteristics made it highly appealing for routine work. The same group reported two other studies with routine LIBS applications, but without details on the multivariate models [100,101].

Martín *et al.* [102] reported a study in which LIBS was applied for the first time to wood-based materials where preservatives containing metals had to be determined. They applied PLS-1 block and PLS-2 block (because of the interdependence of the analytes) to multiplicative scattered-corrected data (a data pretreatment option of most use when diffuse radiation is employed to obtain spectra). They authors studied the loadings of a PCA decomposition to identify the main chemical features that grouped samples. Unfortunately, they did not extend the study to the PLS factors. However, they analysed the regression coefficients to determine the most important variables for some predictive models.

4.7.4 Miscellanea

Stosch *et al.* [103] compared isotope dilution techniques with a novel surface-enhanced Raman scattering (SERS) method to determine creatinine in human serum. The procedure involved the use of isotopically labelled C and N and PLS (87 calibration spectra covering the full range of neat creatinine).

Adams and Allen [104] combined a variable selection algorithm with PLS in order to quantify nine metals in certified geological materials using X-ray fluorescence. It was found that PLS models outperformed the MLR approaches.

Lemberge *et al.* [105] applied PLS regression to archaeological vessels to quantify six metals in the form of their oxides using electron-probe X-ray microanalysis (EPXMA) (180 samples, of which 25 constituted the calibration set) and micro-X-ray fluorescence (53 samples; all samples were required to set the calibration model). The effects of the matrix effects on the models were also studied. Later, the same group used EPXMA to predict the concentration of several analytes in archaeological glass vessels [106]. They developed a so-called M-robust estimator that was combined with PLS in order to avoid the influence of outlying samples and noise.

Resano *et al.* [107] used LA-ICP-MS to fingerprint diamonds from different deposits. Nine elements were selected and the samples were classified by different methods. Different standardisation methods were studied and PLS outperformed other pattern recognition methods. In this particular application, the main aim was not predictive, in the typical sense, but discriminant (this is also termed discriminant-PLS).

Wagner *et al.* [108] developed a PLS model to predict the amount of protein adsorbed in a metallic basis using time-of-flight SIMS (TOF-SIMS) spectra. Study of the multivariate models yielded insight into the surface chemistry and the mechanism for protein resistance of the coatings. The same group reported two other similar studies with satisfactory results [109,110].

References

1. H. Wold, Soft modelling by latent variables: the non-linear iterative partial least squares approach, in J. Gani (ed.), *Papers in Honour of M. S. Bartlett: Perspectives in Probability and Statistics*, Academic Press, London, 1975.
2. H. Wold, Soft modelling: the basic design and some extensions, in K. G. Jöreskog and H. Wold (eds), *Systems Under Indirect Observation, Causality–Structure–Prediction*, Vol II, North-Holland, Amsterdam, 1982, pp. 1–53.
3. I. S. Helland, Some theoretical aspects of partial least squares regression, *Chemom. Intell. Lab. Syst.*, 58, 2001, 97–107.
4. H. Martens and T. Næs, *Multivariate Calibration*, Wiley, Chichester, 1989.
5. S. Wold, M. Sjöstrom and L. Eriksson, PLS-regression, a basic tool for chemometrics, *Chemom. Intell. Lab. Syst.*, 58, 2001, 109–130.
6. S. Wold, P. Geladi, K. Esbensen and J. Öhman, Multi-way principal component and PLS-analysis, *J. Chemom.*, 1, 1987, 41–56.
7. P. Geladi and B. R. Kowalski, *Anal. Chim. Acta*, 185, 1986, 1–17.

8. M. Otto, *Chemometrics*, Wiley-VCH, Weinheim, 2007.
9. S. Wold, N. Kettaneh-Wold and B. Skagerberg, Nonlinear PLS modelling, *Chemom. Intell. Lab. Syst.*, 7, 1989, 53–65.
10. A. Wang, T. Isaksson and B. R. Kowalski, New approach for distance measurement in locally weighted regression, *Anal. Chem.*, 66, 1994, 249–260.
11. T. Naes, T. Isaksson and B. Kowalski, Locally weighted regression and scatter correction for near-infrared reflectance data, *Anal. Chem.*, 62, 1990, 664–673.
12. M. Felipe-Sotelo, M. J. Cal-Prieto, J. Ferré, R. Boqué, J. M. Andrade and A. Carlosena, Linear PLS regression to cope with interferences of major concomitants in the determination of antimony by ETAAS, *J. Anal. At. Spectrom.*, 21, 2006, 61–68.
13. M. Felipe-Sotelo, J. M. Andrade, A. Carlosena and D. Prada, Partial least squares multivariate regression as an alternative to handle interferences of Fe on the determination of trace Cr in water by electrothermal atomic absorption spectrometry, *Anal. Chem.*, 75, 2003, 5254–5261.
14. J. M. Andrade, M. S. Sánchez and L. A. Sarabia, Applicability of high-absorbance MIR spectroscopy in industrial quality control of reformed gasolines, *Chemom. Intell. Lab. Syst.*, 46, 1999, 41–55.
15. S. P. Jacobson and A. Hagman, Chemical composition analysis of carrageenans by infrared spectroscopy using partial least squares and neural networks, *Anal. Chim. Acta*, 284, 1993, 137–147.
16. L. Hadjiiski, P. Geladi and Ph. Hopke, A comparison of modelling nonlinear systems with artificial neural networks and partial least squares, *Chemom. Intell. Lab. Syst.*, 49, 1999, 91–103.
17. M. Blanco, J. Coello, H. Iturriaga, S. Maspoch and J. Pagés, NIR calibration in non-linear systems by different PLS approaches and artificial neural networks, *Chemom. Intell. Lab. Syst.*, 50, 2000, 75–82.
18. R. Brereton, Introduction to multivariate calibration in analytical chemistry, *Analyst*, 125, 2000, 2125–2154.
19. R. DiFoggio, Guidelines for applying chemometrics to spectra: feasibility and error propagation, *Anal. Chem.*, 54(3), 2000, 94A–113A.
20. F. Laborda, J. Medrano and J. R. Castillo, Estimation of the quantification uncertainty from flow injection and liquid chromatography transient signals in inductively coupled plasma mass spectrometry, *Spectrochim. Acta, Part B*, 59(6), 2004, 857–870.
21. M. J. Cal-Prieto, A. Carlosena, J. M. Andrade, S. Muniategui, P. López-Mahía and D. Prada, Study of chemical modifiers for the direct determination of antimony in soils and sediments by ultrasonic slurry sampling-ETAAS with D_2 compensation, *At. Spectrosc.*, 21(3), 2000, 93–99.
22. K. R. Beebe, P. J. Pell and M. B. Seasholtz, *Chemometrics, a Practical Guide*, Wiley, Chichester, 1998.
23. R. Brereton, *Applied Chemometrics for Scientists*, Wiley, Chichester, 2007.
24. T. Næs, T. Isaksson, T. Fearn and T. Davies, *A User-friendly Guide to Multivariate Calibration and Classification*, NIR Publications, Chichester, 2002.

25. A. Savitzky and M. J. E. Golay, Smoothing and differentiation of data by simplified least squares procedures, *Anal. Chem.*, 36, 1964, 1627–1639.
26. J. Riordon, E. Zubritsky, A. Newman, 10 seminal papers, *Anal. Chem.*, 27, 2000, 324A–329A.
27. J. Steinier, Y. Termonia and J. Deltour, Smoothing and differentiation of data by simplified least square procedure, *Anal. Chem.*, 44, 1972, 1906–1909.
28. K. Esbensen, T. Midtgaard and S. Schönkopf, *Multivariate Analysis in Practice*, CAMO, Oslo, 1994.
29. T. Næs and H. Martens, Principal component regression in NIR analysis: viewpoints, background details and selection of components, *J. Chemom.*, 2(2), 1988, 155–167.
30. N. M. Faber, A closer look at the bias–variance trade-off in multivariate calibration, *J. Chemom.*, 13, 1999, 185–192.
31. S. Lanteri, Full validation procedures for feature selection in classification and regression problems, *Chemom. Intell. Lab. Syst.*, 15, 1992, 159–169.
32. M. Forina, S. Lanteri, R. Boggia and E. Bertran, Double cross full validation, *Quim. Anal.*, 12, 1993, 128–135.
33. M. C. Denham, Choosing the number of factors in partial least squares regression: estimating and minimizing the mean squared error of prediction, *J. Chemom.*, 14, 2000, 351–361.
34. S. Wiklund, D. Nilsson, L. Eriksson, M. Sjöström, S. Wold and K. Faber, A randomisation test for PLS component selection, *J. Chemom.*, 21, 2007, 427–439.
35. S. Gourvénec, J. A. Fernández-Pierna, D. L. Massart and D. N. Rutledge, An evaluation of the PoLiSh smoothed regression and the Monte Carlo cross-validation for the determination of the complexity of a PLS model, *Chemom. Intell. Lab. Syst.*, 68, 2003, 41–51.
36. D. N. Rutledge, A. Barros and I. Delgadillo, PoLiSh-smoothed partial least squares regression, *Anal. Chim. Acta.*, 446, 2001, 281–296.
37. R. Todeschini, V. Consonni, A. Mauri and M. Pavan, Detecting 'bad' regression models: multicriteria fitness functions in regression analysis, *Anal. Chim. Acta*, 515, 2004, 199–208.
38. M. P. Gómez-Carracedo, J. M. Andrade, D. N. Rutledge and N. M. Faber, Selecting the optimum number of partial least squares components for the calibration of attenuated total reflectance-mid-infrared spectra of undesigned kerosene samples, *Anal. Chim. Acta*, 585, 2007, 253–265.
39. S. Wold, Cross-validatory estimation of the number of components in factor and principal component models, *Technometrics*, 24, 1978, 397–405.
40. N. M. Faber and R. Rajkó, How to avoid over-fitting in multivariate calibration: the conventional validation approach and an alternative, *Anal. Chim. Acta*, 595, 2007, 98–106.
41. H. van der Voet, Comparing the predictive accuracy of models using a simple randomisation test, *Chemom. Intell. Lab. Syst.*, 25, 1994, 313–323.
42. H. van der Voet, Corrigendum to 'Comparing the predictive accuracy of models using a simple randomisation test', *Chemom. Intell. Lab. Syst.*, 28, 1995, 315–315.

43. D. N. Rutledge and A. S. Barros, Durbin–Watson statistic as a morphological estimator of information content, *Anal. Chim. Acta*, 454(2), 2002, 277–295.
44. D. Veltkamp and D. Gentry, *PLS-2 Block Modeling (User's Manual, v. 3. 1)*, Center for Process Analytical Chemistry, Washington, DC, 1988.
45. K. Esbensen, T. Midtgaard and S. Schönkopf, *Multivariate Analysis in Practice*, CAMO, Oslo, 1994.
46. H. Hotelling, The generalisation of Student's ratio, *Ann. Math. Statist.*, 2, 1931, 360–378.
47. B. M. Wise and N. B. Gallagher, *PLS Toolbox v 1.5*, Eigenvector Technologies, Manson, WA, 1995.
48. F. McLennan and B. R. Kowalski, *Process Analytical Chemistry*, Blackie Academic (Chapman & Hall), London, 1996.
49. D. L. Massart, B. G. M. Vandeginste, S. de Jong, P. J. Lewi and J. Smeyers-Verbke, *Handbook of Chemometrics and Qualimetrics, Part A*, Elsevier, Amsterdam, 1997.
50. N. M. Faber, F. H. Schreutelkamp and H. W. Vedder, Estimation of prediction uncertainty for a multivariate calibration model, *Spectrosc. Eur.*, 16(101), 2004, 17–20.
51. K. Faber and B. R. Kowalski, Improved prediction error estimates for multivariate calibration by correcting for the measurement error in the reference values, *Appl. Spectrosc.*, 51(5), 1997, 660–665.
52. J. A. Fernández Pierna, L. Jin, F. Wahl, N. M. Faber and D. L. Massart, Estimation of partial least squares regression prediction uncertainty when the reference values carry a sizeable measurement error, *Chemom. Intell. Lab. Syst.*, 65, 2003, 281–291.
53. M. J. Griffiths and S. L. R. Ellison, A simple numerical method of estimating the contribution of reference value uncertainties to sample-specific uncertainties in multivariate regression, *Chemom. Intell. Lab. Syst.*, 83, 2006, 133–138.
54. T. Fearn, Assessing calibrations: SEP, RPD, RER and R2, *NIR News*, 13, 2002, 12–20.
55. A. C. Olivieri, N. M. Faber, J. Ferré, R. Boqué, J. H. Kalivas and H. Mark, Uncertainty estimation and figures of merit for multivariate calibration (IUPAC Technical Report), *Pure Appl. Chem.*, 78, 2006, 633–661.
56. R. Boqué, A. Maroto, J. Riu and F. X. Rius, Validation of analytical methods, *Grasas Aceites*, 53, 2002, 128–143.
57. *ISO 3534-1:1993. Statistics – Vocabulary and Symbols – Part 1: Probability and General Statistical Terms*, International Organization for Standardization, Geneva, 1993.
58. A. M. C. Davies and T. Fearn, Back to basics: calibration statistics, *Spectrosc. Eur.*, 18, 2006, 31–32.
59. *American Society for Testing and Materials (ASTM) Practice E1655-00. Standard Practices for Infrared Multivariate Quantitative Analysis, ASTM Annual Book of Standards*, Vol. 03.06, ASTM, West Conshohocken, PA, 2001, pp. 573–600.

60. *ISO Guide 99:1993. International Vocabulary of Basic and General Terms in Metrology (VIM)*, International Organization for Standardization, Geneva, 1993.
61. *ISO/IEC 17025:2005, General Requirements for the Competence of Testing and Calibration Laboratories*, International Organization for Standardization, Geneva, 2005.
62. K. Faber and B. R. Kowalski, Prediction error in least squares regression: further critique on the deviation used in The Unscrambler, *Chemom. Intell. Lab. Syst.*, 34, 1996, 283–292.
63. N. M. Faber and R. Bro, Standard error of prediction for multiway PLS. 1. Background and a simulation study, *Chemom. Intell. Lab. Syst.*, 61, 2002, 133–149.
64. N. M. Faber, X.-H. Song and P. K. Hopke, Prediction intervals for partial least squares regression, *Trends Anal. Chem.*, 22, 2003, 330–334.
65. A. Höskuldsson, PLS regression methods, *J. Chemom.*, 2, 1988, 211–228.
66. H. Van der Voet, Pseudo-degrees of freedom for complex predictive models: the example of partial least squares, *J. Chemom.*, 13, 1999, 195–208.
67. J. Ferré and N. M. Faber, Calculation of net analyte signal for multivariate calibration, *Chemom. Intell. Lab. Syst.*, 69, 2003, 123–136.
68. A. Lorber, K. Faber and B. R. Kowalski, Net analyte signal calculation in multivariate calibration, *Anal. Chem.*, 69, 1997, 1620–1626.
69. K. Faber, A. Lorber and B. R. Kowalski, Analytical figures of merit for tensorial calibration, *J. Chemom.*, 11, 1997, 419–461.
70. *ISO 11843-1:1997: Capability of Detection. Part 1: Terms and Definitions*, International Organization for Standardixation, Geneva, 1997.
71. L. A. Currie, Nomenclature in evaluation of analytical methods including detection and quantification capabilities, *Pure Appl. Chem.*, 67, 1995, 1699–1723.
72. M. C. Ortiz, L. A. Sarabia, A. Herrero, M. S. Sánchez, M. B. Sanz, M. E. Rueda, D. Giménez and M. E. Meléndez, Capability of detection of an analytical method evaluating false positive and false negative (ISO 11843) with partial least squares, *Chemom. Intell. Lab. Syst.*, 69, 2003, 21–33.
73. EURACHEM, *The Fitness for Purpose of Analytical Methods. A Laboratory Guide to Method Validation and Related Topics*, EURACHEM, Teddington, 1998.
74. E. M. M. Flores, J. N. G. Paniz, A. P. F. Saidelles, E. I. Müller and A. B. da Costa, Direct cadmium determination in sediment samples by flame atomic absorption spectrometry using multivariate calibration procedures, *J. Anal. At. Spectrom.*, 18, 2003, 769–774.
75. D. C. Baxter and J. Öhman, Multi-component standard additions and partial least squares modelling, a multivariate calibration approach to the resolution of spectral interferences in graphite furnace atomic absorption spectrometry, *Spectrochim. Acta, Part B*, 45(4–5), 1990, 481–491.
76. D. C. Baxter, W. Frech and I. Berglund, Use of partial least squares modelling to compesate for spectral interferences in electrothermal atomic

absorption spectrometry with continuum source background correction, *J. Anal. At. Spectrom.*, 6, 1991, 109–114.

77. M. Felipe-Sotelo, J. M. Andrade, A. Carlosena and D. Prada, Partial least squares multivariate regression as an alternative to handle interferences of Fe on the determination of trace Cr in water by electrothermal atomic absorption spectrometry, *Anal. Chem.*, 75(19), 2003, 5254–5261.
78. M. Felipe-Sotelo, M. J. Cal-Prieto, M. P. Gómez-Carracedo, J. M. Andrade, A. Carlosena and D. Prada, Handling complex effects in slurry-sampling-electrothermal atomic absorption spectrometry by multivariate calibration, *Anal. Chim. Acta*, 571, 2006, 315–323.
79. M. Grotti, Improving the analytical performance of inductively coupled plasma optical emission spectrometry by multivariate analysis techniques, *Ann. Chim. (Rome)*, 94, 2004, 1–15.
80. M. Glick, K. R. Brushwyler and G. M. Hieftje, Multivariate calibration of a photodiode array spectrometer for atomic emission spectroscopy, *Appl. Spectrosc.*, 45, 1991, 328–333.
81. J. C. Ivaldi, D. Tracy, T. W. Barnard and W. Slavin, Multivariate methods for interpretation of emission spectra from the inductively coupled plasma, *Spectrochim. Acta, Part B*, 47, 1992, 1361–1371.
82. K. Danzer and M. Wagner, Multisignal calibration in optical emission spectroscopy, *Fresenius' J. Anal. Chem.*, 346, 1993, 520–524.
83. K. Danzer and K. Venth, Multisignal calibration in spark- and ICP-OES, *Fresenius' J. Anal. Chem.*, 350, 1994, 339–343.
84. E. H. Van Veen, S. Bosch and M. T. C. De Loos-Vollebregt, Kalman filter approach to inductively coupled plasma atomic-emission-spectrometry, *Spectrochim. Acta, Part B*, 49, 1994, 829–846.
85. E. H. Van Veen and M. T. C. De Loos-Vollebregt, Application of mathematical procedures to background correction and multivariate analysis in inductively coupled plasma-optical emission spectrometry, *Spectrochim. Acta, Part B*, 53(5), 1998, 639–669.
86. D. A. Sadler and D. Littlejohn, Use of multiple emission lines and principal component regression for quantitative analysis in inductively coupled plasma atomic emission spectrometry with charge coupled device detection, *J. Anal. At. Spectrom.*, 11, 1996, 1105–1112.
87. K. Venth, K. Danzer, G. Kundermann and K.-H. Blaufuss, Multisignal evaluation in ICP MS, determination of trace elements in molybdenum–zirconium alloys, *Fresenius' J. Anal. Chem.*, 354(7–8), 1996, 811–817.
88. M. F. Pimentel, B. De Narros Neto, M. C. Ugulino de Araujo and C. Pasquini, Simultaneous multielemental determination using a low-resolution inductively coupled plasma spectrometer/diode array detection system, *Spectrochim. Acta, Part B*, 52(14), 1997, 2151–2161.
89. M. Rupprecht and T. Probst, Development of a method for the systematic use of bilinear multivariate calibration methods for the correction of interferences in inductively coupled plasma-mass spectrometry, *Anal. Chim. Acta*, 358(3), 1998, 205–225.

90. L. Moberg, K. Pettersson, I. Gustavsson and B. Karlberg, Determination of cadmium in fly ash and metal allow reference materials by inductively coupled plasma mass spectrometry and chemometrics, *J. Anal. At. Spectrom.*, 14(7), 1999, 1055–1059.
91. D. M. Haaland, W. B. Chambers, M. R. Keenan and D. K. Melgaard, Multi-window classical least-squares multivariate calibration methods for quantitative ICP-AES analyses, *Appl. Spectrosc.*, 54(9), 2000, 1291–1302.
92. M. L. Griffiths, D. Svozil, P. J. Worsfold, S. Denham and E. H. Evans, Comparison of traditional and multivariate calibration techniques applied to complex matrices using inductively coupled plasma atomic emission spectroscopy, *J. Anal. At. Spectrom.*, 15, 2000, 967–972.
93. M. L. Griffiths, D. Svozil, P. Worsfold, S. Denham and E. H. Evans, Variable reduction algorithm for atomic emission spectra: application to multivariate calibration and quantitative analysis of industrial samples, *J. Anal. At. Spectrom.*, 17, 2002, 800–812.
94. P. Zhang, D. Littlejohn and P. Neal, Mathematical prediction and correction of interferences for optimisation of line selection in inductively coupled plasma optical emission spectrometry, *Spectrochim. Acta, Part B*, 48(12), 1993, 1517–1535.
95. P. Zhang and D. Littlejohn, Peak purity assessment by matrix projection for spectral line selection and background correction in inductively coupled plasma optical emission spectrometry, *Spectrochim. Acta, Part B*, 50(10), 1995, 1263–1279.
96. S. Luan, H. Pang and R. S. Houk, Application of generalized standard additions method to inductively coupled plasma atomic emission spectroscopy with an echelle spectrometer and segmented-array charge-coupled detectors, *Spectrochim. Acta, Part B*, 50(8), 1995, 791–801.
97. B. L. Sharp, A. S. Bashammakh, Ch. M. Thung, J. Skilling and M. Baxter, Bayesian analysis of inductively coupled plasma mass spectra in the range 46–88 Daltons derived from biological materials, *J. Anal. At. Spectrom.*, 17, 2002, 459–468.
98. J. Amador-Hernández, L. E. García-Ayuso, J. M. Fernández-Romero and M. D. Luque de Castro, Partial least squares regression for problem solving in precious metal analysis by laser induced breakdown spectrometry, *J. Anal. At. Spectrom.*, 15, 2000, 587–593.
99. A. Jurado-López and M. D. Luque de Castro, An atypical interlaboratory assay: looking for an updated hallmark (jewelry) method, *Anal. Bioanal. Chem.*, 372, 2002, 109–114.
100. J. L. Luque-García, R. Soto-Ayala and M. D. Luque de Castro, Determination of the major elements in homogeneous and heterogeneous samples by tandem laser-induced breakdown spectroscopy, partial least squares regression, *Microchem. J.*, 73, 2002, 355–362.
101. A. Jurado-López and M. D. Luque de Castro, Laser-induced breakdown spectrometry in jewellery industry, Part II, quantitative characterisation of goldfilled interface, *Talanta*, 59, 2003, 409–415.

102. M. Martín, N. Labbé, T. G. Rials and S. D. Wullschleger, Analysis of preservative-treated wood by multivariate analysis of laser-induced breakdown spectroscopy spectra, *Spectrochim. Acta, Part B*, 60, 2005, 1179–1185.
103. R. Stosch, A. Henrion, D. Schiel and B. Guettler, Surface-enhanced Raman scattering based approach for quantitative determination of creatinine in human serum, *Anal. Chem.*, 77, 2005, 7386–7392.
104. J. M. Adams and J. R. Allen, Quantitative X-ray fluorescence analysis of geological matrices using PLS regression, *Analyst*, 123, 1998, 537–541.
105. P. Lemberge, I. De Raedt, K. H. Janssens, F. Wei and P. J. Van Espen, Quantitative analysis of 16–17th century archaeological glass vessels using PLS regression of EXPMA and μ-XRF data, *J. Chemom.*, 14, 2000, 751–763.
106. S. Serneels, C. Croux, P. Filzmoser and P. J. Van Espen, Partial robust M-regression, *Chemom. Intell. Lab. Syst.*, 79, 2005, 55–64.
107. M. Resano, F. Vanhaecke, D. Hutsebaut, K. De Corte and K. Monees, Posibilities of laser ablation-inductively coupled plasma-mass spectrometry for diamong fingerprinting, *J. Anal. At. Spectrom.*, 18, 2003, 1238–1242.
108. M. S. Wagner, S. Pasche, D. G. Castner and M. Textor, Characterisation of poly(L-lysine)-graft-poly(ethylene glycol) assembled monolayers on niobium pentoxide substrates using time-of-flight secondary ion mass spectrometry and multivariate analysis, *Anal. Chem.*, 76, 2004, 1483–1492.
109. M. S. Wagner, M. Shen, T. A. Horbett and D. G. Castner, Quantitative analysis of binary adsorbed protein films by time-of-flight secondary ion mass spectrometry, *J. Biomed. Materials Res.*, 64A(1), 2003, 1–11.
110. M. Shen, M. S. Wagner, D. G. Castner and T. A. Horbett, Multivariate surface analysis of plasma-deposited tetraglyme for reduction of protein adsorption and monocyte adhesion, *Langmuir*, 19(5), 2003, 1692–1699.

CHAPTER 5

Multivariate Regression using Artificial Neural Networks

JOSE MANUEL ANDRADE-GARDA,[a] ALATZNE CARLOSENA-ZUBIETA,[a] MARÍA PAZ GÓMEZ-CARRACEDO[a] AND MARCOS GESTAL-POSE[b]

[a] Department of Analytical Chemistry, University of A Coruña, A Coruña, Spain; [b] Department of Information and Communications Technologies, University of A Coruña, A Coruña, Spain

5.1 Introduction

This chapter introduces a relatively recent regression methodology based on a set of so-called '*natural computation*' methods. In contrast to classical programming, they work with rules rather than with well-defined and fixed algorithms and came into routine use during the 1990s and gained some popularity. After an initial starting '*rush*', they are now applied with some moderation and, mainly, to handle complex data sets (*e.g.* spectra with strong nonlinearities). Of the many natural computation techniques, *artificial neural networks* (ANNs) stand out, particularly their application in carrying out multivariate regressions. As they constitute a promising way to cope with complex spectral problems (both molecular and atomic), they are introduced here.

During the last two or three decades, chemists became used to the application of computers to control their instruments, develop analytical methods, analyse data and, consequently, to apply different statistical methods to explore multivariate correlations between one or more output(s) (*e.g.* concentration of an analyte) and a set of input variables (*e.g.* atomic intensities, absorbances).

RSC Analytical Spectroscopy Monographs No. 10
Basic Chemometric Techniques in Atomic Spectroscopy
Edited by Jose Manuel Andrade-Garda

Published by the Royal Society of Chemistry, www.rsc.org

On the other hand, the huge efforts made by atomic spectroscopists to resolve interferences and optimise the instrumental measurement devices to increase accuracy and precision led to a point where many of the difficulties that have to be solved nowadays cannot be described by simple univariate, linear regression methods (Chapter 1 gives an extensive review of some typical problems shown by several atomic techniques). Sometimes such problems cannot even be addressed by multivariate regression methods based on linear relationships, as is the case for the regression methods described in the previous two chapters.

Typical situations where linear regression methods can fail involve spectral nonlinearities. Although partial least squares (PLS) has shown important capabilities to handle nonlinearities (see, *e.g.*, [1–5]) and yields surprisingly good results, sometimes this is at the cost of increasing the dimensionality of the model (*i.e.* the number of factors or latent variables considered in the model), which may hamper robustness and model stability. As discussed in Section 4.1, a possible solution to handle nonlinear relationships might consist in using polynomial inner relationships. However, the problem would then be to find the best polynomial. Other possibilities exist to model nonlinearities, such as the use of nonlinear forms of principal component regression (PCR) and PLS, increasing the spectral data set with squared terms (see [6] for a simple, classical paper), splines and multiplicative regression [7], to cite just a few. It is clear that analytical chemists would like a regression method that is able to cope with nonlinearities whenever they are present (or when there are sound reasons to hypothesise their presence) in the data set. Some methods have already been developed in other scientific areas (typically, computer science) to solve problems with conceptually similar difficulties; namely nonlinear behaviours and a large amount of random noise [8,9]. Artificial intelligence is a working field within computer science that tries to discover and describe how the human brain works and how can it be simulated by means of informatic systems (including software and hardware). This discipline developed strongly in recent years and it has proved its efficiency in many scientific fields [10,11]. Of the different possibilities deployed in the computer science field, ANNs constitute a good means to tackle such problems because they work intrinsically with nonlinear functions. Some other major general benefits of the ANNs are summarised briefly below.

They are flexible, in the sense that they do not require an underlying mathematical model (which almost always has some initial postulates and constraints to treat the data). They accept any relation between the predictors (spectra) and the predictand(s) [concentration(s) of the analyte(s)], regardless being linear or not (this might also represent a problem; see Section 5.7). Further, they can work with noisy data or random variance and find rules where they are not obvious for humans or other modelling methods. They are robust in the sense that they support small spectral deviations from the calibration data set, which indeed are common in spectrometric applications, *e.g.* different noise distributions and random fluctuations in the signal caused by particle scattering. Also important for spectroscopists, ANNs have relevant generalisation capabilities. This means that they extract relevant information from known data sets (*learning patterns*) to yield good predictions when new unknown samples are introduced into the model. Here, 'good' should be interpreted as 'good enough to fit the purpose', which should be stated before the modelling steps.

Nonetheless, the use of ANNs is somewhat controversial. Although it is true that they can achieve really good results either in classification or regression, it is also true that they imply a great deal of computer work because many studies are required to obtain a reliable model. We strongly recommend a classical paper in which Kateman underlined several important difficulties that may arise when searching for a satisfactory model based on neural networks [12]. Although the paper is not recent, it remains valid. In addition, it has been reported fairly frequently (see the last section in this chapter reviewing published applications) that ANNs do not always outperform other classical methods, typically PLS. In spite of this, it is acknowledged that ANNs can yield good results in situations where other methods failed.

A daily-life example, although not chemical, is the use of ANNs to configure dynamically the antennas of a satellite to avoid interferences on the receivers. ANNs presented the best solutions and faster than other traditional calculation methods [13].

Interestingly, the historical development of ANNs has suffered from highly enthusiastic epochs, inactive years and, now, a powerful revival. We think that, probably, some more time is needed in the analytical chemistry arena to settle discussions, analyse the pros and cons and address two important (in our opinion) main issues:

1. to develop simplified ways of searching for the best neural network (important developments have been done by combining ANNs with genetic algorithms; interested readers should consult the literature [14–18]) and making them available to a broad range of users;
2. to formulate some sort of '*application rules*', *i.e.* general statements that may guide the user on deciding when to apply ANNs instead of other regression methods, such as PLS and PCR.

We would like to refer here to a brief resume of the history of ANNs given by Jain *et al.* [19]. ANN research underwent three main periods of extensive activity. The first was during the 1940s, when they were developed, after McCulloch and Pitts' pioneering work [20]. The second was from 1960 to 1969, when Rosemblatt developed a form of neural net termed perceptron [21], which improved several previous ideas. Then, strong criticisms from some workers emphasising the limitations of a simple perceptron extinguished the initial enthusiasm and made the interest in neural networks almost disappear. It was in early 1980s when Hopfield published his work [22] and with the development of the back-propagation learning algorithm for multilayer perceptrons that interest in ANNs was boosted again (this is the learning paradigm most widely used at present and it will be explained later). The back-propagation algorithm was first proposed by Werbos in the context of regression [23], although it was popularised by Rumelhart and McClelland [24].

5.2 Neurons and Artificial Networks

Artificial neural networks (ANNs) emulate some human brain characteristics, such as the ability to derive conclusions from fuzzy input data, the capacity to memorise patterns and a high potential to relate facts (samples). If we examine carefully those human (animal) abilities, they share a common basis: they cannot be expressed through a classical well-defined algorithm; rather, they are based on a common characteristic: experience. Humans can solve situations according to their accumulated experience, rather on a conscious and strict reasoning procedure.

It is commonly argued that ANNs can implement solutions to problems that are difficult or impossible to solve in a classical algorithmic way. To visualise this idea, consider this hypothetical situation: you want to classify a set of old objects you have kept in your store room for years. As you are a good scientist, you try to measure as many characteristics as possible (maybe to sell the objects in batches) and since you like chemometrics, you applied a typical classification technique (let us say, PCA or cluster analysis), but you did not obtain satisfactory results. Nevertheless, you can see yourself that some objects could form a 'more or less' (in the ambiguous human language) definite group. Why were those chemometric techniques so unsuccessful? Probably because most multivariate chemometric techniques work with strict equations. For instance, cluster analysis defines similarity between two samples according to some distance measurement, but there are many mathematical forms to define a distance (all them correct!) and they do not consider fuzzy definitions (that broken, rusty device that belonged to your grandfather was, after all, an old bicycle model). If you translate this trivial example to your working field and the very many problems you can suffer, it becomes immediately clear that the newest multivariate regression techniques should be versatile and should be based on new, different approaches to handling problems.

ANNs try to mimic in a simplified form the way in which the brain works, as this is the most perfect example we have of a system that is capable of acquiring knowledge by means of experience. Hence, as the brain itself, an ANN is constituted of an interconnected set of basic units of processing, each called a neuron or process element.

Figure 5.1 shows a simplified view of the main structure of a biological neuron compared with an artificial neuron. A biological neuron (and its artificial counterpart) needs an electrical signal (which, in the artificial neuron, is substituted by a mathematical function) to start functioning. This is called the *input function*. All input signals are combined (how this can be performed is explained in Section 5.3) to obtain a signal which is processed (treated) within the neuron (this is called the *activation function*) and, finally, an output signal is obtained which leaves the neuron (this is called *output function*) (see Figure 5.2). Because of the many similarities, computer science uses the same jargon as biologists (Figure 5.1):

1. An input, x_i, represents the signal which come from either the outer world (of the neuron) or from other neuron(s). The signal is captured by the dendrites.

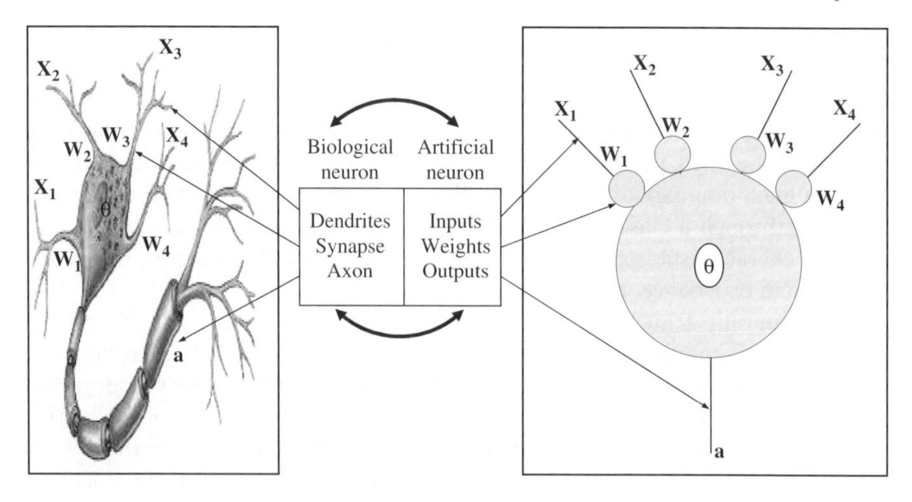

Figure 5.1 Conceptual relationships between biological and artificial neurons.

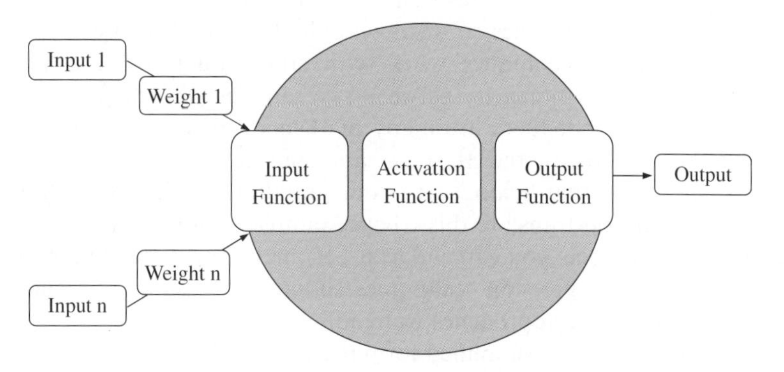

Figure 5.2 Conceptual operations performed in an artificial neuron.

2. A weight, w_i, denotes the intensity of the synapse that connects two neurons; therefore x_i and w_i are numerical values (scalars).
3. θ is a threshold value that a neuron must receive to be activated. Biologically, this happens inside the cell, but the artificial neuron needs a dedicated function (equation). This threshold value (and its related mathematical function) is called bias.
4. The signal that leaves the neuron, a, can typically be transmitted to other neuron(s) or be the final result of the computing process.

An ANN is a set of interconnected neurons (also termed *nodes*, *cells*, *units* or *process elements*) distributed in a specific arrangement, usually termed an *architecture*. In general, neurons are organised in layers (see Figure 5.3).The most common neural nets, the feed-forward nets, are fully connected, *i.e.* each node is connected to all the nodes in the next layer. The information we want to enter in

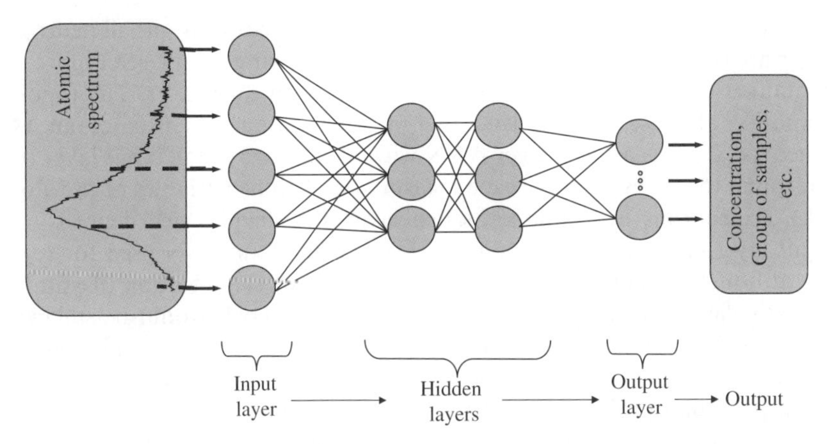

Figure 5.3 Internal organisation of an artificial neural network. In general, there is a neuron per original variable in the input layer of neurons and all neurons are interconnected. The number of neurons in the output layer depends on the particular application (see text for details).

the ANN (*e.g.* the spectra) is given to the '*input layer*', which is the set of neurons that receive directly the information from the external world of the net. This layer is almost always depicted on the left-hand side of graphical representations (Figure 5.3).

In general, there is an input neuron per each original variable (*e.g.* a neuron per wavelength in molecular spectra or a neuron per time variable in atomic spectrometry). The input signals are transferred to one or several '*hidden layers*' (they are internal to the network and have no direct contact with the outer surroundings). The number of hidden layers can be any positive value although, usually, the number is kept low: in Analytical Chemistry no more than one layer is recommended in most situations. Graphically, the hidden layers are drawn between the input layer and the final line of neurons. The terminology used to count the number of hidden layers may vary, but commonly it is just the number of intermediate layers between the first (input) and the last (output) layer. Finally, the last arrangement of neurons is called the output layer and is drawn on the right-hand side of the diagrams (Figure 5.3). How many neurons should be placed in the output layer will depend on the problem at hand and how you want to represent it [8,25,26]. If you want to perform a regression, usual choice is to set up just a neuron (which will yield the value predicted by the whole ANN). If you are classifying several types of mine ores according to a suite of metals measured by ICP, it would be wise to set up one neuron per class of tools.

The neural networks where information flows from the input to the output layer are frequently termed '*feed-forward*' ANNs and they are by far the most often employed type in Analytical Chemistry; they are considered here 'by default', so this term will not be mentioned again for brevity.

As mentioned above, the topology, architecture or taxonomy of an ANN (they are synonymous) defines its physical structure; *i.e.* how many neurons have been set by the user, their connections (in most applications all neurons in a layer are connected to all the neurons of the following layers; see Figure 5.3), the number of layers and the number of neurons in each layer. The topology summarises all these characteristics in a simple notation. For instance, an ANN defined as 22/6/1 is a three-layer ANN with 22 neurons in the input layer (they correspond to 22 absorbances each measured at 22 time variables of the atomic peak), six neurons in a unique hidden layer and, finally, one neuron in the output layer. This would be the typical structure of an ANN used for regression, although we could also use it for some sort of classification (if you have only two classes of objects, a neuron in the output layer would be sufficient). Therefore, it is important to explain clearly what the ANN is being used for.

Unfortunately, there are no rules to decide in advance which topology will solve a particular problem best. This means that many assays have to be done, varying the number of layers and the number of neurons per layer. The good news is that, as pointed out in the paragraph above, the numbers of neurons in the input and output layers are almost immediately defined by the number of variables you have and your purpose (regression or classification). The other parameters will be set after empirical studies. These are considered in some detail in subsequent sections of this chapter. Often experience will help, but remember that each problem is different and so are the spectral artefacts, noise, *etc.* Do not immediately accept an ANN you have seen published somewhere as the best one for you because, almost certainly, it will need some further optimisation.

In this context, two recent studies tried to simplify the optimisation process of the ANNs by developing new optimisation schemes [27,28], and Boozarjomehry and Svrcek [29] even presented a procedure to design a neural network automatically. Although the last paper is rather technical, the first sections give an easy-to-follow perspective of some new trends in designing ANNs. We should mention that many recent methodologies to optimise an ANN 'automatically' include procedures for extensive searches among a wealth of possible solutions and often some form of genetic algorithms is used [14,15,17,18]. We will not consider this option here in order to maintain the text at an introductory level.

In following sections, the basis of the main components of a neuron, how it works and, finally, how a set of neurons are connected to yield an ANN are presented. A short description of the most common rules by which ANNs learn is also given (focused on the error back-propagation learning scheme). We also concentrate on how ANNs can be applied to perform regression tasks and, finally, a review of published papers dealing with applications to atomic spectrometry, most of them reported recently, is presented.

5.3 Basic Elements of the Neuron

5.3.1 Input Function

As shown in Figures 5.1 and 5.3, a neuron usually receives several simultaneous inputs (except for the neurons placed at the input layer, which only receive a signal). Each input (x_1, x_2, . . . , x_n) is a continuous variable (absorbance, concentration of an element, *etc.*) which becomes weighted by a synaptic weight or, simply, weight (w_{1i}, w_{2i}, . . . , w_{ni}, respectively, where w_{ji} represents the weight associated to the input signal *i* on neuron *j*). The function of the weight in the artificial neuron is similar to the synaptic function in the biological neuron, *i.e.* to strengthen some signals (it is though that human learning implies repeated strengthening of some neural connections) and/or to cancel out some others. A weight can be positive, so it promotes the signal, reinforcing its transmission, or negative, which inhibits the signal and, therefore, weakens the transmission of that particular input signal. To sum up, the mission of the weights is to adjust the influence of the input values in the neurons (see Figure 5.4). Then, all weighted input signals are summed so that the overall net input to each neuron can be expressed as

$$Net_j = \sum_{i=1}^{q} w_{ji}x_i + \theta_j \tag{5.1}$$

where θ_j is the bias of the neuron, which can be treated as a general weight of the neuron (like a threshold). The weights can be initialised randomly (most frequently) or derived from some previous calculations. Anyway, they have to be optimised during model development and the corresponding optimisation process is called training (or learning), and this will be discussed in Section 5.4.

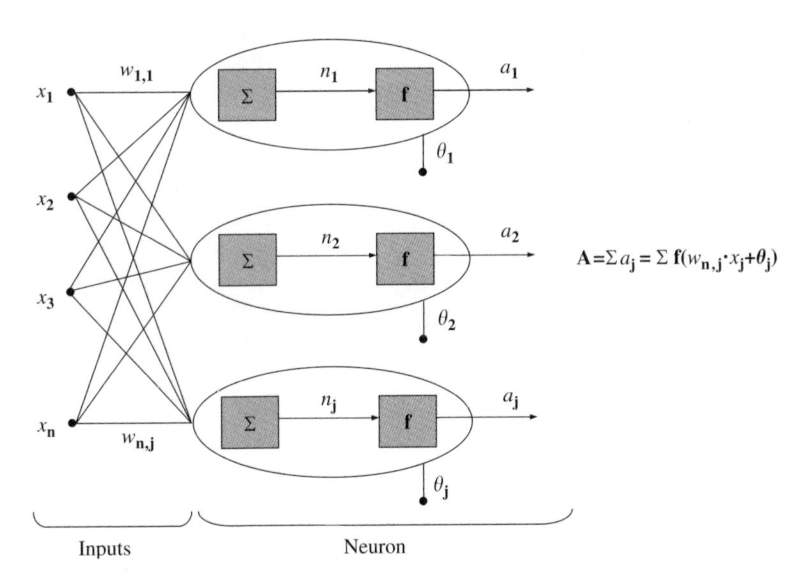

Figure 5.4 Simplified view of the mathematical operations performed in a neuron. See text for a detailed explanation of all terms.

5.3.2 Activation and Transfer Functions

A biological neuron can be active (excited) or inactive (not excited). Similarly, the artificial neurons can also have different activation status. Some neurons can be programmed to have only two states (active/inactive) as the biological ones, but others can take any value within a certain range. The final output or response of a neuron (let us call it a) is determined by its transfer function, f, which operates on the net signal (Net_j) received by the neuron. Hence the overall output of a neuron can be summarised as:

$$a_j = f_j(Net_j)$$

or

$$a_j = f_j\left(\sum_{i=1}^{q} w_{ji}x_i + \theta_j\right) \tag{5.2}$$

The numerical value of a_j determines whether the neuron is active or not. The bias, θ_j, should also be optimised during training [8]. The activation function, a_j, ranges currently from 0 to 1 or from -1 to $+1$ (depending on the mathematical transfer function, f). When a_j is 0 or -1 the neuron is totally inactive, whereas a value of $+1$ denotes that it is totally active.

The transfer function can be either a linear or a nonlinear function and this has to be decided depending on the problem that the ANN has to solve. The usual functions are the linear function, the threshold (or step) function, the Gaussian function, the sigmoid function and the hyperbolic tangent function (see Table 5.1). Each activation function has advantages and disadvantages and, disappointingly, no general rules exist to select one instead of another to solve a particular problem. As was explained for PLS (Chapter 4), the only '*golden rule*' is to make tests to seek out the most suitable option for your own problem [7]. Despite this, the hyperbolic tangent function is the option most frequently applied, mainly when a continuous range of values is needed (*e.g.* for calibration). Table 5.1 shows that the hyperbolic tangent function yields a sigmoid plot, very similar to the classical logarithmic sigmoid. The main difference is that the former yields output values between -1 and $+1$ whereas the latter gives values between 0 and 1. Note that these are normalised values and so to obtain the 'true' concentration values a final operation should rescale them back to the original concentration units.

5.3.3 Output Function

The output function determines the value that is transferred to the neurons linked to a particular one. In the same way as each biological neuron receives a lot of inputs (maybe one per dendrite) but outputs only a signal through the

Table 5.1 Activation functions currently employed in artificial neurons, where n represents the overall net input to the neuron and a denotes the result of the activation function

Activation function	*Input values*	*Output*	*Typical icon*	*Acronym*
Hard limit	$n < 0$ $n \geq 0$	$a = 0$ $a = 1$	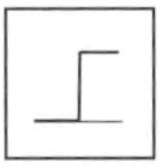	*hardlim*
Symmetric hard limit	$n < 0$ $n \geq 0$	$a = -1$ $a = +1$		*hardlims*
Positive linear	$n < 0$ $0 \leq n$	$a = 0$ $a = n$		*poslin*
Linear		$a = n$	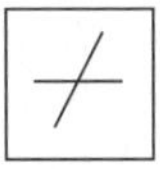	*purelin*
Saturated linear	$n < 0$ $0 \leq n \leq 1$ $n > 1$	$a = 0$ $a = n$ $a = 1$		*satlin*
Symmetric saturating linear	$n < -1$ $-1 \leq n \leq 1$ $n > 1$	$a = -1$ $a = n$ $a - +1$		*satlins*
Log-sigmoid	$n \geq 0$	$a = \dfrac{1}{1 + e^{-n}}$	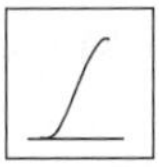	*logsig*
Hyperbolic tangent sigmoid	$-1 \leq n \leq 1$	$a = \dfrac{e^{n} - e^{-n}}{e^{n} + e^{-n}}$	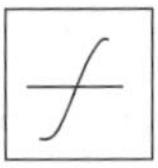	*tansig*
Competitive	Neuron with n maximum Other neurons	$a = 1$ $a = 0$	C	*compet*

axon, the artificial neuron processes all inputs and gives a final signal. Although the combination of the input signals has already been made by the input and the result processed by the activation functions, we can process the signal of the activation function slightly further by means of a post-process function. For instance, if you are classifying samples in two groups, it would be advantageous to set that while the value obtained by the activation function is lower than a certain threshold (*e.g.* 0.50), no signal exits the neuron. This example would require a so-called binary output function, which yields a 1 if the value of the activation function is greater than a predefined threshold value and 0 otherwise. This function is termed '*hard limit*' in Table 5.1; sometimes it is also called the Heaviside function or the step function and it represents an 'all-or-none' behaviour.

Going back to the main issue of this book, multivariate calibration, the most common situation is to accept the value of the activation function without further processing. In this case, the output function has no effect and it just transfers the value to the output (this can be considered as an identity function). Recall that the final response of the ANN has to be scaled back to obtain concentration units.

5.3.4 Raw Data Preprocessing

It has been experienced that ANNs work best when they handle numerical values restricted to a small scale, which should be common for all variables. To achieve this, an initial preprocessing of the atomic spectra (raw data) is required. Autoscaling, mean centring or 0–1 scaling can be recommended for this purpose. Preprocessing avoids undue effects caused by the different scales and ranges of the initial variables. As already discussed in previous chapters (see, *e.g.*, Chapter 4), different attempts have to be made to look for the best scaling option, although the [0 . . . 1] and [−1 . . . +1] options are frequently used in ANNs.

Note that before scaling the training data set, one has to be aware of the absorbances that might occur on the validation set (in general, on the unknowns). This is so because the [0 . . . 1] and [−1 . . . +1] ranges should bracket not only the absorbances of the calibration samples but also those that might occur on the validation set and future unknowns.

Despagne and Massart [7] gave some additional general rules. If sigmoid or hyperbolic tangent transfer functions are used, the [−1 . . . +1] range was recommended for the spectral set of variables. The Y value (variable to be predicted) should also be scaled when nonlinear functions are selected for the output layer. Hence the target values of the calibration set (*i.e.* the values we want to predict) should be scaled [0.2 . . . 0.8] or [−0.8 . . . +0.8] when a sigmoid or a hyperbolic tangent transfer function is used, respectively. This would prevent the ANN from yielding values on the flat regions of the mathematical functions and, therefore, improve the predictions.

5.4 Training an Artificial Neural Network

Once an architecture for an ANN has been decided, it is time to develop the model and see how it works. Setting an architecture involves deciding on the input, activation and output functions. In Section 5.4.1 it will be discussed that training an ANN consists of a process by which both the weights and the biases are modified iteratively until satisfactory predictions are given by the net (it is said that the ANN '*learnt*'). Since ANNs do not have a predefined model because the mathematical equations to obtain a solution are not strictly predefined (unlike, *e.g.*, MLR; where a set of regression coefficients is calculated to minimise a criterion), a sequential iterative optimisation process guided by an error measure (usually the mean square error, MSE) has to be followed to minimise the global prediction error yielded by the net.

In this section, the process by which an ANN learns is summarised. To simplify (although without lack of generality), explanations will be given for a network with an input layer, a hidden layer and an output layer. Finally, some notes will be presented for networks implying more than one hidden layer. In all cases, i denotes the number of variables in the input vector (1, 2, . . . , I; *i.e.* the atomic spectral variables); j represents the number of neurons in the hidden layer (1, 2, . . . , j); and k is the number of neurons in the output layer (1, 2, . . . , k).

As for any calibration method, we need a reliable set of standards where the atomic spectra have been measured and the concentration(s) of the analyte(s) is (are) known. Here 'reliable' is used *sensu stricto*, which means that the standards should be as similar as possible to the unknown samples. This can be justified easily by recalling that ANNs extract knowledge from a training set to, later, predict unknowns. If and only if the standards resemble the unknowns closely can the ANNs be expected to extract the correct knowledge and predict future samples properly. As for MLR, PCR and PLS, this issue is of paramount importance.

The previously scaled atomic spectrum of a standard (technically, it is called a training pattern) enters the net throughout the input layer (a variable per node). Thus, an input neuron receives, simply, the information corresponding to a predictor variable and transmits it to each neuron of the hidden layer (see Figures 5.1, 5.3 and 5.4). The overall net input at neuron j of the hidden layer is given by eqn (5.3), which corresponds to eqn (5.1) above:

$$n_j^0 = \sum_{i=1}^{i} w_{ji}^0 x_i + \theta_j^0 \tag{5.3}$$

w_{ji}^0 is the weight that connects neuron i of the input layer with neuron j at the first hidden layer, x_i is the ith absorbance (intensity, m/z ratio, *etc.*) of the atomic spectrum, θ_j^0 is the *bias* of neuron j in the hidden layer and the upper index (^{0}) represents the layer to which every parameter belongs, in this case the hidden layer.

As was explained above, each neuron of the hidden layer has one response (a_j^0), calculated with the activation function (f^0):

$$a_j^0 = f^0\left(n_j^0\right) \tag{5.4}$$

The j outputs, all a_j^0 values of the neurons of the hidden layer, are now the inputs to the next layer (here throughout this explanation the output layer). Of course, they need to be weighted again, although using connection weights other than those above. This process can be repeated again through the different hidden layers of the ANN. The net input signal for a neuron in the output layer is described by the equation

$$n_k^s = \sum_{j=1}^{j} w_{kj}^s a_j^0 + \theta_k^s \tag{5.5}$$

where the superscript s indicates the output layer, w_{kj}^{s} is the weight which joins neuron j at the hidden layer with neuron k at the output layer, θ_{k}^{s} is the *bias* of the kth neuron in the output layer and n_{k}^{s} is the net input to neuron k of the output layer.

As in our example here there is only a neuron at the exit layer (we are considering only calibration), the activation function yields a value that is the final response of the net to our input spectrum (recall that the output function of the neuron at the output layer for calibration purposes is just the identity function):

$$a_{k}^{s} = f^{s}\left(n_{k}^{s}\right) = \text{response} \tag{5.6}$$

Undoubtedly, we do not expect the ANN to predict exactly the concentration of the analyte we used to prepare the standard (at least in these initial steps), so we can calculate the difference among the target (t) and predicted value (a):

$$\delta_{k} = \left(t_{k} - a_{k}^{s}\right) \tag{5.7}$$

Also, since we have to pass all atomic calibration spectra (let us say we have m standards) through the ANN, we have an overall prediction error, ep. Usually, the average prediction error (or mean square error, MSE) is calculated as

$$ep^{2} = \frac{1}{m}\sum_{m=1}^{m}(\delta_{m})^{2} \tag{5.8}$$

To obtain a successful learning process, the ANN has to update all weights and biases to minimise the overall average quadratic error.

5.4.1 Learning Mechanisms

Provided that the information entering the ANN flows through the net in order to obtain a signal that has to be related to a property (*e.g.* concentration of an analyte), the ANN needs a formal way to calculate the correct output (concentration) for each input vector (atomic spectrum) contained in the calibration set. This calls for some systematic way to change the weights and biases in an iterative process termed learning, training or '*conditioning*'. The set of atomic spectra on which this process is based is called the learning (training or '*conditioning*') data set. Each atomic spectrum has a corresponding concentration value for the analyte of interest (although more than one analyte can be predicted).

In most applications, the input, activation and output functions are the same for all neurons and they do not change during the training process. In other words, learning is the process by which an ANN modifies its weights and bias terms in response to the input information (spectra and concentration values). As for the biological systems, training involves the 'destruction', modification and 'creation' of connections between the neurons. This corresponds to the reinforcement of some weights (which acquire high values) and the cancellation of some others (with values around zero). The learning process will finish either when all weights remain stable after several iterations (sometimes this is called convergence) or when an error threshold is reached (more usual) [30–32].

It follows that the most important aspect to train an ANN is to determine how the weights are modified in an iterative, automatic process; that is, the criterion to change the weights. How this is done determines the different learning (training) algorithms. Diverse learning methods exist [30,32] of which some typical examples are cited first, then we will concentrate on the most common one.

Generally, the ANN training methods can be classified as unsupervised and supervised methods. In *unsupervised learning*, the ANN is not informed of the errors it is committing because it does not know whether the output for a particular input is correct or not. In Analytical Chemistry, this learning mode is used mainly for pattern recognition purposes to find groups of samples (or variables). Within unsupervised learning, two learning procedures can be mentioned:

1. The so-called '*Hebbian learning rule*' (to honour Canadian neuropsychologist Donald Hebb, who proposed it in 1949) specifies how much the weight of the connection between two neurons should increase or decrease as a function of their activation. The rule states that the weight should increase whether the two interconnected neurons are active simultaneously (otherwise the weight should decrease). This rule is not used too frequently nowadays because it works well provided that all the input patterns are orthogonal or uncorrelated.
2. '*Competitive learning*' or '*comparative learning*'. Here, different nodes compete among them to offer the output, so only one neuron at the competitive layer (either at the output or hidden layers) will be activated. That is, only one neuron will have an output different from zero. The 'winning' node will be that for which the weights are most similar to the input sample. Learning consists of increasing the connections of the winning neuron and weakening the other connections, so that the weights of the winner will be very close to the values of the input sample (*e.g.* atomic spectrum). Kohonen neural networks (also termed self-organizing maps, SOMs), can be classified here and they are used fairly frequently nowadays. Some references to atomic spectrometry can be mentioned, although they will not be discussed in detail because they deal with applications beyond our objectives (see, *e.g.*, [33–36]). Just to summarise, these neural networks are highly recommended for performing classification and/or pattern recognition tasks. This means that they are applied once the data have been obtained and the raw data matrix (samples times variables) is ready. Several recent examples are presented in Section 5.11.

In *supervised learning*, the training process is controlled by an external agent (supervisor), which analyses the final output of the ANN and, if it does not coincide with the 'true' or 'target' value, makes the weights continue to change. This control will remain until a minimum error (set in advance by the chemist) is reached. This is the training mode most commonly employed in Analytical Chemistry whenever a calibration model has to be obtained. Note the quotation marks around *true*, which refers to the concentration(s) value(s) associated with each atomic spectrum by the analyst before computations start. Several learning procedures are possible here, including the following [30–32]:

1. '*Reinforcement learning*': the network knows whether its output is correct or not but there is no measure of the error. The weights will increase after positive behaviours or decrease (be punished) after negative behaviours. Hence positive behaviours (or behaviours reinforced positively) are learnt, whereas negatively reinforced behaviours are punished.
2. '*Stochastic learning*' consists in changing the weights randomly and evaluating how well the ANN works after each change. In general, this procedure is time consuming and computing demanding. Distributions of probability can be used.
3. '*Error correction learning*' (currently known as *back-propagation*). This learning mode compares the output(s) of the ANN with the true concentration value(s) associated in advance with each spectrum. The error derived from such a comparison will control the ANN training. In the most straightforward way, the errors can be used to adjust the top weights directly by means of a predefined algorithm. Error-correction learning algorithms attempt to minimise the overall ANN error on each iteration.

5.4.2 Evolution of the Weights

A suite of initial weights is required to initiate the net before learning starts. They can be selected after preliminary studies although, usually, they are assigned a random value (between -1 and $+1$). Two possibilities exist to back-propagate the error:

1. Continuous update or '*pattern mode*' [37], where changes in the weights are made immediately after a value is predicted for each standard (pattern). This means that after each iteration or epoch all weights changed m times, being m the number of standards (patterns) in the calibration set.
2. Periodic update or '*batch mode*'. Here, the weights are not updated until all standards have passed across the net. All partial changes are summed for each weight and, before the next iteration starts, the weights are changed accordingly [37]. This mode resembles very much the use of the RMSEC to select the best dimensionality of a PLS model; it is computationally fast (even though it requires a great deal of computer memory – although this is not too relevant for most applications on modern personal computers) and it allows for large changes in the weights, which allows convergence to be reached slightly faster.

In simple terms, new weights are computed as $weight_{new} = (learning\ rate \times error) + (momentum \times weight_{old})$. The error was defined above as the difference between the predicted and the true values; sometimes, the true value is also referred to as target value [12,38,39]. The learning rate is a factor that can be modified by the analyst in order to control the velocity of the training process (high values yield faster trainings, but if they are too large, the weights fluctuate widely and convergence cannot be reached). The *momentum* is an optional constant that smoothes the fluctuations that occur on the weights (low values prevent large fluctuations on the weights and facilitate convergence, therefore a too low momentum makes the training process slow). It is a constant which can be any number between 0 and 1. When the momentum constant is zero, the change in the weights is based solely on the proportion of the overall error that is attributed to that weight. When the momentum constant is 1, the weight changes as in the previous interaction (the proportion of the overall error is ignored) [40]. In general, the learning rate is more influential than the momentum constant on the optimisation process [8].

5.5 Error Back-propagation Artificial Neural Networks

Error back-propagation artificial neural networks (BPNs) constitute, without doubt, the paradigm most widely applied to develop ANNs nowadays [39,41,42]. It constitutes the most common choice thanks to its recognised capability to solve a wide range of problems related to, mainly, regression and supervised classification. The name refers to feed-forward ANNs which are trained using an error back-propagation algorithm. BPNs were developed in the 1960s [23], although they were not used widely until the end of the 1980s. The main reason was the lack of an effective learning algorithm to study problems of high dimension [43].

BPNs 'learn' how to relate a predefined set of atomic spectra in the calibration set with their associated target values (*e.g.* concentrations) using a supervised two-phase cycle. We will refer to the example presented in Figure 5.3 although with a unique neuron in the output layer. There, an atomic spectrum entered the net at the first layer of neurons (input layer), and the signals flow throughout the connections (weighted by the weights and biases) layer after layer until an output is obtained. The output is then compared with the 'true' (reference or target) concentration and an error is derived for each sample. Here the second phase starts. The errors are then transmitted backwards from the output layer to each unit in the hidden layer that contributes directly to the output [12,44,45]. However, each unit in the hidden layer receives only a portion of the total error signal, based roughly on the relative contribution the unit made to the original output. This 'back-propagation of the error' is repeated layer by layer, until each unit in the network has received an error signal that describes its relative contribution to the total error.

The overall process is repeated many times until the ANN converges to a point where the weights do not change and a model is obtained that predicts (classifies) correctly all the standards [38]. Frequently, convergence is not reached and a preset threshold error is allowed for. Figure 5.5 presents a general scheme with some more details of the back-propagation process.

5.6 When to Stop Learning

The simplest way to stop the iterative process by which the weights and biases are updated is to fix in advance a threshold error. When the average error [considering the calibration standards, see eqn (5.8)] is lower than the threshold value, the net stops training and the last set of weights and biases are stored [41]. In some publications it is said that the ANN converged.

From a pragmatic viewpoint, we must say that it is not unusual that some trials do not reach convergence and, sometimes, the errors fluctuate strongly, without a definite pattern. This may be caused by (among other possibilities) a wrong architecture, a too large learning rate, the existence of outlying samples, wrong scaling of the spectra or, simply, the random nature of the initial weights which might cause the net to be stacked in a local (inadequate) minimum. Another rough criterion to stop learning is to cease the training process after a given number of iterations have been performed (*e.g.* 10 000). This acts also as a warning criterion to avoid the net being stacked or to avoid too long computation times (probably because the ANN is not good enough).

A useful approach consists in monitoring the average errors of the calibration and control data sets (see below for more details on the latter set) in the same plot, although the availability of this graph depends on the software that can be used. Charts such as that shown in Figure 5.6 are typically obtained [27,46]. This approach is used very frequently and it requires three data sets; namely one for calibration, one for testing or controlling (*i.e.* checking the models while the ANN is being optimised [9]) and one for validation. Svozil *et al.* [37] termed this procedure '*early stopping*', which clearly underlines that it is used to avoid overfitting. The main idea is to stop training exactly when the net has extracted enough knowledge to predict both the calibration and, more importantly, the test data sets. Note that the test data set is used after the ANN has been trained to verify its performance. Analogously to PLS dimensionality (see Chapter 4), the more extensive the ANN learning, the lower is the average training error and the worse the predictions of the control and testing sets and, thus, of the unknowns. The reason is that a too long learning period makes the ANN memorise too particular spectral details of the calibration spectra, which may correspond to noise, random fluctuations, spectral artefacts, *etc.* Unfortunately, such specific details contribute only to fitness and not to prediction, *i.e.* an overfitted model is obtained. As in the previous chapter regarding PLS, it is necessary to reach a compromise between calibration error (underfitting), training time and prediction error (overfitting). A simple way is to use charts such as that in Figure 5.6, where a vertical line signals where a satisfactory compromise was obtained.

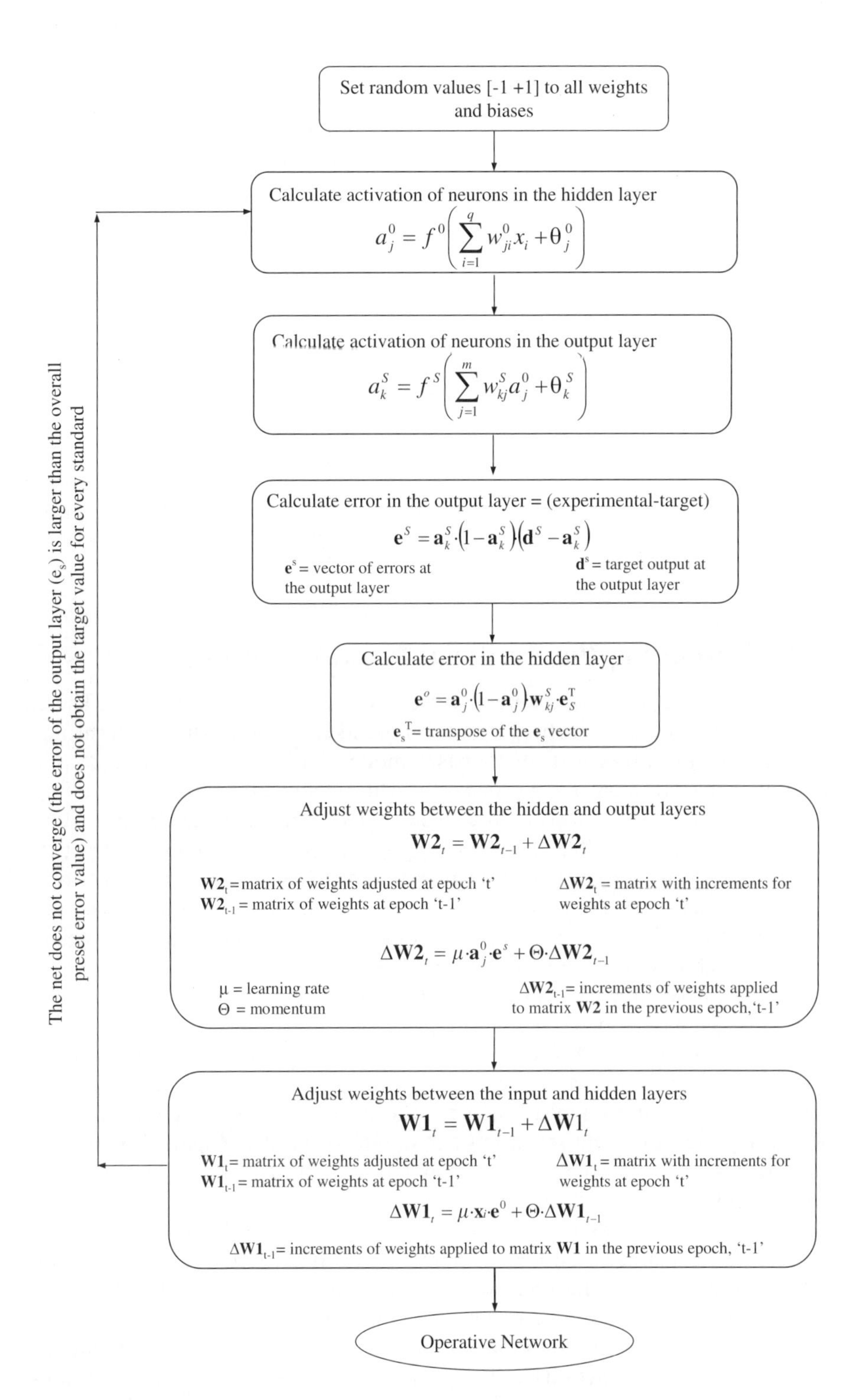

Figure 5.5 Simplified view of the process by which the prediction error of an artificial neural network is back-propagated from the output to the input layers. See text for more details.

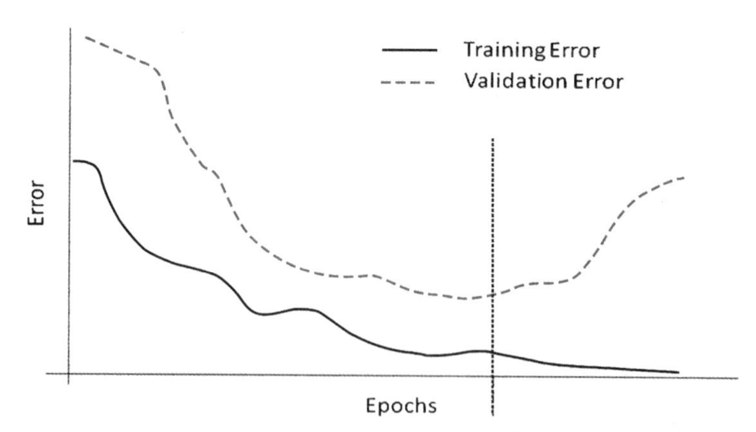

Figure 5.6 Example of a control chart depicting the training and validation errors as a function of learning epochs (cycles). The vertical line denotes the compromise between the modelling and prediction abilities.

5.7 Validating the Artificial Neural Network

Underfitting and overfitting were discussed in some detail in Section 4.4. There, it was explained that overfitting is much more likely to occur than underfitting. The same can be argued for ANN-based models, although one should keep in mind that they are very much prone to overfit. In effect, the intrinsic behaviour of the ANNs leads them to predict the (limited number of) calibration standards as closely as possible. Therefore, if any unexpected phenomenon appears on the unknown samples that we want to predict (*e.g.* a spectral artefact or a new component), it may well happen that the net does not predict those samples properly. This is in fact also a problem with any classical (univariate) calibration, but it is exacerbated when ANNs are used since they really try to 'memorise' what they have seen (the calibration set), although, unfortunately, new samples may be predicted wrongly [47].

Hence validating the ANN model with a set of new samples is really critical. As we saw in PLS, we would need a set of samples (validation set) where the atomic spectra had been measured and whose concentrations are known to us. Regarding the three data sets needed to develop a model, Zupan [9] suggested that each set should contain approximately one-third of the overall spectra, with a tendency for the training set to be the smallest and the validation set to be the largest. The control set (which is used to control how the network evolves) has often been ignored as many users assign its task to the validation set, but according to the philosophy of learning and testing, the validation samples should in no way be involved in obtaining the model.

The validation spectra are input to the net, the concentrations are predicted and, finally, the predictions are compared with the reference (target) concentrations. Of course, neither the weights nor the biases are modified in this stage. The objective is to obtain models which predict the validation samples with a minimum average prediction error. In PLS this was the RMSEP (root mean square error of prediction), and although of course it can also be used here, the MSE (mean square error) is used more frequently. A typical goal is to obtain similar average errors for the calibration and validation samples, although in order to decide whether a model is valid or not fitness-for-purpose is a key factor as the analyst needs to know in advance the level of error he/she can accept. If the calibration and validation errors are monitored in chart plots such as Figure 5.6, they will help the analyst to make his/her final decision.

Zorriassatine and Tannock [25] also discussed the number of standards. Although there is no consensus on how to set the number of samples in each data set, the general trend is to include samples containing all relevant sources of variation of both random and non-random patterns. Experiments have shown that using small training and validation data sets may produce undesirable results (bad models when used to predict true unknowns; see also [7]). In this respect, a good experimental design will be of help not only to bracket our working space in a really structured way but also to organise the number of samples to be prepared. To minimise the possibility of chance correlations on the weights, Livingstone *et al.* [48] considered the ratio of the number of standards to network connections. Values such as 1.8 resulted in a network that would memorise the data (too few samples), whereas a value of 2.2 contained an insufficient number of connections for training to be successful (too few nodes/connections). This strongly coincides with the Vapnik–Cernonenkis dimension [7], which, for a one-hidden-layer network, suggests that the number of samples should be twice the number of weights in the ANN. Although this value may be flexible, the ratio of the number of samples to the number of adjustable parameters should be kept as high as possible and close to 2. If this cannot be reached, a simplification of the architecture of the ANN would be a good alternative. In this respect, Walczak recommended not to use more than one hidden layer composed of 3–5 neurons [49].

5.8 Limitations of the Neural Networks

There is no doubt about the useful capabilities of the ANNs to yield good solutions in many different applications. They are specially suited to obtain a solution when noisy or strong nonlinearities are contained into the data. Nevertheless, they are not trivial to apply and have disadvantages (as with any other regression method).

Kateman [12] considered that ANNs are not really based on theoretical models and, therefore, they lack of a structured and systematic development. Their implementation and optimisation are based on a trial-and-error approach. This is labour demanding and time consuming and there are always doubts about whether the selected model was 'the best' one we could find.

A typical drawback of neural nets is that they need a relatively high number of samples. Nevertheless, good PLS model development also needs as many samples as possible and three data sets [calibration, test (or control) and validation]. This problem (difficulty may be a better word) leads us to another typical drawback, namely flexibility (although this was mentioned as a possible advantage at the beginning of this chapter). ANNs are so powerful that they can really learn or memorise easily the calibration set but predict badly unknown samples. This is especially true with small calibration sets and, therefore, ANNs can yield overfitted models which are useless for practical purposes. Extrapolation can also be cumbersome and therefore it is not recommended [7].

One of the most common criticisms of ANNs is that they cannot be interpreted at all. They act like 'black-boxes' where raw data sets enter the box and final predictions are obtained. Many efforts are being made to improve this issue and most common options consider the size of the weights. The higher the weights, the more important the associated variables are (strictly, the connected neurons, which can be traced back to the input layer). Despite this, Féraud and Clérot [50] explained that the nonlinear nature of the hidden neurons (hidden layer) can make the size of the weights misleading in some situations. Although this is probably true for regression models, the study of the weights is of interest in many pattern recognition applications (mainly when Kohonen SOMs are employed).

Disappointingly, the internal structure of the net (*i.e.* the values of the weights and biases) is not always visible to the analyst, mainly when regression problems are considered. In contrast, the weights for the different layers of Kohonen SOMs can be visualised fairly easily and they were used to derive general interpretations [51]. Some attempts have been made also to gain knowledge on how the BPN organises the net [52]. Also, recent approaches specifically looked for ways to interpret how an ANN functions by establishing rules of the 'if . . . then' type. Although they seem promising, they are still under development [53].

Although ANNs can, in theory, model any relation between predictors and predictands, it was found that common regression methods such as PLS can outperform ANN solutions when linear or slightly nonlinear problems are considered [1–5]. In fact, although ANNs can model linear relationships, they require a long training time since a nonlinear technique is applied to linear data. Despite, ideally, for a perfectly linear and noise-free data set, the ANN performance tends asymptotically towards the linear model performance, in practical situations ANNs can reach a performance qualitatively similar to that of linear methods. Therefore, it seems not too reasonable to apply them before simpler alternatives have been considered.

5.9 Relationships with Other Regression Methods

ANNs can be related to some other classical regression methods. Here we will summarise the excellent introductory work of Næs *et al.* [54] and Despagne and Massart [7], in particular those relationships among ANNs and the three regression methods considered in this book. Although Kateman stated that the relationships are loose because of their widely different natures [12], we considered it of interest to present some reflections in order to unify all chapters of this book. We will focus on some general, practical aspects rather on the more formal ones.

When the simplest technique considered in this book, multiple linear regression, was presented as an immediate generalisation of the Lambert–Beer–Bouguer law (see Chapter 3), it was shown that it can be expressed as $y = a + b_1x_1 + b_2x_2 + \ldots + b_nx_n + e$, where y is the concentration of the analyte to be predicted, the bs are the regression coefficients, the xs are the atomic absorbances, intensities, *etc*., at times $1, \ldots, n$ of the atomic peak and e is the error term. This equation is equivalent to eqn (5.2) in Section 5.2.2, and therefore MLR might be considered as an ANN without any hidden layer with the transfer function being the identity function (*i.e.* the activation function is not weighted). Alternatively, it is possible to include hidden layers provided that linear transfer functions are used in both the hidden and output layers. In this case, the two successive linear combinations performed in the ANN are equivalent to a single MLR regression [7]. However, the MLR equation is fitted to the experimental data through the classical criterion of least squares (minimisation of the sum of the squared residuals). This involves a matrix inversion, which often yields severe problems, as discussed. Matrix inversion is a one-step calculation. In contrast, ANNs require an iterative process (with as many as 100 000 iterations) where the coefficients are adjusted sequentially until a minimum error is obtained. In fact, if we use an ANN to calibrate a system where only linear relationships exist between the spectral data and the concentrations, the ANN solution converges to an MLR model.

In PCR and PLS, a set of abstract factors were used to describe the spectral data and, then, relate them to the concentrations. These factors can be visualised easily as combinations of original variables (see Chapters 3 and 4). The spectra of the standards can be projected into the new abstract factors space, regressed against the concentrations and studied further. For simplicity, assume that both the spectra and the concentrations were mean centred (as is usually done). Equation (5.2) can be seen also as a PCR or PLS model when all transfer functions are linear [54]. The weights between the input and hidden layers are equivalent to the spectral loadings on the different factors and the activation function within the nodes of the hidden layer can act as the scores in PCR and PLS. As for MLR, the main difference is how the regression parameters are obtained. In ANNs, they are obtained after an iterative training process where the overall error is minimised. In PCR and PLS, we have serious restrictions in order to give as relevant scores for prediction as possible, for example, scores orthogonality and maximisation of the covariance between the **X**- and the **Y**-blocks (the cornerstone of the PLS algorithm). The lack of restrictions for ANNs may explain why they are so sensitive to overfitting whereas, in contrast, the severe restrictions for PCR and PLS facilitate the chemical interpretation of the models [54].

Often, the original spectra that we have recorded have many variables and so the number of input nodes in any ANN would be too large, with a subsequent increase in the number of weights and, finally, in the time required to optimise it and an urgent need for more standards. In order to simplify the architecture of the net, reduce the risk of overfitting and optimise the ANN faster, it is common practice to perform a preliminary PCA or PLS regression on the raw spectral data. Once the number of factors has been decided, the corresponding scores are input to the net and, finally, the ANN is developed. In doing so, Despagne and Massart [7] recommended not to preprocess the original data. The use of PCA or PLS ensures that only a few input neurons are used (one neuron per factor) and that only relevant information is given to the net. Outliers can also be removed easily before they go on into the net. A note of caution is relevant [7]: as PCA is a linear projection method, it cannot preserve the structure of a nonlinear data set (as would happen whenever we have a strong nonlinear behaviour in our spectra). This means that we should not restrict the number of PCs to the very first ones but try to cope with those which consider minor sources of variance (which might, presumably, be related to nonlinear problems). As a general rule of thumb (but neither definitive nor applicable to all cases), 15 PCs should suffice to describe the data set. The same reasoning can be applied to PLS, even though it is more powerful to handle nonlinearities and, probably, they should be modelled with fewer factors. We again face the well-known iterative approach so much needed to develop a satisfactory model.

5.10 Worked Example

In order to present a worked case study, we will consider here the same data set as employed in Chapter 4 to develop the PLS example. As explained in Section 5.8, an important limitation of the ANNs is that they do not provide graphical data to analyse their behaviour and, to some extent, one has to rely only on the numerical final outputs (*i.e.* the values predicted for the testing and, eventually, the validation sets). Although a strict protocol to develop a satisfactory neural network model cannot be given easily, several major steps can be underlined. Nevertheless, recall that developing a predictive ANN model requires an iterative process and, therefore, some steps might involve going back and testing again what you had set previously.

Probably the first issue to be decided upon is the (tentative) architecture of the net. This implies setting the number of neurons in the input layer, the number of hidden layers (and number of neurons on each) and the number of neurons in the output layer. Usually, the input and output neurons are established according to the type of problem to be addressed, although the user can take several approaches to codify the input and output data in the initial stages of the study. In the present example, each atomic peak was defined by 108 spectral variables (*i.e.* atomic absorbances at different times of the atomic peak). Although this is not very many compared with molecular spectrometry, where more than 1000 variables can be handled, it is a fairly large number to use as training data for a neural network. Not only can the training time be a problem when such a large number of spectral variables is managed but also, most importantly, do we have a so large number of calibration standards? In general we do not have so many spectra (standards) so we should consider two strategies:

1. To develop our calibration data set using an experimental design (see Chapter 2) in order to be, hopefully, reasonably sure that all the experimental domain is represented by the standard solutions to be prepared.
2. To apply chemometric tools to reduce the number of input variables used to characterise the samples without losing relevant information. In Chapter 3, a powerful tool to carry out such a reduction in the dimensionality of the original data was described in some detail: principal components analysis (PCA). In effect, if we first perform a PCA of the atomic peaks of the calibration set, we can consider that the scores of the first A principal components (PCs) alone represent the calibration standards accurately. Hence, instead of using the 108 original absorbance variables to train the ANN, only a reduced number of PC scores can be employed. This number (A) has to be optimised, although a reasonable value to start with can be somewhere around 10 (in our case 10 PCs explained 99.92% of the initial variance, which seemed good enough).

Subsequently, we fixed the number of neurons in the input layer as the number of PC scores (here we started with 10). Regarding the hidden layer, we followed current suggestions [49] and a reduced number of neurons were assayed, from 2 to 6. The number of neurons in the output layer is often simpler to set as we are interested in performing a regression between a set of input variables and a variable to be predicted (*e.g.* concentration of an analyte) and a unique neuron would be sufficient.

Before training the net, the transfer functions of the neurons must be established. Here, different assays can be made (as detailed in the previous sections), but most often the hyperbolic tangent function ('*tansig*' function in Table 5.1) is selected for the hidden layer. We set the linear transfer function ('*purelin*' in Table 5.1) for the output layer. In all cases the output function was the identity function (*i.e.* no further operations were made on the net signal given by the transfer function).

Finally, we have to set a stopping criterion, which may be either the number of iterations of the net (sometimes this is also termed '*number of epochs*', which corresponds to number of times the net is trained with the calibration data set and a corresponding update of the weights is performed) or a maximum average global error that we would accept for the prediction capabilities of the net. In many programs the latter is the '*mean square error*' (MSE), defined as $\text{MSE}=\Sigma(\hat{y} - y_{\text{reference}})^2/n$ ($\hat{y}$ = concentration of the analyte predicted by the net for a sample; n = number of samples). Most often both criteria are given simultaneously to prevent the net from being stacked in a local minimum and save time. Another option consists in ceasing the training phase when the error in the validation data starts to increase, to avoid overfitting (this is called early stopping). Depending on the software that can be handled, the learning rate can also be modified. If it is too low, the net will evolve too slowly and it may become trapped in a local solution (not the global one). If it is too large, the net will change rapidly but it may also show a random behaviour (bad and good predictions in successive iterations). The average error, learning rate and number of iterations are usually fixed after some preliminary trials. In addition, they can be employed to test whether the scores (input data) should be scaled further. We tested the net using non-scaled scores and scores scaled 0–1 (*i.e.* dividing all scores by their absolute maximum value), as is common when the '*tansig*' transfer function is selected. In this example, we obtained best results with the scaled scores.

To summarise, the first net we trained was characterised by:

- topology: 10 neurons (input layer)/2 neurons (hidden layer)/1 neuron (output layer);
- transfer functions: *tansig* (hidden layer) and *purelin* (output layer);
- learning rate, 0.0001; maximum number of epochs (iterations), 300 000; maximum acceptable mean square error, 25 (53 calibration standards);
- data pretreatment: normalisation 0–1 (division by the absolute maximum score).

Figure 5.7 depicts how the calibration error of the net (calculated using the calibration set) evolved as a function of the number of epochs. It is obvious

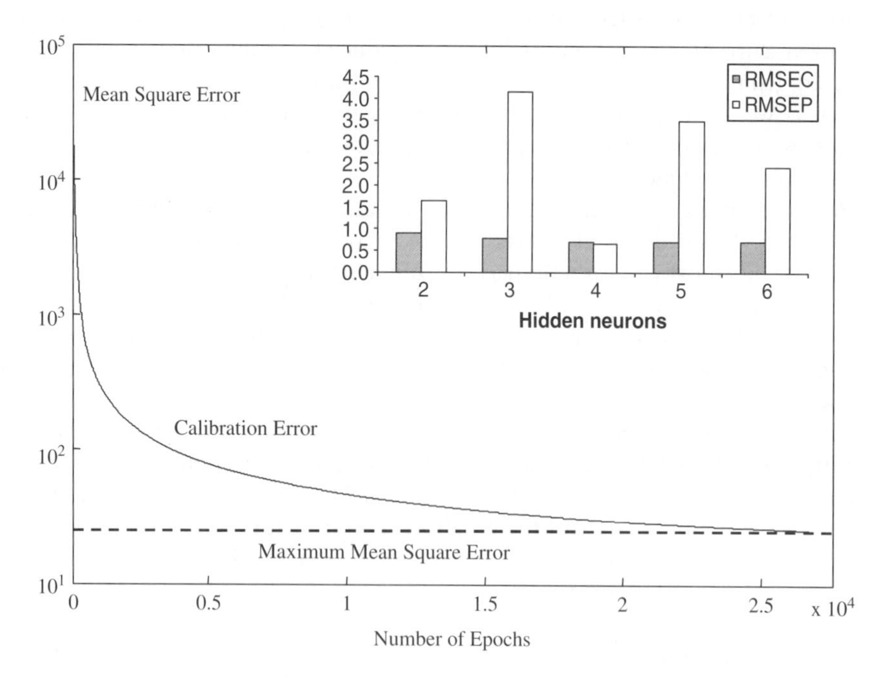

Figure 5.7 Evolution of the overall calibration error as a function of the number of epochs the ANN is trained. The inset shows the prediction capabilities of the net as a function of the number of neurons in the hidden layer.

that the net learns very fast and that it converges nicely to the maximum mean square error that we wanted to accept. Note that the MSE calibration error will decrease continuously since the ANN tends to memorise perfectly the calibration standards (which is not very useful to predict new samples). As for PLS, we must make a compromise between the calibration and the prediction abilities. Then, more trials have to be made, changing the parameters mentioned above in order to find an ANN with satisfactory prediction capabilities (considering the test set). Different numbers of hidden neurons should be attempted – sometimes even the number of hidden layers can be increased – but tests can also be made changing other parameters.

As the MSE is slightly inconvenient to evaluate the prediction capabilities of the ANN models, two more user-friendly statistics can be defined: the root mean square error of calibration (RMSEC) and the root mean square error of prediction (RMSEP). They are defined in the same way as for PLS (see Chapter 4, Section 4.5.6): RMSEC(P) = $[\Sigma(\hat{y} - y_{\text{reference}})^2/n]^{1/2}$ [$\hat{y}$ = concentration of the analyte predicted by the net for a sample; n=number of calibration(validation) samples]. Note that in contrast to what had been defined for PLS, the RMSEC here does not take into account the degrees of freedom of the model.

The inset in Figure 5.7 shows that both the RMSEC and RMSEP reach a minimum when four neurons are considered in the hidden layer. This was verified running more tests and it was concluded that this should be an optimum topology for this particular example. The best RMSEP corresponded to an ANN which yielded a 0.49 $\mu g\, ml^{-1}$ error. In addition, it was tested whether a larger or smaller number of PC scores might improve the predictions. Two extreme situations were studied considering 5 and 15 scores per sample. Improvements were not observed and, as a consequence, the net was not changed further.

It is worth comparing briefly the PLS (Chapter 4) and ANN models. The ANN selected finally uses four neurons in the hidden layer, which is exactly the same number of latent variables as selected for PLS, a situation reported fairly frequently when PLS and ANN models perform similarly. The RMSEC and RMSEP were slightly higher for PLS, 1.4 and 1.5 $\mu g\, ml^{-1}$, respectively, and they were outperformed by the ANN (0.7 and 0.5 $\mu g\, ml^{-1}$, respectively). The best predictive capabilities of the neural network might be attributed to the presence of some sort of spectral nonlinearities in the calibration set and/or some spectral behaviour not easy to account for by the PLS linear models.

Unfortunately, the random initialisation of the weights of the ANNs during the training phase may lead to neural networks with slightly different prediction capabilities, even when no parameter changed between successive trials. Therefore, we strongly recommend spending some time running the net several times in order to verify its performance, even when an ANN seemed satisfactory enough. As an example, the RMSEPs calculated for seven different trainings of an ANN with the same architecture were 0.5, 3.1, 1.7, 3.0, 0.9, 0.7 and 1.5 $\mu g\, ml^{-1}$. In two different runs, the predictive results were not very good. Two trainings led to good predictive results, similar to the corresponding PLS results (for which the RMSEPs were around 1.5 $\mu g\, ml^{-1}$). The fact that this particular topology yielded excellent results on three different occasions (errors lower than 1 $\mu g\,ml^{-1}$) reinforced the idea that its good predictions did not occur just by chance.

5.11 Examples of Practical Applications

Although ANNs are a rather recent technique, the number of applications in the atomic spectrometry field is relatively important and most studies were published during the last decade. Most applications consider only classification tasks to assess different geographical origins, brands, final qualities, *etc.*, *i.e.* pattern recognition applications. This shows that ANNs constitute a trustworthy and interesting way to solve some difficult problems. Unfortunately, far fewer applications dealing with multivariate calibration by ANNs were found and, accordingly, it was decided to present both types of applications. Probably, the number of papers reporting on the use of multivariate regression by ANNs will increase further, mainly to cope with complex problems which could not be solved satisfactorily by other regression means. In the following paragraphs different analytical atomic techniques and ANN approaches are reviewed and some major ideas presented by the authors are summarised.

Some of the pioneering studies published by several reputed authors in the chemometrics field [55] employed Kohonen neural networks to diagnose calibration problems related to the use of AAS spectral lines. As they focused on classifying potential calibration lines, they used Kohonen neural networks to perform a sort of pattern recognition. Often Kohonen nets (which were outlined briefly in Section 5.4.1) are best suited to perform classification tasks, whereas error back-propagation feed-forwards (BPNs) are preferred for calibration purposes [56].

5.11.1 Flame Atomic Absorption and Atomic Emission Spectrometry (FAAS and FAES)

López-Molinero *et al.* [57] developed a chemical volatilisation method to determinate Sb in poly(ethylene terephthalate) using FAAS. An ANN with a back-propagation error algorithm was applied to explore the surface response in order to evaluate whether any other optimum apart from the one previously proposed could be found.

Frías *et al.* [33] determined 11 elements in wines using FAAS (K, Na, Ca, Mg, Fe, Cu, Zn, Mn, Sr) and FAES (Li and Rb). Using cluster analyses and Kohonen SOMs they revealed differences in wines according to their geographical origin and the ripening state of the grapes. Further, they also found that error back-propagation artificial neural networks showed better prediction ability than stepwise linear discriminant analyses.

Hernández-Caraballo *et al.* [58] determined Cu, Zn and Fe concentrations in Cocuy de Penca clear brandy by FAAS. They applied linear and quadratic discriminant analysis and ANNs trained with an error back-propagation algorithm to estimate if the different beverages could be distinguished. They employed the concentrations of the elements in the final product to assess the geographical location of the manufacturers and the presence or absence of sugar in the end product. Various ANNs, comprising a linear function in the input layer, a sigmoid function in the hidden layer(s) and a hyperbolic tangent function in the output layer, were evaluated.

Padín *et al.* [34] characterised Galician quality brand potatoes using FAAS and FAES. A multilayer feed forward ANN, a self-organizing map with adaptative neighbourhood and their combination were used to classify potatoes according to their original brand.

5.11.2 Electrothermal Atomic Absorption Spectrometry (ETAAS)

Of the small number of publications dealing with the calibration issue, a few can be cited. Hernández-Caraballo *et al.* [59] applied BPNs and aqueous standards to develop calibration models capable of enlarging the (typically) short linear dynamic range of Cd curves. They employed a univariate approach, since only those absorbances at the maxima of the atomic peaks were regressed against Cd concentrations. Unfortunately, they did not apply the model to real samples and they did not consider that the atomic maxima could appear at slightly different wavelengths. In a later study [60], they compared BPNs with other mathematical functions, with the best results for the BPNs.

Recently, Felipe-Sotelo *et al.* [61] developed an ANN multivariate calibration model capable of handling complex interfering effects on Sb when soil, sediments and fly ash samples were analysed by slurry sampling–ETAAS (SS–ETAAS). Sometimes, spectral and chemical interferences cannot be totally resolved in slurries using chemical modifiers, ashing programmes, *etc.*, because of the absence of a sample pretreatment step to eliminate/reduce the sample matrix. Hence the molecular absorption signal is so high and structured that background correctors cannot be totally effective. In addition, alternative wavelengths may not be a solution due to their low sensitivity when trace levels are measured. To circumvent all these problems, the authors employed current PLS, second-order polynomial PLS and ANNs to develop predictive models on experimentally designed calibration sets. Validation with five certified reference materials (CRMs) showed that the main limitations of the models were related to the SS–ETAAS technique; *i.e.* the mass/volume ratio and the low content of analyte in some solid matrices (which forced the introduction of too much sample matrix into the atomiser). Both PLS and ANNs gave good results since they could handle severe problems such as peak displacement, peak enhancement/depletion and peak tailing. Nevertheless, PLS was preferred because the loadings and correlation coefficients could be interpreted chemically.

5.11.3 Inductively Coupled Plasma (ICP-OES and ICP-AES)

Van Veen and De Loos-Vollebregt [62] reviewed various chemometric procedures which had been developed during the last decade for ICP-OES analysis. In these procedures, data reduction by techniques such as digital filtering, numerical derivatives, Fourier transforms, correlation methods, expert systems, fuzzy logic, neural networks, PCA, PLS, projection methods, Kalman filtering, MLR and generalised standard additions were discussed.

Schierle and Otto [63] used a two-layer perceptron with error back-propagation for quantitative analysis in ICP-AES. Also, Schierle *et al.* [64] used a simple neural network [the bidirectional associative memory (BAM)] for qualitative and semiquantitative analysis in ICP-AES.

In the multivariate calibration field, Khayatzadeh *et al.* [65] compared ANNs with PLS to determine U, Ta, Mn, Zr and W by ICP in the presence of spectral interferences. For ANN modelling, a PCA preprocessing was found to be very effective and, therefore, the scores of the first five dominant PCs were input to an ANN. The network had a linear transfer function on both the hidden and output layers. They used only 20 samples for training.

An ANN was used by Magallanes *et al.* [66] to optimise hydride generation–inductively coupled plasma atomic emission spectrometry (HG–ICP-AES) coupling for the determination of Ge at trace levels.

Catasus *et al.* [67] studied two types of neural networks: traditional multilayer perceptron neural networks and generalised regression neural networks (GRNNs) to correct for nonlinear matrix effects and long-term signal drift in ICP-AES.

The recoveries of several rare earth elements in leachates obtained from apatite concentrates were determined by Jorjani *et al.* [68] for ICP-AES and ICP-MS. A neural network model was used to predict the effects of operational variables on the La, Ce, Y and Nd recoveries in the leaching process. The neural network employed was a feed-forward one.

In recent years, a number of new methods for the correction of spectral interferences in ICP have been reported. Zhang and Ma [69] reviewed the characteristic advantages and also the weak points of the more efficient and updated methods such as multiple linear regression, Kalman filter, generalised standard additions method, factor analysis, on-line intelligence background correction, ANNs, Fourier transform deconvolution, wavelet transform and computer spectrum stripping for correcting spectral interferences.

Derks *et al.* [70] employed ANNs to cancel out noise in ICP. The results of neural networks (an Adaline network and a multi-layer feed-forward network) were compared with the more conventional Kalman filter.

Moreno *et al.* [71] studied the metal content of different classes of wines by ICP-OES (Al, Ba, Cu, Fe, Mn, Sr, Zn, Ca, K, Na and Mg) and ETAAS (Ni and Pb). They employed 'probabilistic neural networks' for the classification of the two classes of wines.

Similarly, Alcázar *et al.* [72] characterised beer samples according to their mineral content measured by ICP-AES (Zn, P, B, Mn, Fe, Mg, Al, Sr, Ca, Ba,

Na and K) employing different pattern recognition techniques [PCA, linear discriminant analysis (LDA) and ANNs]. ANNs were trained by an error back-propagation algorithm and they were found to be very efficient for classifying and discriminating food products.

Álvarez *et al.* [73] compared the performance of LDA and ANNs to classify different classes of wines. Metal concentrations (Ca, Mg, Sr, Ba, K, Na, P, Fe, Al, Mn, Cu and Zn) were selected as chemical descriptors for discrimination because of their correlation with soil nature, geographical origin and grape variety. Both LDA and ANNs led to a perfect separation of classes, especially when multi-layer perceptron nets were trained by error back-propagation.

Sun *et al.* [74] employed a 3-layer artificial neural network model with a back-propagation error algorithm to classify wine samples in 6 different regions based on the measurements of trace amounts of B, V, Mn, Zn, Fe, Al, Cu, Sr, Ba, Rb, Na, P, Ca, Mg and K by ICP-OES. Similarly,

Balbinot *et al.* [36] classified Antarctic algae by applying Kohonen neural networks to a data set composed of 14 elements determined by ICP-OES.

Samecka-Cymerman *et al.* [75] measured the concentrations of Al, Be, Ca, Cd, Co, Cr, Cu, Fe, K, Mg, Mn, Ni, Pb and Zn in the aquatic bryophytes *Fontinalis antipyretica*, *Platyhypnidium riparioides* and *Scapania undulata* using ICP and ETAAS. A self-organizing feature map or Kohonen network was used to classify the bryophytes in terms of the concentrations of elements, in order to identify clusters within the examined sampling sites and to identify relationships between similar classes. PCA was also applied. Both techniques yielded distinct groups of aquatic bryophytes growing in streams flowing through different types of rock, groups which differed significantly in the concentrations of certain elements.

Fernández-Cáceres *et al.* [76] determined metals (Al, Ba, Ca, Cu, Fe, K, Mg, Mn, Na, Sr, Ti and Zn) in 46 tea samples by ICP-AES. PCA, LDA and ANNs were applied to differentiate the tea types. LDA and ANNs provided the best results in the classification of tea varieties and its relation to geographical origin. Analogously, Anderson and Smith [77] employed different pattern recognition methods (PCA, discriminant function analyses and neural network modelling) to differentiate the geographical growing regions of coffee beans.

Zhang *et al.* [78] analysed the metal contents of serum samples by ICP-AES (Fe, Ca, Mg, Cr, Cu, P, Zn and Sr) to diagnose cancer. BAM was compared with multi-layer feed-forward neural networks (error back-propagation). The BAM method was validated with independent prediction samples using the cross-validation method. The best results were obtained using BAM networks.

5.11.4 Inductively Coupled Plasma Mass Spectrometry (ICP-MS)

Waddell *et al.* [79] investigated the applicability of ICP-MS data obtained from the analysis of ecstasy tablets to provide linkage information from seizure to seizure. The data generated were analysed using different pattern recognition techniques, namely PCA, hierarchical clustering and ANNs.

Nadal *et al.* [35] determined the concentrations of As, Cd, Cr, Hg, Mn, Pb and V in soil and chard samples collected in various industrial sites in Tarragona (Spain). Concentrations in soils were determined by ICP-MS. In chard samples, the levels of As, Cd, Hg, Mn and Pb were measured using ICP-MS, whereas Cr and V concentrations were determined by ETAAS. A Kohonen SOM showed differences in metal concentrations according to the geographical origin of the samples.

Pérez-Trujillo *et al.* [80] determined the contents of 39 trace and ultratrace elements, among them 16 rare earth elements, in 153 wines from the Canary Islands (Spain). Back-propagation ANNs were employed to classify the set of wines according to the island of origin.

Thorrold *et al.* [81] used LDA and ANN models to classify fish as a function of their natal estuary. They investigated the ability of trace element and isotopic signatures measured in otoliths to record the nursery areas of juvenile (young-of-the-year) weakfish *Cynoscion regalis* from the east coast of the USA.

5.11.5 X-ray Fluorescence (XRF)

Kowalska and Urbanski [82] applied an ANN to compute calibration models for two XRF gauges. The performance of the ANN calibration was compared with that of a PLS model. It was found that the ANN calibration model exhibited better prediction ability in cases when the relationship between input and output variables was nonlinear.

Luo *et al.* [83] used an ANN to perform multivariate calibration in the XRF analysis of geological materials and compared its predictive performance with cross-validated PLS. The ANN model yielded the highest accuracy when a nonlinear relationship between the characteristic X-ray line intensity and the concentration existed. As expected, they also found that the prediction accuracy outside the range of the training set was bad.

Bos and Weber [84] compared two ANN optimisation procedures, namely genetic algorithms and backward error propagation, for quantitative XRF spectrometry. They used already published data from thin-film Fe–Ni–Cr samples for which the genetic algorithm approach performed poorly. They found that (as expected) the larger the training set, the better were the predictions. They also pointed out that samples outside the training set could not be predicted. Neural networks were shown to be useful as empirical calibration models with good quantitation abilities, with sufficient accuracy to compete successfully with various other common calibration procedures.

ANNs were compared with univariate linear regression in order to calibrate an XRF spectrometer to quantify Ti, V, Fe, Ni and Cu in polymetallic ores by Kierzek *et al.* [85].

Luo [86] proposed a kind of neural cluster structure embedded in neural networks. The ANN is based on the error back-propagation learning algorithm. The predictive ability of the neural cluster structure was compared with that of common neural net structures. A comparison of predictability with four neural networks was presented and they were applied to correct for matrix effects in XRF.

Long *et al.* [87] employed an ANN approach for the analysis of low-resolution XRF spectra. Instead of peak analysis and fitting the experimental results to a mathematical function as is common when conventional algorithms are used, the ANN method takes the spectrum as a whole, comparing its shape with the patterns learnt during the training period of the network.

Self-organizing ANNs (Kohonen neural nets) were employed for classifying different steels [88]. Twelve relevant elements were selected for data processing through ANNs.

Boger and Karpas [89] employed neural networks to interpret ion mobility spectrometry (IMS) and XRF spectral data and to derive qualitative and quantitative information from them.

Vigneron *et al.* [90] applied ANNs to analyse XRF spectra to dose uranium and plutonium at various stages of the nuclear fuel cycle.

Hernández-Caraballo *et al.* [91,92] evaluated several classical chemometric methods and ANNs as screening tools for cancer research. They measured the concentrations of Zn, Cu, Fe and Se in blood serum specimens by total reflection XRF spectrometry. The classical chemometric approaches used were PCA and logistic regression. On the other hand, two neural networks were employed for the same task, *viz.*, back-propagation and probabilistic neural networks.

Majcen *et al.* [93] studied linear and nonlinear multivariate analysis in the quality control of industrial titanium dioxide white pigment using XRF spectrometry.

Bos *et al.* [94] compared the performance of ANNs for modelling the Cr–Ni–Fe system in quantitative XRF spectroscopy with the classical Rasberry–Heinrich model and a previously published method applying the linear learning machine in combination with singular value decomposition. They studied whether ANNs were able to model nonlinear relationships, and also their ability to handle non-ideal and noisy data. They used more than 100 steel samples with large variations in composition to calibrate the model. ANNs were found to be robust and to perform better than the other methods.

5.11.6 Secondary Ion Mass Spectrometry (SIMS)

Hutter *et al.* [95] reviewed surface image analysis by SIMS. Detection of chemical phases by classification using neural networks or denoising of scanning-SIMS images by wavelet filtering demonstrated the increased performance of analysis imaging techniques. Analogously, Tyler [96] applied different multivariate statistical techniques (PCA, factor analysis, neural networks and mixture models) to explore spectral images.

Nord and Jacobsson [97] proposed several approaches to interpret ANN models. The results were compared with those derived from a PLS regression model where the contributions of variables were studied. Notably, they employed simple architectures composed of a unique hidden layer. They discovered that the variable contribution term in ANN models is similar to that in PLS models for linear relationships, although this may not be the case for nonlinear relations. In such cases, the proposed algorithms can give additional information about the models.

Sanni *et al.* [98] used neural networks and PCA for classification of adsorbed protein static time-of-flight (TOF) SIMS spectra.

The utility of ANNs as a pattern recognition technique in the field of microbeam analysis was demonstrated by Ro and Linton [99]. Back-propagation neural networks were applied to laser microprobe mass spectra (LAMMS) to determine interparticle variations in molecular components. Self-organizing feature maps (Kohonen neural networks) were employed to extract information on molecular distributions within environmental microparticles imaged in cross-section using SIMS.

References

1. M. Felipe-Sotelo, J. M. Andrade, A. Carlosena and D. Prada, Partial least squares multivariate regression as an alternative to handle interferences of Fe on the determination of trace Cr in water by electrothermal atomic absorption spectrometry, *Anal. Chem.*, 75(19), 2003, 5254–5261.
2. J. M. Andrade, M. S. Sánchez and L. A. Sarabia, Applicability of high-absorbance MIR spectroscopy in industrial quality control of reformed gasolines, *Chemom. Intell. Lab. Syst.*, 46(1), 1999, 41–55.
3. S. P. Jacobsson and A. Hagman, Chemical composition analysis of carrageenans by infrared spectroscopy using partial least squares and neural networks, *Anal. Chim. Acta*, 284(1), 1993, 137–147.
4. L. Hadjiiski, P. Geladi and Ph. Hopke, A comparison of modelling nonlinear systems with artificial neural networks and partial least squares, *Chemom. Intell. Lab. Syst.*, 49(1), 1999, 91–103.
5. M. Blanco, J. Coello, H. Iturriaga, S. Maspoch and J. Pagès, NIR calibration in nonlinear systems by different PLS approaches and artificial neural networks, *Chemom. Intell. Lab. Syst.*, 50(1), 2000, 75–82.
6. S. Sekulic, M. B. Seasholtz, Z. Wang, B. R. Kowalski, S. E. Lee and B. R. Holt, Nonlinear multivariate calibration methods in analytical chemistry, *Anal. Chem.*, 65(19), 1993, 835A–845A.
7. F. Despagne and D. L. Massart, Neural networks in multivariate calibration, *Analyst*, 123(11), 1998, 157R–178R.
8. J. R. M. Smits, W. J. Melssen, L. M. C. Buydens and G. Kateman, Using artificial neural networks for solving chemical problems. Part I: multi-layer feed-forward networks, *Chemom. Intell. Lab. Syst.*, 22(2), 1994, 165–189.

9. J. Zupan, Introduction to artificial neural network (ANN) methods: what they are and how to use them, *Acta Chim. Slov.*, 41(3), 1994, 327–352.
10. A. Pazos Sierra (ed.), *Redes de Neuronas Artificiales y Algoritmos Genéticos*, University of A Coruña, A Coruña, 1996.
11. S. Haykin, *Neural Networks: a Comprehensive Foundation*, 2nd edn., Prentice Hall, Englewood Cliffs, NJ, 1998.
12. G. Kateman, Neural networks in analytical chemistry?, *Chemom. Intell. Lab. Syst.*, 19(2), 1993, 135–142.
13. J. C. Brégains, J. Dorado de la Calle, M. Gestal Pose, J. A. Rodríguez González, F. Ares Pena and A. Pazos Sierra, Avoiding interference in planar arrays through the use of artificial neural networks, *IEEE Antennas Propag. Mag.*, 44(4), 2002, 61–65.
14. M. S. Sánchez, E. Bertran, L. A. Sarabia, M. C. Ortiz, M. Blanco and J. Coello, Quality control decissions with near infrared data, *Chem. Intell. Lab. Syst.*, 53(1–2), 2000, 69–80.
15. M. S. Sánchez and L. A. Sarabia, A stochastic trained neural network for nonparametric hypothesis testing, *Chemom. Intell. Lab. Syst.*, 63(2), 2002, 169–187.
16. R. Leardi, Genetic algorithms in chemometrics and chemistry: a review, *J. Chemom.*, 15(7), 2001, 559–569.
17. M. P. Gómez-Carracedo, M. Gestal, J. Dorado and J. M. Andrade, Chemically driven variable selection by focused multimodal genetic algorithms in mid-IR spectra, *Anal. Bioanal. Chem.*, 389(7–8), 2007, 2331–2342.
18. A. Blanco, M. Delgado and M. C. Pegalajar, A genetic algorithm to obtain the optimal recurrent neural network, *Int. J. Approx. Reasoning*, 23(1), 2000, 61–83.
19. A. K. Jain, J. Mao and K. M. Mohiuddin, Artificial neural networks: a tutorial, *Computer*, 29(3), 1996, 31–34.
20. W. S. McCulloch and W. Pitts, A logical calculus of ideas immanent in nervous activity, *Bull. Math. Biophysics*, 5, 1943, 115–133.
21. R. Rosemblatt, *Principles of Neurodynamics*, Spartan Books, New York, 1962.
22. J. J. Hopfield, Neural networks and physical systems with emergent collective computational abilities, *Proc. Natl. Acad. Sci. USA*, 79(8), 1982, 2554–2558.
23. P. Werbos, Beyond regression: new tools for prediction and analysis in the behavioral sciences, *PhD Thesis*, Department of Applied Mathematics, Harvard University, Cambridge, MA, 1974.
24. D. E. Rumelhart and J. L. McClelland, *Parallel Distributed Processing: Exploration in the Microstructure of Cognition*, MIT Press, Cambridge, MA, 1986.
25. F. Zorriassatine and J. D. T. Tannock, A review of neural networks for statistical process control, *J. Intell. Manuf.*, 9(3), 1998, 209–224.
26. M. J. Adams, *Chemometrics in Analytical Spectroscopy*, Royal Society of Chemistry, Cambridge, 1995.

27. Z. Boger, Selection of the quasi-optimal inputs in chemometric modelling by artificial neural networks analysis, *Anal. Chim. Acta*, 490(1–2), 2003, 31–40.
28. Z. Boger, R. Weber, Finding an optimal artificial neural network topology in real-life modeling, presented at the ICSC Symposium on Neural Computation, article No. 1, 1403/109, 2000.
29. R. B. Boozarjomehry and W. Y. Svrcek, Automatic design of neural network structures, *Comput. Chem. Eng.*, 25(7–8), 2001, 1075–1088.
30. C. M. Bishop, *Pattern Recognition and Machine Learning*, Springer, Berlin, 2007.
31. C. M. Bishop, *Neural Networks for Pattern Recognition*, Oxford University Press, Oxford, 1995.
32. T. Mitchell, *Machine Learning*, McGraw Hill, New York, 1997.
33. S. Frías, J. E. Conde, M. A. Rodríguez, V. Dohnal and J. Pérez-Trujillo, Metallic content of wines from the Canary islands (Spain). Application of artificial neural networks to the data analysis, *Nahrung/Food*, 46(5), 2002, 370–375.
34. P. M. Padín, R. M. Peña, S. García, C. Herrero, R. Iglesias and S. Barro, Characterization of Galician (N. W. Spain) quality brand potatoes: a comparison study of several pattern recognition techniques, *Analyst*, 126(1), 2001, 97–103.
35. M. Nadal, M. Schuhmacher and J. L. Domingo, Metal pollution of soils and vegetation in an area with petrochemical industry, *Sci. Total Environ.*, 321(1–3), 2004, 59–69.
36. L. Balbinot, P. Smichowski, S. Farias, M. A. Z. Arruda, C. Vodopivez and R. J. Poppi, Classification of Antarctic algae by applying Kohonen neural network with 14 elements determined by inductively coupled plasma optical emission spectrometry, *Spectrochim. Acta, Part B*, 60(5), 2005, 725–730.
37. D. Svozil, V. Kvasnicka and J. Pospichal, Introduction to multi-layer feed-forward neural networks, *Chemom. Intell. Lab. Syst.*, 39(1), 1997, 43–62.
38. R. J. Erb, Introduction to back-propagation neural network computation, *Pharma. Res.*, 10(2), 1993, 165–170.
39. S. A. Kalogirou, Artificial intelligence for the modelling and control of combustion processes: a review, *Prog. Energy Combust. Sci.*, 29(6), 2003, 515–566.
40. Z. Ramadan, P. K. Hopke, M. J. Johnson and K. M. Scow, Application of PLS and back-propagation neural networks for the estimation of soil properties, *Chemom. Intell. Lab. Syst.*, 75(1), 2005, 23–30.
41. V. M. López Fandiño, Análisis de componentes principales no lineales mediante redes neuronales artificiales de propagación hacia atrás: aplicaciones del modelo de Kramer, *PhD Thesis*, Instituto Químico de Sarriá, Ramon Llull University, Barcelona, 1997.
42. G. G. Andersson and P. Kaufmann, Development of a generalized neural network, *Chemom. Intell. Lab. Syst.*, 50(1), 2000, 101–105.

43. P. A. Jansson, Neural networks: an overview, *Anal. Chem.*, 63(6), 1991, 357A–362A.
44. L. Burke and J. P. Ignizio, A practical overview of neural networks, *J. Intell. Manuf.*, 8(3), 1997, 157–165.
45. J. Gasteiger and J. Zupan, Neural networks in chemistry, *Angew. Chem. Int. Ed. Engl.*, 32, 1993, 5403–527.
46. J. E. Dayhoff and J. M. DeLeo, Artificial neural netowrks. Opening the black box, *Cancer, Suppl.*, 91(8), 2001, 1615–1635.
47. E. Richards, C. Bessant and S. Saini, Optimisation of a neural network model for calibration of voltametric data, *Chemom. Intell. Lab. Syst.*, 61(1–2), 2002, 35–49.
48. D. J. Livingstone, D. T. Manallack and I. V. Tetko, Data modelling with neural networks: advantages and limitations, *J. Comp.-Aided Mol. Des.*, 11(2), 1997, 135–142.
49. B. Walczak, Modelling methods based on SNVs and related techniques, Plenary Lecture presented at the IIIrd Workshop in Chemometrics, Burgos, Spain, 15–16 September 2008.
50. R. Féraud and F. Clérot, A methodology to explain neural network classification, *Neural Networks*, 15(2), 2002, 237–246.
51. A. M. Fonseca, J. L. Vizcaya, J. Aires-de-Sousa and A. M. Lobo, Geographical classification of crude oils by Kohonen self-organizing maps, *Anal. Chim. Acta*, 556(2), 2006, 374–382.
52. C. Ruckebusch, L. Duponchel and J.-P. Huvenne, Interpretation and improvement of an artificial neural network MIR calibration, *Chemom. Intell. Lab. Syst.*, 62(2), 2002, 189–198.
53. J. R. Rabuñal, J. Dorado, A. Pazos, J. Pereira and D. Rivero, A new approach to the extraction of ANN rules and to their generalisation capacity throught GP, *Neural Comput.*, 16(7), 2004, 1483–1523.
54. T. Næs, K. Kvaal, T. Isaksson and C. Miller, Artificial neural networks in multivariate calibration, *J. Near Infrared Spectrosc.*, 1(1), 1993, 1–11.
55. Y. Vander Heyden, P. Vankeerberghen, M. Novic, J. Zupan and D. L. Massart, The application of Kohonen neural networks to diagnose calibration problems in atomic absorption spectrometry, *Talanta*, 51(3), 2000, 455–466.
56. J. M. Andrade, M. J. Cal-Prieto, M. P. Gómez-Carracedo, A. Carlosena and D. Prada, A tutorial on multivariate calibration in atomic spectrometry techniques, *J. Anal. At. Spectrom.*, 23(1), 2008, 15–28.
57. A. López-Molinero, P. Calatayud, D. Sipiera, R. Falcón, D. Liñán and J. R. Castillo, Determination of antimony in poly(ethylene terephthalate) by volatile bromide generation flame atomic absorption spectrometry, *Microchim. Acta*, 158(3–4), 2007, 247–253.
58. E. A. Hernández-Caraballo, R. M. Ávila-Gómez, T. Capote, F. Rivas and A. G. Pérez, Classification of Venezuelan spirituous beverages by means of discrimimant analysis and artificial neural networks based on their Zn, Cu and Fe concentrations, *Talanta*, 60(6), 2003, 1259–1267.

59. E. A. Hernández-Caraballo, R. M. Avila-Gómez, F. Rivas, M. Burguera and J. L. Burguera, Increasing the working calibration range by means of artificial neural networks for the determination of cadmium by graphite furnace atomic absorption spectrometry, *Talanta*, 63(2), 2004, 425–431.
60. E. A. Hernández-Caraballo, F. Rivas and R. M. Ávila de Hernández, Evaluation of a generalized regression artificial neural network for extending cadmium's working calibration range in graphite furnace atomic absorption spectrometry, *Anal. Bioanal. Chem.*, 381(3), 2005, 788–794.
61. M. Felipe-Sotelo, M. J. Cal-Prieto, M. P. Gómez-Carracedo, J. M. Andrade, A. Carlosena and D. Prada, Handling complex effects in slurry-sampling–electrothermal atomic absorption spectrometry by multivariate calibration, *Anal. Chim. Acta*, 571(2), 2006, 315–323.
62. E. H. Van Veen and M. T. C. De Loos-Vollebregt, Application of mathematical procedures to background correction and multivariate analysis in inductively coupled plasma-optical emission spectrometry, *Spectrochim. Acta, Part B*, 53(5), 1998, 639–669.
63. C. Schierle and M. Otto, Comparison of a neural network with multiple linear regression for quantitative analysis in ICP-atomic emission spectroscopy, *Fresenius' J. Anal. Chem.*, 344(4–5), 1992, 190–194.
64. C. Schierle, M. Otto and W. Wegscheider, A neural network approach to qualitative analysis in inductively coupled plasma-atomic emission spectroscopy (ICP-AES), *Fresenius' J. Anal. Chem.*, 343(7), 1992, 561–565.
65. M. Khayatzadeh Mahani, M. Chaloosi, M. Ghanadi Maragheh, A. R. Khanchi and D. Afzali, Comparison of artificial neural networks with partial least squares regression for simultaneous determination by ICP-AES, *Chin. J. Chem.*, 25(11), 2007, 1658–1662.
66. J. F. Magallanes, P. Smichowski and J. Marrero, Optimisation and empirical modeling of HG-ICP-AES analytical technique through artificial neural networks, *J. Chem. Inf. Comput. Sci.*, 41(3), 2001, 824–829.
67. M. Catasus, W. Branagh and E. D. Salin, Improved calibration for inductively coupled plasma-atomic emission spectrometry using generalized regression neural networks, *Appl. Spectrosc.*, 49(6), 1995, 798–807.
68. E. Jorjani, A. H. Bagherieh, Sh. Mesroghli and Ch. Chelgani, Prediction of yttrium, lanthanum, cerium and neodymium leaching recovery from apatite concentrate using artificial neural networks, *J. Univ. Sci. Technol. Beijing*, 15(4), 2008, 367–374.
69. Z. Zhang and X. Ma, Methods for correction of spectral interferences in inductively coupled plasma atomic emission spectrometry, *Curr. Top. Anal. Chem.*, 3, 2002, 105–123.
70. E. P. P. A. Derks, B. A. Pauly, J. Jonkers, E. A. H. Timmermans and L. M. C. Buydens, Adaptive noise cancellation on inductively coupled plasma spectroscopy, *Chemom. Intell. Lab. Syst.*, 39(2), 1997, 143–160.

71. I. M. Moreno, D. González-Weller, V. Gutierrez, M. Marino, A. M. Cameán, A. G. González and A. Hardisson, Differentiation of two Canary DO red wines according to their metal content from inductively coupled plasma optical emision spectrometry and graphite furnace atomic absorption spectrometry by using probabilistic neural networks, *Talanta*, 72(1), 2007, 263–268.
72. A. Alcázar, F. Pablos, M. J. Martín and A. G. González, Multivariate characterisation of beers according to their mineral content, *Talanta*, 57(1), 2002, 45–52.
73. M. Álvarez, I. M. Moreno, A. Jos, A. M. Cameán and A. G. González, Differentiation of two Andalusian DO 'fino' wines according to their metal content from ICP-OES by using supervised pattern recognition methods, *Microchem. J.*, 87(1), 2007, 72–76.
74. L. X. Sun, K. Danzer and G. Thiel, Classification of wine samples by means of artificial neural networks and discrimination analytical methods, *Fresenius' J. Anal. Chem.*, 359(2), 1997, 143–149.
75. A. Samecka-Cymerman, A. Stankiewicz, K. Kolon and A. J. Kempers, Self-organizing feature map (neural networks) as a tool in classification of the relations between chemical composition of aquatic bryophytes and types of streambeds in the Tatra national park in Poland, *Chemosphere*, 67(5), 2007, 954–960.
76. P. L. Fernández-Cáceres, M. J. Martín, F. Pablos and A. G. González, Differentiation of tea (*Camellia sinensis*) varieties and their geographical origin according to their metal content, *J. Agric. Food Chem.*, 49(10), 2001, 4775–4779.
77. K. A. Anderson and B. W. Smith, Chemical profiling to differenciate geographic growing origins of coffee, *J. Agric. Food Chem.*, 50(7), 2002, 2068–2075.
78. Z. Zhang, H. Zhuo, S. Liu and P. D. B. Harrington, Classification of cancer patients based on elemental contents of serums using bi-directional associative memory networks, *Anal. Chim. Acta*, 436(2), 2001, 281–291.
79. R. J. H. Waddell, N. NicDaéid and D. Littlejohn, Classification of ecstasy tablets using trace metal analysis with the application of chemometric procedures and artificial neural networks algorithms, *Analyst*, 129(3), 2004, 235–240.
80. J.-P. Pérez-Trujillo, M. Barbaste and B. Medina, Chemometric study of bottled wines with denomination of origin from the Canary Islands (Spain) based on ultra-trace elemental content determined by ICP-MS, *Anal. Lett.*, 36(3), 2003, 679–697.
81. S. R. Thorrold, C. M. Jones, P. K. Swart and T. E. Targett, Accurate classification of juvenile weakfish *Cynoscion regalis* to estuarine nursery areas based on chemical signatures in otoliths, *Mar. Ecol. Prog. Ser.*, 173, 1998, 253–265.
82. E. Kowalska and P. Urbanski, XRF full-spectrum calibration technique using artificial neural network, *Pol Nukleonika*, 42(4), 1997, 879–887.

83. L. Luo, C. Guo, G. Ma and A. Ji, Choice of optimum model parameters in artificial neural networks and application to X-ray fluorescence analysis, *X-Ray Spectrom.*, 26(1), 1997, 15–22.
84. M. Bos and H. T. Weber, Comparison of the training of neural networks for quantitative X-ray fluorescence spectrometry by a genetic algorithm and backward error propagation, *Anal. Chim. Acta*, 247(1), 1991, 97–105.
85. J. Kierzek, A. Kierzek and B. Malozewska-Bucko, Neural networks based calibration in X-ray fluorescence analysis of polymetallic ores, *Pol. Nukleonika*, 40(3), 1995, 133–140.
86. L. Luo, Predictability comparison of four neural network structures for correcting matrix effects in X-ray fluorescence spectrometry, *J. Trace Microprobe Tech.*, 18(3), 2000, 349–360.
87. X. Long, N. Huang, T. Li, F. He and X. Peng, An artificial neural network analysis of low-resolution X-ray fluorescence spectra, *Adv. X-Ray Anal.*, 40, 1998, 307–314.
88. J. F. Magallanes and C. Vázquez, Automatic classification of steels by processing energy-dispersive X-ray spectra with artificial neural networks, *J. Chem. Inf. Comput. Sci.*, 38(4), 1998, 605–609.
89. Z. Boger and Z. Karpas, Application of neural networks for interpretation of ion mobility and X-ray fluorescence spectra, *Anal. Chim. Acta*, 292(3), 1994, 243–251.
90. V. Vigneron, A. C. Simon, R. Junca and J. M. Martinez, Neural techniques applied to analysis of X-ray fluorescence spectra. Example of determination of uranium, *Analusis*, 24(9–10), 1996, 37–41.
91. E. A. Hernández-Caraballo, F. Rivas, A. G. Pérez and L. M. Marcó-Parra, Evaluation of chemometric techniques and artificial neural networks for cancer screening using Cu, Fe, Se and Zn concentrations in blood serum, *Anal. Chim. Acta*, 533(2), 2005, 161–168.
92. E. A. Hernández-Caraballo and L. M. Marcó-Parra, Direct analysis of blood serum by total reflection X-ray fluorescence spectrometry and application of an artificial neural network approach for cancer diagnosis, *Spectrochim. Acta, Part B*, 58(12), 2003, 2205–2213.
93. N. Majcen, F. X. Rius and J. Zupan, Linear and nonlinear multivariate analysis in the quality control of industrial titanium dioxide white pigment, *Anal. Chim. Acta*, 348(1–3), 1997, 87–100.
94. A. Bos, M. Bos and W. E. Van der Linden, Artificial neural network as a multivariate calibration tool: modelling the Fe–Cr–Ni system in X-ray fluorescence spectroscopy, *Anal. Chim. Acta*, 277(2), 1993, 289–295.
95. H. Hutter, Ch. Brunner, St. Nikolov, Ch. Mittermayer and M. Grasserbauer, Imaging surface spectroscopy for two- and three-dimensional characterisation of materials, *Fresenius' J. Anal. Chem.*, 355(5–6), 1996, 585–590.
96. B. Tyler, Interpretation of TOF-SIMS images: multivariate and univariate approaches to image de-noising, image segmentation and compound identification, *Appl. Surf. Sci.*, 203–204, 2003, 825–831.

97. L. I. Nord and S. P. Jacobsson, A novel method for examination of the variable contribution to computational neural network models, *Chemom. Intell. Lab. Syst.*, 44(1–2), 1998, 153–160.
98. O. D. Sanni, M. S. Wagner, D. Briggs, D. G. Castner and J. C. Vickerman, Classification of adsorbed protein static TOF-SIMS spectra by principal component analysis and neural networks, *Surf. Interface Anal.*, 33(9), 2002, 715–728.
99. C.-U. Ro and R. W. Linton, New directions in microprobe mass spectrometry: molecular microanalysis using neural networks, *Microbeam Anal.*, 1(2), 1992, 75–87.

Subject Index

Note: Figures and Tables are indicated by *italic page numbers*.

Abbreviations: ANN = artificial neural network; CLS = classical least squares; ILS = inverse least squares; PCR= principal component regression; PLS = partial least squares